6000 04450

WITHDRAWN

T0035893

Representations of Space and Time

Representations of SPACE AND TIME

Donna J. Peuquet

THE GUILFORD PRESS
New York   London

© 2002 The Guilford Press
A Division of Guilford Publications, Inc.
72 Spring Street, New York, NY 10012
www.guilford.com

All rights reserved

No part of this book may be reproduced, translated, stored in a
retrieval system, or transmitted, in any form or by any means,
electronic, mechanical, photocopying, microfilming, recording,
or otherwise, without written permission from the Publisher.

Printed in the United States of America

This book is printed on acid-free paper.

Last digit is print number: 9 8 7 6 5 4 3 2 1

Library of Congress Cataloging-in-Publication Data

Peuquet, Donna J.
    Representations of space and time / Donna J. Peuquet.
      p.   cm.
    Includes bibliographical references (p.) and index.
    ISBN 1-57230-773-0 (hardcover)
    1. Geographical perception. 2. Space and time. 3. Information
storage and retrieval systems—Geography.   I. Title.
    G71.5 .P48   2002
    304.2′3—dc21   2002005303

UNIVERSITY OF HERTFORDSHIRE
HATFIELD CAMPUS LRC
HATFIELD AL10 9AD

BIB

CLASS 304.23 PEU

LOCATION MAIN TERM OWL

BARCODE 6000404450

PREFACE

This book is about the representation of geographic space and the dynamics that occur within that space. Historically, research on spatiotemporal representation has been very fragmented, and there are two different directions from which the topic has been addressed: the philosophical and cognitive perspective on how we understand ourselves and how we react to our surroundings in an everyday sense, and the natural science perspective on how we understand the true nature of the world around us. The former focuses on informal spatiotemporal knowledge as represented and utilized by humans, and the latter on the development of formal representations of earth-related phenomena. Both are obviously important and interrelated components of how we learn and gain understanding about the world around us.

More recently, problems encountered in geographic information science (GIScience), as well as geographic information systems (GIS) as computer-based tools for handling and analyzing geographic data, have sparked much interest in the issue of geographic representation from both the cognitive and the natural science perspectives. From the cognitive perspective, the issue is how to represent both computer databases and graphic visualizations of geographic data in ways that are more "intuitive" to the user. From the perspective of scientific analysis, the issue is how to provide, using the very large and heterogeneous data sets now available, consistent and coherent representational techniques that preserve the information content of the data and that also can be used in a wide variety of analytical and modeling contexts. Part of this issue involves the representation of space–time *dynamics*—something that historically has been ignored in GIS but has always been integrated in process models devoted to and designed for certain specific tasks. GIS data representation's static view of the world means that dynamic models cannot be easily integrated into GIS.

I first grappled with how to integrate philosophical and cognitive principles into geographic database representation in 1987 when I wrote a journal article on the topic. I attempted to delineate a combined perspective in which more human-centered computer-based representations could be developed than had historically been the case. It proved to be an extremely difficult task within the space limitations of the article simply because of the many aspects involved. From that initial beginning, I continued to develop my ideas in a series of other articles and seminars. It wasn't until I had the good fortune of being granted sabbatical leave and a Guggenheim Fellowship from the spring of 1999 to the fall of 2001 that I was able to devote the time I needed to weave together the many threads and complete the writing of this book.

I believe it is by integrating cognitive principles as well as ideas from other disciplines that deal with spatial and space–time issues that we can (and need to) make a leap forward. Yet there is a careful balance that needs to be struck: finding new and better approaches to representation should not mean abandoning what has been learned in the field of GIScience over the last 30 to 40 years. I have tried look very broadly across a number of disciplines and examine how things can fit together in an integrated theory. It is reassuring that, in my explorations for the book, I found we need not go in a different direction, but rather that there are striking parallels and similarities in the ideas and principles among the fields I looked at. From a GIScience standpoint, I believe the aim should be a more comprehensive view—in essence, one that fills in the gaps and shows how the principles relate on several levels.

The aim of this book, therefore, is to examine both the cognitive and philosophical as well as the formal representation perspectives in order to reveal their commonalities and links, and to thereby provide insight for a unified representational framework. In the discussion I compare concepts from a diverse array of disciplines, including philosophy, cognitive science, linguistics, mathematics, artificial intelligence, database management systems, cartography, and GIScience. I draw on and combine these concepts into a set of general principles and an overall framework for representing geographic phenomena.

The book is divided into two parts. In the first I examine some fundamental philosophical issues concerning the nature of both space and time, and the nature of spatial knowledge. Following that I discuss a number of theoretical approaches, from various relevant disciplines, to how geographic space is cognitively represented. In the second part, I address computer-based geographic representations.

Chapter 1 provides a brief historical overview of computer-based geographic data processing, the current state of the art, and what potential there is for a different, cognitively informed approach to computer modeling. Chapters 2 through 7 explore theories of how humans acquire, store, and use spatial knowledge. Again, I use a historical perspective to show (1) the commonalities and linkages among the theories of various fields and (2) the evolution of these theories over time, in their view of learning and vision—as well as other forms of sensory perception—and in their view of how maps, language, and other symbolic forms can convey knowledge. Chapters 8, 9, and 10 act as a bridge between the two parts of the book. Chapter 8 draws together into a single framework the common threads from various views on specific aspects of how people acquire and store spatial knowledge. Before examining how this framework and the various theories regarding human knowledge representation, interaction, and learning can be productively applied in a computing context, Chapters 9 and 10 discuss current challenges and opportunities regarding computer storage and use of geographic data and higher-level information. In the remainder of Part II, Chapter 11 provides a review of computer-based geographic data representation techniques as they have developed up to now. Chapter 12 provides a broader view of computer-based representation approaches from a historical and more abstract perspective, which links it with the framework developed in Chapter 8. Chapter 12 also covers recent work that outlines a new, integrative database representation approach. Chapters 13 and 14, respectively, examine how cognitive principles can be, and are being, applied in interacting with computers and computer databases, and how they can be used for implementing advanced geographic databases. Chapter 15 provides some brief closing comments.

This book is intended for researchers and professionals in both computer and noncomputer aspects of modeling geographic space and is designed to be used as a text in advanced-level courses on models of geographic space, especially cognitive models, and/or data models for GIS. It could serve as the basis of a two-course sequence utilizing the material in Parts I and II, respectively.

ACKNOWLEDGMENTS

The general ideas in this book were influenced by many mentors, colleagues, and friends. It is impossible to adequately acknowledge all of the traditions and people that in many ways contribute to a work such as this. All that can be done is to mention some of them and hope that the rest will know who they are and that their help along the way is deeply appreciated. Those directly involved with bringing the book to fruition include my department head, Roger Downs, and the dean of my college, John Dutton, who allowed me to take necessary time off from my other duties at Penn State University. The generous support of the Guggenheim Foundation gave me the time not only to complete the book, but also to investigate some new directions in the use of advanced computer-based representations that would not have been possible otherwise. I am indebted to Martien Molenaar and Menno-Jan Kraak of the International Institute for Aerospace Survey and Earth Science (ITC) in Enschede, The Netherlands, for inviting me to the ITC. In addition to a shorter trip to the ITC in the fall of 1999, I worked with Menno-Jan Kraak and his colleagues on new methods of database and visual representation of complex, dynamic geographic data from February through May of 2000. This resulted in material in Chapter 7 on how to use visualization to aid creative thinking and discovery in the exploration of geographic data. Menno-Jan also provided comments on the full and earlier drafts of the manuscript. Alan MacEachren helped in the area of visualization in our many discussions over the years, particularly in helping me recognize connections between visual and database forms of representation. Barbara Tversky provided valuable feedback as I was delving into the cognitive psychology literature, and she provided comments on the entire manuscript. Duane Marble also provided comments on the entire manuscript. In addition, I am indebted to those who, as graduate students, participated in the seminars where we together discussed and pondered much of the material in this book in its formative stages and who gave me their reactions to earlier versions of the manuscript, particularly Elizabeth Wentz, Trudy Suchan, Nik Huffman, Liujin Qian, Jeremy Mennis, and Diansheng Guo.

There are many people without whom this book could not have been published. Seymour Weingarten, Editor in Chief of The Guilford Press, remembered me and my work from 8 years earlier and was willing to take on this project, and Kristal Hawkins, Geography Editor, ushered things through the process. Craig Williams, at Penn State, managed to interpret my drawings and employed his considerable computer illustration and GIS software skills to produce many of the graphics.

CONTENTS

PART II
The Computer as a Tool
for Storing and Acquiring Spatial Knowledge

CHAPTER 1

INTRODUCTION

The representation and use of geographic information influences almost every aspect of our personal, cultural, social, and economic lives. It is therefore not surprising that the study of representing geographic space has a long and varied history. By geographic space, I mean the space of the environment around us, the greater world in which we live. This is distinguished from other kinds of space on the basis of scale and its implications from the perspective of personal experience. For example, personal space (i.e., that in our immediate proximity) and tabletop space (i.e., that laid out before us on a surface) encompass things that we can take in with our senses at that moment in time. We can see, hear, smell, and often touch things in space on this scale and move them around. Astronomical space involves distances so vast and other forces, such that not only are we unable to directly experience most of the elements within it, but also the basic ways in which things interact are different from what we humans normally experience. Our world (i.e., our geographic environment) is also a very dynamic place. Things change *in* space *over* time, inseparably intertwining the representation of space and time as they relate to our experiential world.

As I just mentioned, geographic space is the space of our environment as we step outside our house or apartment. We can directly experience things in geographic space, but the scale is such that we cannot do so all at once. We may also be able to see an entire city at once from an airplane, but to "see" an entire country such as the United States requires a map. Standing on a street corner, we can easily see the front of a building before us, but to see what is behind it, we would need to walk around the building. To touch the building, we would need to walk up to it. Most objects in geographic space (i.e., buildings, trees, lakes, mountains, etc.) are of such size that we certainly cannot pick them up and move them in any normal sense.

The study of representing geographic space can be traced back to ancient philosophy and geography. These two fields also correspond to the two distinct perspectives from which the topic of geographic data representation has been addressed historically. The first perspective is cognitive and deals with how we humans store and use geographic knowledge on an everyday basis. The understanding of how we view geographic space is a means of understanding ourselves and our interactions with the experiential world. The second perspective deals with developing formalized and quantified models of space, and is the realm of scientific analysis. Although curiosity plays a large part in both of these perspectives, they are also driven by the practical need to maximize our ability to survive in our environment.

From a cognitive perspective, we use "mental maps" to drive to work, to decide where to live, and, in general, to understand our surroundings. The term "mental map" has been used over the past 30 years or so as a metaphor for our conceptual models of geographic space, that is, how we mentally store and use knowledge about our environment, although there has been discussion in the literature about accepting this metaphor literally. These "mental maps" represent extremely large and complex collections of information, yet their acquisition and use are so automatic that for the most part we do not consciously think about those processes as they are happening within our own minds. Certainly, when people are asked *how* the sum total of their geographic knowledge is arranged conceptually, they cannot give a real answer. We can only gain glimmerings of this arrangement by asking people to draw, for example, their mental map of the United States, their home town, or how they get to work. What quickly emerges is that the specific representation used for performing a given spatial task varies for different types of tasks, as do the types of information employed. Representations may also differ considerably between individuals, yet we manage to communicate spatial information on a routine basis.

From an analytical and modeling perspective, the manner in which geographic information is represented and used is a central issue of geography and, indeed, any field that studies phenomena on, over, or under the surface of the earth. A representational scheme is required for, and is in fact inextricably linked with, the process of spatial analysis and the modeling of geographic phenomena. For example, in routing problems, spatial information is typically represented as links between places denoted as points. In market area studies, a continuous surface is divided into Thiessen polygons around point locations. For other purposes, these same places may be represented as polygonal objects defined locationally by explicit boundaries. The selection of information to be represented and the representational scheme for storing the information is thus often driven by the analytical technique to be utilized. The inherent structure within the information, such as interrelationships among various elements, must also be preserved, because the specific intent of analysis is to derive information about this inherent structure. In other words, the results of any analysis can be greatly influenced by how the phenomena under study are viewed. Nevertheless, the suitability of any representational scheme for a use with a given analytical technique is often determined intuitively. More systematic methods, with more predictable outcomes, are required for dealing with large, multifaceted problems.

With the development of computing technology and the availability of large, shared databases in digital form, geographic information systems (GIS) and related geographic software (e.g., automated mapping, routing, and navigation systems) have become almost ubiquitous, with a wide variety of everyday and scientific uses. GIS have now also become essential for deriving information from the massive amounts of available geographic data. The term "geographic information system" is used to denote any integrated computer software package that is specifically designed for use with geographic data and performs a comprehensive range of data-handling tasks, including data input, storage, retrieval, and output, in addition to a variety of statistical and analytical processes. Although research on GIS began within geography, it is now part of a broader interdisciplinary field in its own right—GIScience—that includes aspects of computer science, psychology, cartography, surveying, statistics and mathematics, as well as geography.

The initial GIS of the late 1960s and early 1970s were developed as technological solutions to urgent needs for handling larger and more complex geographic data sets than could be handled via traditional manual methods.[1] The data representation scheme either translated the points, lines, and areas portrayed on the traditional (and static) map into digital form or was grid-based, in order to be compatible with standard computer output (initially the line printer) and input devices. These representation schemes became known as the vector and raster formats, respectively. Subsequent development of GIS continued to be centered on incremental, ad hoc improvements of previous approaches. For example, there was much debate in the 1970s and throughout most of the 1980s as to which of the traditional data structures, raster or vector, was the best approach for large, multipurpose systems (Chrisman 1974, 1987; Faust 1998; Peuquet 1979, 1988a). Techniques and concepts from other fields such as database management systems (DBMS) and image processing have also been directly applied in individual implementations (Abel 1988; Guptill and

Stonebraker 1992; Meier and Ilg 1986; Samet, Rosenfeld, Shaffer, and Webber 1984; van Oosterom and Vijlbrief 1991; Waugh and Healey 1987). These efforts and arguments for or against various options remained focused on storage/retrieval efficiency and other technical issues. The issue of how to represent space–time dynamics in GIS had been ignored until the late 1980s, in part because of a dominant static paradigm inherited from traditional cartographic representation and a lack of temporal data.

At present, a series of global-scale problems are being viewed with increasing urgency: The tropical rain forests are rapidly being cut down, yet the demands of rapidly growing populations in these regions are difficult to meet; the end of the Cold War and the resultant emergence of new democracies have made international relations and the world economy more unpredictable. Such problems involve complex natural and human systems that are also often intertwined. These complex systems are certainly very difficult to understand and predict. Because of the spatial data-handling and analytical potential of GIS, their use is critical as an enabling technology for addressing these problems. GIS have managed over time to become practical tools for dealing with spatial problems at all scales. Given commercially available satellite imagery with resolutions as fine as 1 meter and establishment of spatial data clearinghouses (National Research Council 1999; Sui and Goodchild 2001), with necessary data becoming increasingly available, GIS, mapping systems, and related types of software have now become almost ubiquitous. They serve everyday purposes such as route planning for automobile trips, checking the weather, and history and geography instruction in primary and secondary education. Mapping systems with global positioning systems (GPS) available in cars and trucks help tourists find their way and facilitate deliveries and emergency services.

The widespread availability of many different kinds of observational data at multiple scales and, in many cases, complete spatial coverage that documents change over time, have taken us significantly beyond the traditional, static paper map in the vicarious exploration of our world, and, indeed, of synthetic worlds. Computing technology allows users the capability to create their own dynamic maps instantly; to combine these with photographic imagery, video clips, and other types of graphics with text and sound; and to change them interactively at will.

Nevertheless, because progress has historically relied on a fragmented gathering of approaches inherited from cartography, imposed by hardware, or borrowed from other computer-related fields, we are faced with the current situation in which (1) increased functionality has characteristically been accompanied by increased conceptual complexity, making GIS progressively more nonintuitive for the user; (2) the initial promise of true analytical capability within an integrated data-handling environment has still not been fully realized almost 30 years after the first GIS became operational; and (3) the large, shared, multiscale, heterogeneous databases now being developed to address urgent social, economic and environmental problems are so much more vast and exhaustive than ever before that the computer must also be used to aid learning—in filtering, selecting, and interpreting the raw data— before any further analysis can be carried out.

The lack of a cohesive or integrative GIS theory has been recognized by the GIS community for more than 10 years (Anselin and Getis 1992; Goodchild 1990; Peuquet 1988a). Michael Goodchild, whose keynote address at the 1990 Spatial Data Handling Conference in Zurich brought widespread attention to the need for a science of geographic information as a separate intellectual field of study, coined the term "geographic information science." An expanded version of this paper, published subsequently in the *International Journal of Geographical Information Systems* (now called the *International Journal of Geographical Information Science*), posed a number of questions, that go beyond the immediate technology, but are central to GIS. It asked

> how to model time-dependent geographical data; how to capture, store and process three-dimensional geographical data; how to model data for geographical distributions draped over surfaces embedded in three dimensions; how to explore such data, for example, what exploratory metaphors are useful; and how to evaluate the geographical perspective on information and processes relative to more conventional perspectives? (Goodchild 1992, p. 41)

Since then, geographic information science (GIScience) has emerged as an interdisciplinary field that includes aspects of geography, computer science, mathematics, psychology, and philosophy, as well as other fields. GIScience has also progressed from what in hindsight can be viewed as an emphasis on solving immediate technical problems to focus on a more unified conceptual approach and broader, more abstract intellectual issues.

Certainly, this is attributable to a maturation of the field. Nevertheless, a central and unifying issue (perhaps *the* most central and unifying issue)— how data and information are represented and used—had its first glimmerings in the very early days of GIS, as evidenced in several papers presented at the 1977 Harvard Advanced Study Symposium on Topological Data Structures (Chrisman 1978; Peuquet 1978; Sinton 1978), and also a common theme in the Goodchild list mentioned earlier. One of the important emphases of this list, how to deal with multidimensional data, remains one of the primary technical challenges today, particularly that of multiple observations over time. A paper published at about the same time that GIScience, as such, first emerged (Peuquet 1988a) was an explicit effort to provide a high-level, conceptual, representational framework. This was later extended to incorporate representation in the temporal as well as spatial dimensions (Peuquet 1994). Helen Couclelis (1993) provided valuable insight by emphasizing the pervasive nature of spatial and space–time representation as a truly interdisciplinary issue, with applications in both the everyday world and science.

Aside from being the product of pure intellectual curiosity, continuing research in a theory of geographic representation has not forgotten the practical context of its application within GIS. Today, concerns about the user and how people cognitively represent geographic information, learn, and solve spatial problems are drawing much attention to the issue of representation. It is now realized, retrospectively, that the focus in GIS for 30 years has been driven by faster and more flexible ways to represent geographic data within the computer, and the human user has been forced to accommodate to the computer representation of the data and capabilities. As already mentioned, many of the representational methods were taken directly from cartography. Perhaps this was one reason why the capabilities and characteristics of the user were often considered to be secondary, in addition to overcoming undeniable technical problems to advance functionality. After all, historically, cartography was developed with the human user of the map in mind. The appropriateness of the techniques borrowed from cartography, or the way in which they were translated into a computing context from a cognitive perspective, was, however, never really questioned until recently.

Due in large part to the complexity and nonintuitiveness of current systems, we are now encountering problems beyond the historical focus on increasing flexibility and efficiency of geographic representations. In an everyday context, in-vehicle navigation systems have proven distracting to the driver, even with voice commands. More "intuitive" representations, both in terms of database and graphical displays, are needed, so that these systems can be quickly learned and understood. This requires that these computer-based representations more effectively take human cognition into account, including focus of attention.

In addition to "traditional" database manipulation, analysis, and visualization tasks, GIS (and GIS users!) also now need to filter through vast amounts of data to find patterns and associations. Although there are already some statistical techniques for doing this, such as Baysean classification, the degree of sophistication needed in this capability can only be fully realized by learning from the more heuristic, rule-based techniques used by people in finding patterns. In more general terms, GIS need to be more in-line with human cognition, so that the potential of computers can be better utilized by people as a tool to help in dealing with everyday spatial tasks and as an aid in solving spatial problems. The goal here is to combine the speed, perfect recall, and tireless computational power of modern computing technology with the capabilities of humans to recognize complex visual patterns, as well as intuition and trained judgment. Therefore, from both everyday and scientific perspectives of geographic representation, environmental cognition has become a major component of GIScience, particularly over the past 10 years (Mark, Freksa, Hirtle, Lloyd, and Tversky 1999).

The topic of environmental cognition, itself a component of the interdisciplinary field of cognitive science, has aspects that today span a broad range of disciplines, including philosophy, psychology, linguistics, anthropology, neuroscience, artificial intelligence, and geography. Nevertheless, it is precisely the complex and interdisciplinary nature of this topic that resulted in a fragmentary approach to its study. Environmental cognition does not fall into any single traditional cognitive field, but rather lies among (or between) all of them. Each of these fields has historically considered environmental cognition only from its own specialized aspect until relatively recently, with the advent of cognitive science as an interdisciplinary field over the past 30 years. As a result, the literature concerned with environmental cognition is still widely scattered and does not leave us with any generalized view or integrated theoretical framework from which to draw.

This book therefore examines concepts and approaches for representing space across a range of disciplines in order to bring thoughts together on a number of fundamental questions that relate to computer representation of geographic information: What is special about geographic space and geographic knowledge as opposed to other types of knowledge? How do people perceive their environment? What role do these perceptions play in learning? What are the commonalities and differences in the ways that people view the world and solve spatial problems? What can be applied from these insights in a computing context to help the usability of GIS and extend their capabilities as a tool for gaining geographic information? What should and should not be applied to provide an optimal mix of human and computing capabilities in a cooperative, man–machine environment?

The ultimate goal is to find commonalities and linkages that can be developed into a conceptually coherent framework, moving toward a general theory that encompasses currently fragmented ideas and principles. It is important to note, however, that there should be no expectation that a single, all-encompassing computer database representation can be derived, nor that all representations can be reduced to a few common, atomistic elements. Certainly, the issue of geographic representation, particularly considering both its cognitive and database/knowledge base model aspects, is far too complex (and contains too much context-specific variation) for the derivation of any single, all-encompassing, representation. Such an atomistic approach presumes not only a simplicity of structure but also a mechanistic behavior that cannot be justified.

The reader must also be cautioned that this does not constitute an exercise in simply finding invariances among multiple representations. To do so would assume that all representations are fundamentally alike, that Einstein's view of the world is no different than Newton's or that of a child finding the way home from school. "What is the nature of spatiotemporal representation?" therefore remains a hopelessly unconstrained question. Rather, we must look at a specific representation and ask, "What is it about the nature of this specific view of space–time that makes it valid and useful?" By asking this same question over an array of views from different contexts, we gain insight into their linkages.

It is my hope that the general representational insights thereby derived from what is acknowledged to be a very complex topic can find common anchor points and a general structure of relationships within the complexity, show the range of possibilities, and lead to the right questions, as the age of cyberspace matures as a realm of geographic representation and analysis.

NOTE

1. The first GIS, operational in 1968, was designed by Environment Canada in order to handle a national natural resources inventory.

PART I

THEORIES OF WORLD KNOWLEDGE REPRESENTATION

We see [the world] as being outside ourselves, although it is only a mental representation of what we experience inside ourselves. . . . Time and space thus lose that coarser meaning which is the only one everyday experience takes into account.

—RENÉ MAGRITTE (in Scutenaire 1948, p. 83)

CHAPTER 2

REPRESENTATION
VERSUS REALITY

WHAT ARE "SPACE" AND "TIME"?

Space and time are among the most fundamental of notions. They provide a basis for ordering all modes of thought and belief. In the history of human thought, space and time were so basic that they were regarded as the source of the world in ancient mythological, religious, and philosophical systems, including *Chaos* and *Kronos* in ancient Greek mythology, *akasa* and *kala* in Indian philosophy, and *Zurvan* in early Zoroastrianism (Akhundov 1986).

We are constantly reminded of the importance of space and time in modern everyday life when we use such expressions in ordinary language as "Everything has its *place*," or "To which one are your referring—this one *here* or the one over *there*?" The words *place, here,* and *there* are references to a conceptual framework of knowledge about the world. In short, things occur or exist in relation to space and time. They are also among the most fundamental elements in a number of sciences, particularly in physics and geography. In physics, space and time are basic unifying concepts, since most physical notions are introduced by means of operational rules employing spatial distances and/or temporal intervals. Space is certainly a basic unifying disciplinary element within geography, which has been called the science of *distance* and *space* (Bartels 1982). Among other terms also mentioned as basic in geography (place, region, and pattern in particular), space is an implied component of their meanings. Time is important in geography as the study of processes over space and time. Nevertheless, the nature of both space and time remains an open issue as the topic of many articles and books (the current one included) in a variety of fields.

The dictionary definition of the term "space" certainly reveals a significant

11

level of confusion, as in *Webster's New Twentieth Century Dictionary*, second edition,

1. distance extending without limit in all directions; that which is thought of as a boundless, continuous expanse extending in all directions or in three directions, within which all material things are contained
2. distance, interval, or area between or within things; extent; room; as "leave a wide *space* between rows"
3. (enough) area or room for some purpose . . .

which accords "space" a total of *12* separately enumerated definitions.

For the term "time," the situation is even worse. In the same dictionary, "time" is variously defined as

1. the period between two events or during which something exists, happens, or acts: measured or measurable interval
2. a period of history, characterized by a given social structure, set of customs, etc.; as, medieval *times*

   .
   .
   .

12. a point in duration; a moment; an instant
13. a precise instant, second, minute, hour, day, week, month, or year, determined by clock or calendar; as the *time* of the accident

    .
    .
    .

19. indefinite, unlimited duration in which things are considered as happening in the past, present, or future; every moment there has ever been or ever will be

among a total of *29* definitions! The notions of space and time are also very closely connected. To occur is to take *place*. In other words, to exist is to have being within both *space* and *time*. Spatial expressions are also frequently used in everyday language to express temporal notions: "He's *over the hill*," "We're getting *close* to Christmas." This entanglement of thing, space, and time adds to the difficulty of analyzing these concepts.

VIEWS OF SPACE AND TIME: IDEAS FROM EARLY MYTH TO MODERN SCIENCE

In order to bring these two fundamental concepts and their complexities into clearer focus, it is necessary to begin (as much as possible) at the beginning and trace the evolution of thought. Modern, everyday scientific and philosophical concepts of space and time can be clearly traced to primitive mythological and religious notions. Ancient Greek philosophy is rooted in these notions, yet it developed a distinctive world view. These same notions can also be discerned in the clashes between different philosophical and scientific currents throughout the following centuries. This applies to Descartes' extensional notion, and Kant's subjective, a priori concept of space and time, which is discussed later in this chapter. Certainly, the primitive and classical concepts underwent significant changes as they were incorporated into different systems of philosophy and natural science, and came to function in different sociocultural environments. Nevertheless, the changes in the concepts of space and time were far from a complete evolutionary metamorphosis. The constancy of a few basic, parallel threads throughout this long historical evolution of thought is striking.

The earliest notions in early Western mythology on the nature of space and time can be described in general terms as a progression of the world from Chaos to Cosmos. Chaos, the initial state of the mythological universe, is the boundless abyss, infinite space. In Hesiod's *Theogony* (Hesiod 1999), his fifth-century B.C. rendition of an ancient Greek creation myth, time is considered so important that it is personified, along with Earth as a god. Gaia (Earth) appears and eventually gives birth to Kronos (Father Time). Order is gradually imposed through a multigenerational succession of battles between good and evil deities. Cosmos is the final state of order. Like the developing Greek state, this Cosmos consists of both political and natural components (Aveni 1989). Although the final state is one of order, there is no notion of the world as a unified whole, with an overall order. Rather, there is relative order within a multiplicity of unconnected pieces, or territories, and discrete events. In other words, space and time are discontinuous, although the story does have an overall forward-moving evolution. The idea of a connected world develops in later mythologies, as exemplified in the works of Homer.

The frequent occurrence of the cyclical view of time seems to be a reflection of the close association to nature and its rhythms in the everyday life of early cultures. Hesiod's *Works and Days* (Hesiod 1999) shows that for the early Greeks, time on an everyday scale consisted of the ordered rhythm of human activities within the seasons of the year and its corresponding repeated cycle of sensible events: bird migrations, planting time, and so on (Aveni 1989). The notion of cyclic time also appears repeatedly in the mythology and religion of other cultures. The Mayans believed that history would repeat itself every 260 years. This is a period of time they called the *lamat*, which was the fundamental element of their calendar (Coveney and Highfield 1990).

Before the rise of Christianity, the cyclical view of time dominated Western thought. Only the Hebrews and the Zoroastrians, and the occasional writer such as Seneca, had previously thought of time as progressive and nonrepeating. For the Hebrews, a linear view of time was a natural adjunct to the emphasis on the eventual deliverance and salvation of Israel. For the Christians, it was the emphasis on the final salvation of the world. Notions of cyclical and linear time continue to coexist. Time is measured by a repeating sequence of days and months that derive from ancient Babylonian and Hellenistic astrological beliefs.

In the work of Homer, an ordering of events can be seen, with the sense of time being continuous and open-ended, proceeding out of the past, through the present, and into an open-ended future. Space is also seen as continuous and connected. With the emergence of these notions of time and space, it is possible, for example, to retrace the voyage described in Homer's *Odyssey*. Both the idea of the Cosmos and the evolution of the concept correspond to the gradual development and expansion of sociopolitical organization into unified structures (*polis*, state, empire).

Greek thinkers became increasingly interested in mathematical and physical problems. One such question concerned the divisibility of matter and (continuous) space. Anaxagoras introduced the concept of infinite divisibility into natural philosophy: "For of the small there is no smallest, but always a smaller. (For that which is cannot be cut away to nothing.)" (Sider 1981, p. 54). This thesis served as the basis of early Greek continuous mathematics and the foundation of the scientific doctrine of continuous space and time. A differing vein of thought that developed from this, atomism, reduced everything to infinitely separable (and separate) particles—bodies adrift in space, with space itself being the container of these objects—the Void. Atomism has earlier roots in Pythagoreanism. The Pythagoreans thought of Cosmos as a harmonious unity of such basic opposites as the limited and the unlimited, representing the origin of the notion that space and time have two aspects: On the one hand, as the Void and infinite, they are the receptacle of objects, like a boundless box; on the other hand, they are the order of these objects and processes (Akhundov 1986).

Related to the idea of minimal units of space and time, the atomists also held that the smallest, physically indivisible, units of matter combine with empty space to form the total of all physical objects. These units are made up of what atomists termed *amera*, which are truly indivisible, have no parts, never exist in a free state, and serve as the criterion of mathematical indivisibility.

In examining the overall view of the atomists, there are two spaces: continuous physical space as receptacle (the Void), and mathematically discrete space in which *amera* serve as the standard for measuring matter within space. In ancient Indian doctrines, among others, this same notion occurs with *akasa* (boundless space) on one hand, and *dish*, the space of geometric figures in which direction and position are defined on the other, the latter being a space that allows metric relations (Akhundov 1986). In accordance with the atomism of space, Democritus assumed the atomistic nature of time and motion. Epicurus later developed these ideas into an integrated system, treating the properties of mechanical motion in the context of discrete space and time. The notion of discrete space and time dominated until Galileo and Newton advanced their theories.

Plato's concepts of time derive from the atomists, yet for the first time a distinction was made between reality and human understanding of it. His work *Timeus* begins with the distinction between Being and Becoming, in which Being (the here and now) is a fundamental world "apprehended by intelligence and reason is always in the same state," while the world of Becoming (the realm of time) "is conceived by opinion with the help of sensation and without reason, [and] is always in a process of becoming and perishing and never really is" (1949 §28, p. 12). The Idea (the world of ideas) is identical to Being. This world possesses time all at once as a durationless "now." This world always "is," in its wholeness.

Plato does not deny the reality of past or present time but merely says that it is inappropriate to divide time into past, present, and future with respect to the world of Ideas. To understand the distinction between Being and Becoming, he used the analogy of a journey (Becoming) and its destination (Being), claiming that only the latter is real. It also follows from this distinction that the physical world has only a secondary reality, and that observations of the real world and empirical experimentation are therefore irrelevant in the search for knowledge (Coveney and Highfield 1990). Plato regarded time as the moving image of eternity. He also indicated the numerical nature of time, which is measured through the revolution of the heavens; that is, time is ordered by the laws of the Cosmos (Akhundov 1986).

Plato's beliefs dominated the Golden Age of Greek thought, around 400 B.C., and this scheme was later extended in Newton's system. But whereas Plato regarded time as the moving image of eternity, Newton viewed relative time as the moving empirical image of absolute time. This absolute time, under conditions of instantaneous action at a distance, tends to acquire the character of static time, or eternity.

Aristotle, a pupil of Plato, valued observation more than his teacher and advocated a close interplay between observation and belief. Aristotle's spatiotemporal notions are relational and as such are contrasted with the Void of Democritus. Aristotle's space is a system of relations between material objects, which means that location in space is a *property* of material objects, in contrast to Democritus' Void as the *receptacle* of objects.

Aristotle, who rejected the notion of atoms (because nature does not take leaps) and the Void (because nature abhors a vacuum), instead developed his famous conception of space as *topos*, that is, space as place:

> Nevertheless, it clearly seems to be a fact that place "is." First, there is displacement. Where now there is water, there will be air when the water has gone (as out of a vessel); and then again some other body will occupy the same place. The place, therefore, seems to be different from all the bodies which successively displace one another. That "in" which the air is now, is that "in" which the water was before. Consequently, the place was clearly something; that is, the location was different from the bodies which, by passing into and out of it, changed places. (1961, IV, §208b, p. 58)

Place, which exists together with objects, and all objects are located in a place, thereby becomes a necessary *condition* for the existence of any object. This conception of space agrees with the atomists' notion of the Void as a receptacle, except for Aristotle; *topos* is filled not with discontinuous atomic matter but with continuous matter (i.e., *topos* is voidless). Aristotle's space is also finite and limited; not a boundless box. He viewed time as also being continuous. To Aristotle, the moment, or "now," was not an element of discontinuity but, more appropriately, was described as a linking element that continualizes temporal duration. He also subscribed to the Greek distinction in his day of successive, measurable time (*chronos*) and that of episodes, including seasons: time with value to people relative to goals and the cycles of living (*caraways*). Time, like space, is equally present everywhere. Unlike space, however, time is infinite.

Aristotle's cosmological model consists of a finite space with the earth at its center and was divided into two levels—the earthly and the celestial. These two spheres contain completely different kinds of entities with completely different motions that, in total, organically combined static, dynamic, and cyclic structures of time and the substantial and relational levels in the structure of space. Aristotle tried to resolve absolute versus relative space by denying space an empirical reality and instead giving it a metaphysical reality as pure form, thereby conceptually distinct from matter.

The Aristotelian view of space and time can be summarized as follows: (1) Space is finite; (2) time is infinite; (3) empty space does not exist; (4) space is divided into two levels—the earthly and the celestial—that obey different laws, have different structures, and do not overlap (Akhundov 1986). Some of Aristotle's notions were criticized or refined (perhaps most notably by Ptolemy) in subsequent centuries, but they remained substantially unchanged and unquestioned for almost 2,000 years, until the time of Newton.

Although his basic assumptions to explain cosmic motion were later proved incorrect, Aristotle provided the necessary logical structure for further scientific inquiry and an organically derived model of time and space that was workable, at least for the cultural context of that day. The metric properties of space were explored in detail by the early Greek geometers to describe and measure the earth. These *static* descriptions became formalized in the axioms and theorems of Euclidean geometry. Theologians in the Middle Ages were encouraged by this power of calculation, but the most attractive feature of the system was its geocentricity, which seemed to offer "scientific" confirmation of Christian anthrocentrism, and the Aristotelian model of the universe was decreed as fact by Church law.

The Renaissance was a time of major socioeconomic change that inspired, and was inspired by, a revival of intellectual curiosity and the spirit of discovery. Unsatisfied with established notions of the world and contemporary interpretations of ancient philosophy, Renaissance thinkers turned directly to the writings of the ancient Greeks. Discovery of the *dynamic* properties of space in time evolved, culminating in a systematic mathematical formalization with the work of Newton. Geographic and cosmographic dogmas collapsed, although not without significant social trauma—the Inquisition being one reaction.

Copernicus developed a purely physical view of how the universe works. Using the notions of the ancient Greeks as a foundation, he constructed a uniform theory of space and time wherein both the earth and the heavens operate according to the same laws. However, he also maintained the Aristotelian view that space is finite (Akhundov 1986).

Renaissance thinkers changed the notions of space and time in two fundamental ways. At the beginning of the Renaissance, the ancient distinction between "earth" and "the heavens" was the official Christian Church doctrine of the day. The Copernican revolution eliminated this distinction and thereby signified two significant advances. First, the notion of a continuous and unending time was put on a firm and measurable scientific basis. Second, the notion that space and time could be studied together as related aspects of a unified Cosmos became generally accepted.

Galileo recognized empty space, but Descartes did not. In his view, the chief attribute of any object is that it occupies space. Moreover, the only things we can be assured of knowing about material things are their measurable, geometric properties. This close association of space with the existence of objects forced him to reject empty space, because he would otherwise have had to admit the existence of nonmaterial objects. Descartes stated:

> That a vacuum in the philosophical sense of the term (that is, a space in which there is absolutely no substance) cannot exist is evident from the fact that the extension of space, or of internal place, does not differ from the extension of body. From the sole fact that a body is extended in length, breadth and depth we rightly conclude that it is a substance; because it is entirely contradictory for that which is nothing to possess extension. And the same must also be concluded about space which is said to be empty; that, since it certainly has extension, there must necessarily also be substance in it. (1983, pp. 46–47)

Descartes, as well as Locke, explicitly advocated an epistemological dualism with the following basic tenets:

1. The world is composed of at least two sets of entities: (external) material things and ideas (i.e., perceptions).
2. These ideas alone are known in consciousness as objects, whereas material things are only inferred.
3. The inferred material things are always nonidentical to the perceived objects.

For Descartes, the *idea* (*l'idée représentative*) is that which is known, and there is no intrinsic causal relation between material things and the generation of specific ideas. He asserted that the only knowledge we have of material things is that of their substance, duration, order, quantity, and "perhaps" some other general notions, corresponding to Locke's and Galileo's notion of primary qualities. They can be defined as the basic qualities any external object must have in order to be material. To Galileo, these are the properties that are sufficient and necessary for physics, in accordance with his theory that the material world is sufficiently characterized by these "extensional" qualities, such as color, odor, texture, and so on. From Descartes onward, ideas were considered objects of knowledge.

As seen from antiquity forward, through the Middle Ages and the Renaissance, there existed a concept of two times: one more absolute and linear, the other more relative and cyclical. Descartes espoused the notion of two times, but these were seen through the lens of theology (as were his ideas on space and matter). Descartes attributed absolute time to the external world and external objects. As such, this time is measurable, but is also eternal and attributable to God. Relative time, according to Descartes, is that time directly experienced by man as a mode of thought. In the *Meditations on First Philosophy* (Descartes 1996), Descartes also advanced the notion that time is discrete, explaining this as a demonstration of God's omnipotence. God directly intervenes with each successive discrete instant and such interference is the cause of both the continuing existence of each individual and the cause for the entire diversity of natural objects (Akhundov 1986).

Newton later developed a natural–philosophical revision of the Cartesian concepts of space, time, motion, and matter. In his *Principia*, he presents a theory of motion of material objects based on his notion of absolute space and time:

> Absolute, true, and mathematical time, of itself and from its own nature always flows equably without relation to anything external and by another name is called duration. . . . Absolute space, in its own nature, without relation to anything external, remains always similar and immovable. (1962, p. 6)

This view has been described as somewhat like the Cheshire cat's grin: that which remains after all substance disappears. It also provided a view of absolute space and time that is more purely a physical description, relying on observation and measurement.

In Newton's view, absolute space and time are the backdrop upon which the dynamics of physical objects can be measured, not as measurable properties intrinsic to physical objects themselves. This stands in sharp contrast to Descartes' rejection of empty space. Space and time were maintained as discrete notions, as they had been previously through history. Because the path of a moving object is *through* space *in* time, his theory connects space, time, and objects in a system of physical laws—Newton's laws of motion. Any event could thereby be regarded as having a distinct and definite position in space and occurring at a particular moment in time. Spatial and temporal distances between events are well-defined. Time thus becomes a type of abstract, universal order that exists by and in itself, regardless of what happens *in* time. Within Newton's space–time framework, the movement of a body changes the position of that body, but it changes neither the framework itself nor the relationship of other objects to that framework.

With this view of absolute and discrete time and space, Newton also had to postulate absolute and universal simultaneity; a "clockwork universe." Every place (and everything that may happen to be contained within space) is connected by the same moment of "now." Leibniz postulated universal time, yet viewed both time and space as relative rather than absolute. He held that space and time do not exist either in their own right as substances or as being identical with matter. Rather, they are properties of material objects, in that space is the order of coexistence and time is the order of successive phenomena. Nevertheless, Newton's view of absolute space and absolute time dominated science until the beginning of the 20th century, when a relative and continuous view was adopted by Einstein as a central theme of his work. Perhaps the primary reason for the dominance of the Newtonian perspective of space and time for approximately 200 years is the revolutionary influence that Newton's *Principia* had on scientific methodology, whereby new standards for rigor of thought based on inductions from experimental evidence within a mathematical framework were introduced virtually overnight. Natural philosophy, which previously had been open to controversy and speculation, was established on a solid foundation and elevated to natural science in the modern sense of the term (Kroes 1988). Newton's methodology and theories were adopted as a single package.

The relative view of space–time continues to dominate modern physics as well as 20th-century science in general. Einstein based his work on Minkowski's revolutionary view of a combined space–time. Minkowski's work in mathematics at the end of the 19th century was characterized by a deliberate application of "geometric intuition" to fields of mathematics beyond geometry and particularly to number theory, in which his major work was *The Geometry of Numbers* [Geometrie der Zahlen] (Minkowski 1953). In his subsequent work in physics, Minkowski applied his visual–geometric approach in pure mathematics to the development of his physics of space–time, wherein time is viewed as an additional dimension or axis in a four-dimensional geometry— that is, x, y, z, and t, with t representing time in a hypercube multivariate coordinate space. He eventually took this even further and ascribed a physical reality to the geometry of space–time. Thus, for Minkowski it was not that physical laws can be equivalently expressed through a mathematical construct, but rather that, in a certain sense, the world *is* a four-dimensional, non-Euclidean manifold (Galison 1985).

The work of Newton marked the shift in a long historical debate from what has been termed the philosophy of space to the science of space. It is also primarily due to the work of Newton that science itself has been based on a mathematical paradigm ever since. This paradigm has become seemingly paradoxical, with many types of "spaces"—curved spaces, n-dimensional spaces, topological spaces, set spaces, and so on—all developed in various contexts with their own self-contained, formalized language designed to describe such spaces.

The field of mathematics has subsequently been accused of divorcing itself from physical reality in developing these self-contained and completely abstract spaces. Lefebvre expressed the problem as follows:

> The proliferation of mathematical theories (topologies) has aggravated the old "problem of knowledge": how were transitions to be made from mathematical spaces (i.e., from the mental capacities of the human species, from logic) to nature in the first place, to practice in the second, and thence to the theory of social life—which also presumably must unfold in space? (1992, p. 3)

The claim is not that mathematics has lost the capability to describe physical reality, but rather that it has lost the focus on doing so. Contained within this is a tension between the Newtonian/positivistic view that it is through the process of measurement and objective *description* that space can be understood, and the Cartesian, cognitively centered view that real knowledge lies in the abstraction of observation—in the *idea*. In this latter view, the representation, and thereby the manipulation of purely abstract spaces, becomes a powerful tool for understanding.

SPACE AND TIME AS A CONTEXT FOR UNDERSTANDING

The Kantian notion of space and time as a backdrop is partially based on that of Newton, yet its perspective is very different. Kant espoused the Cartesian dual nature of reality, with both external (physical reality) and internal (cognitive) components, but he went far beyond that. Kant's space and time are concepts that we possess at birth. This is a response to John Locke's earlier notion that the mind of a neonate is a *tabula rasa*, that we start with a "blank slate" and *all* knowledge is based upon experience. Kant argued that in order for what we know (our cognitive understanding) to have any correspondence to external reality, there must be some concepts that are innate and intuitive. This has become known as his principle of "subjective a priori." In order to perceive objects relative to their location in the external world, there must be some preexisting notion of space that is innate and intuitive. Similarly, a preexisting notion of time is needed in order to perceive changes in location (motion) or in objects: "At the basis of their empirical intuition lies a pure intuition (of space and of time) which is *a priori*" (Kant 1955, §11, p. 36). Time and space become the *forms* of sensory experience; the basis on which the human mind inevitably arranges all knowledge.

In his *Critique of Pure Reason* Kant (1950) points out that it follows from their a priori nature that space and time are

1. *Universal*—the world of phenomena are represented against a single spatial and temporal background. In this conception, geometry serves as the system of synthetic a priori knowledge.
2. *Insuppressible*—we can assume, as a mental exercise, the nonexistence of any or all objects, but not of space or time. Similarly, an individual with no experience (a newborn) would still have an innate sense of time and space.
3. *Necessary*—they are required for all sensory perception and universal application.
4. *Unique*—there is but one space and one time; all so-called spaces are only parts of that one space, and different times are periods of that one time.
5. *Infinite*—because time and space are not, and cannot be, objects in themselves, they have no boundaries or limits.

To summarize, Kant said that the only reality we can know is that which is filtered by our current and previous perceptions cast in the context of an innate sense of space and time. From this, we are also led to the conclusion that what we take to be real cannot be separated from the act of knowing, or in Kant's terms, our known (cognitive) world is a *construction* of thought. He thus continued Descartes' separation of knowing from reality and cast the capability of knowing into an explicitly spatiotemporal context.

The neo-Kantians of the late 19th and early 20th centuries criticized Kant's ideas for a lack of historical perspective and cultural context, particularly as they pertained to scientific knowledge. The fundamental difference in neo-Kantian thought is that the personal and scientific knowledge that Kant assumed to be a completely individual achievement, beginning with innate concepts, was instead acknowledged and extended as a cumulative achievement of mankind (Pos 1958).

According to the Marburg school of neo-Kantianism, the unity of experience is not a given but is rather a task. This task does not establish concepts as absolutes but only provides an orderly process for learning. On an individual level, this provides a framework for cognition. On a scientific level, it provides a guiding principle for inquiry. Within this line of thinking, space and time become the basic modes of order, the modalities within which experiences and knowledge can be referenced and integrated.

The philosopher Ernst Cassirer (1874–1945), of the Marburg school, utilized this interpretation and saw knowledge as the sum of all previous experience. He was one of the first to view the human experience of space and time from an explicitly developmental perspective. Moreover, he gave wide acceptance to the notion that the acquisition of knowledge is a truly dynamic activity of the mind, and that knowledge itself is dynamic. This means that knowledge is not acquired in a strictly cumulative fashion; that from time to time, our basic assumptions must be revised in order to bring new experience into noncontradictory alignment with previous experience.

The theories of Copernicus and Newton, and more recently, Einstein's theory of relativity, are demonstrations of such periodic reexamination. These shifts in our understanding demonstrate how scientific theory develops within a historical and cultural context. The history of how space is viewed is certainly not one of strictly linear accumulation, nor is it one of a succession of sudden and history-shattering new ideas. Instead, it is a history of steady evolution punctuated sporadically with new insight. Through the eons, we can see the development from a strictly local and sense-dependent space to notions of a completely constructed and abstract space. In this respect, the histories of science and philosophy are two aspects of one and the same intellectual process: It is the historical process by which distinctive cultures have evolved (Hartman 1949).

The way any individual sees the world is filtered not just by the sum total of past and current experience. Those experiences themselves incorporate a specific cultural context. This means that not only do all individuals possess their own "world view" but also every area of an individual's knowledge varies in mode of experience and level of knowledge. Nevertheless, the sum total of an individual's knowledge is an interrelated whole. Similarly, each scientific discipline constitutes a unique and distinct perspective, yet all are interrelated. Therefore, only the *process* of knowing is universal. This process is also continuous and open-ended.

The American scientist, philosopher, and logician C. S. Peirce (1839–1914) developed a theory of signs, or semiotics,[1] that anticipated Cassirer's view of human experience. Peirce asserted that all thought and experience is through the use of signs. A sign "is something which stands to somebody for something in some respect or capacity" (in Hartshorne and Weiss, 1931–1958, 2.28).[2] In other words, objects that we perceive in our environment evoke *meaning* based upon our previous experiences. This, as Morris (1938) has pointed out, implies a dual nature for objects in the environment. In a public sense, any object is itself—the reality—and is accessible to one person as well as another through sensory perception. In a private sense, it is an element of individual consciousness and is individually related to a concept.

The work of Kant, and others since, represents a dramatic break with previous understanding of how spatial and temporal notions are formed. The individual is no longer seen as a passive observer of an already existing reality, but instead determines the "shape" of space and time for him- or herself. This is one of the most important developments in modern thought (Wallace 1974). Nevertheless, Werner and Piaget first advanced notions of how people develop an understanding of reality beyond the realm of philosophical speculation, and placed theories of environmental knowledge within a firm scientific framework.

Werner was interested in understanding the process of learning in general, whether in children, adults, or animals, but like Piaget, his experimental work was focused on the initial learning of children. He argued that the development of environmental knowledge in children proceeds from an initial state of a lack of differentiation between the infant and its environment to increased differentiation and articulation (Werner 1948, 1957).

Piaget began with the classic epistemological problem of the nature of reality and whether our perception of that reality could be assumed to be accurate (Piaget and Inhelder 1956). Not only did Piaget maintain and clarify the Cartesian distinction of external and internal space, but he also detailed the causal link between the two in his theory. The Piagetian view holds that knowledge is constructed through action upon, and interaction with, people and things in the external world.

The experimentally based theories of Werner and Piaget validated the Kantian contention that what constitutes the environment for a particular individual is an intellectual construction by that individual, but like Cassirer, they attributed a more central role to sensory input. Piaget's theory, in particular, differs from previous thinking in two very fundamental respects: First, Piagetian theory holds that the acquisition of knowledge is a process of *interaction* with the world and *adaptation* to it; second, space and time as conceptual categories are not innate but must be constructed as part of that interaction and adaptation process. To Piaget, the only innate processes are those that are biologically motivated. These are the sensory–motor abilities that enable learning at the earliest stages.

In contrast to the more expansive perspective of cognitive psychology, the "ecological psychology" of J. J. Gibson (1966, 1979) was concerned more specifically with the *perception* of sensory input, not cognition. Nevertheless, Gibson's views meshed with Piaget's approach concerning the importance of interaction with the environment and how this is (perhaps more strongly) biologically driven. Human beings interact constantly with, and are an inseparable part of, their environment. According to Gibson, we perceive the outside world and the elements in it relative to our capability to interact with those elements. Thus, trees, for example, are *climb-up-able*. Gibson speaks of such interactional opportunities provided by the environment as *affordances*. He furthermore proposed that the key to understanding how humans perceive the world lies in identifying the invariant or essential properties of the world as experienced and defined affordances as being these invariants: "The observer may or may not perceive or attend to the affordance, according to his needs, but the affordance, being invariant, is always there to be perceived" (J. J. Gibson 1979, p. 137).

This idea also led Marr (1982) to what he called the "primal sketch" of the image, and he based the first unified representational framework of the seen world on this key concept. The overall mechanism proposed by Marr is that descriptive primitives are built up from the most detailed level through successive groupings, producing hierarchies of entities and spatial patterns. Although Marr's theory was concerned specifically with how humans process visual input, the overall process has application to learning in general, as we see in later chapters.

Variations in how geographical space is perceived have been studied empirically by behavioral geographers in experiments dealing with spatial cognition and spatial choice. As summarized in the volumes by Golledge and Rushton (1976), and Downs and Stea (1973), spatial behavior has been demonstrated to be a function of perceptual views of individuals in which inexactness and variability prevail. Views of the world vary among individuals and depend on the particular task at hand. Thus, a geomorphologist's view of a mountain would be different from that of a climatologist or a botanist, yet all three would recognize the same entity as a mountain. Space is also viewed differently by the same individual when, say, carefully measuring a garden and planning the spatial allocation of flowers as opposed to enjoying its beauty later in the summer.

Indeed, the world is full of "spaces"—physical, mathematical, geographic, cartographic, social, economic, and today, even cyberspace. The world is also full of "times"—geologic, astrological, seasonal, and so on. Different views of space arise in everyday life and in science because of different levels of abstraction from different viewpoints and modes of thought, depending on the situation. The various ways that space, time, and their properties may appear to individuals are due to differences in attention to detail, the cultural environment of the individual, access to technology, the amount of formal education, and simply the amount of total experience. Differences may also be altered by physical attributes, such as being very short (as in the young) or blind.

Variability in perceived world views and the resulting inexactness in language have long been of central concern in linguistics. Without some commonality among the views of individuals, it would be impossible for people to communicate with one another. How can language be very subjective yet still be intelligible? How does language relate to people's individual world views? These questions have been investigated within an explicitly spatiotemporal context by only a few researchers within the linguistics community, however. Talmy (1983) has cited a number of reasons for this, all of which relate to the unique difficulties of investigating spatiotemporal concepts within language:

1. *Complexity*—the multidimensionality of space–time is a much more complex problem domain.
2. *Spatial terms often perform double duty*—most spatial expressions can also be used in a temporal context (*in* the house, *in* a minute).
3. *Context*—the meaning of spatial and temporal expressions is dependent upon:

   a. *Scale*—on a global scale, State College, Pennsylvania, is *near* Pittsburgh. However, for someone in State College facing the 3-hour drive to Pittsburgh, State College would definitely not be viewed as being near Pittsburgh.
   b. *Frame of reference*—my *left* is always on my left side, no matter which way I turn, but my left is on the right of someone facing me. I may also be facing (i.e., my frontward orientation) toward the east or the north, and so on, depending upon how I turn relative to this absolute, objective referencing system.

All of these complexities, of course, deal with cognitive interpretations of subjective spatial and temporal relationships. Nevertheless, Clark (1973), Talmy (1983), Herskovits (1986), and others have shown that there is indeed a high degree of commonality in spatial and temporal concepts at a fundamental level that transcends individual experience and, indeed, variation in culture and specific languages. For example, all cultures have some notion of relative direction. In Western cultures, geographic direction is often expressed in terms of east, west, north, and south. Ancient Hawaiians expressed direction as mountainward or seaward. In some Native American cultures, it was expressed in terms of upstream or downstream.

Clark (1973) has shown that, at least in English and related languages, the description of time is based on a spatial metaphor that has very specific properties. In general terms, this "temporal space" of everyday speech is linear, ordered, and directional. The time axis, then, like x, y, or z, axes, can be given an arbitrary zero point marking some reference base and is asymmetrical about this zero point, as in the sense of past, present, and future. Thus, time is often described using those spatial terms that do not imply a multidimensional context, such as long–short and near–far: "It has been a long day"; "Spring is still far away." Given the asymmetrical, or directional, view of time, temporal relations are also described in terms of *front–back* relations, which happen to be both linear and directional in their implied orientation: "The children are looking forward to Christmas"; "Let's put that behind us"; "That happened back in the last century." This last pair of expressions also exemplifies what appears to be two movement metaphors within the directionality of time: one in which time (and events contained within it) are moving past the reference base, and another in which the reference base is moving forward (futureward) past stationary events.

Cartography also conveys different views of space. The use of cartographic projections, scales, and colorations, among other devices, to convey a specific view of space within a standardized yet flexible representational framework is a highly developed art. How do maps relate to people's individual world views? Certainly, there must be a correspondence between the map compiler and the map reader, in order to convey an intended message, similar to language. This is also a complex issue, in part because of the dual nature of maps. On the one hand, the map is a formal, spatial *structure*, being composed of an assemblage of points, lines, curves, surfaces, and volumes that can be completely and unambiguously represented in an appropriate coordinate system (Freeman 1973). The components of such a spatial structure and their interrelationships must conform to a specific set of uniform, mathematically defined rules in order to convey the intended message correctly. On the other hand, the map is also a graphic *image*—a natural, visual object characterized by the variation of pattern, lightness and darkness, and possibly color. It may or may not convey meaning, as in the case of an abstract painting.

The study of maps as *objective*, spatial structures has employed a variety of mathematical subfields, including topology, graph theory, lattice theory, and analytic and projective geometry. Much has already been accomplished in the development of unified representational schemes in specific areas. Primary examples include work by Tobler (1961) and Snyder (1984) on map projections, and White (1979, 1984) and Corbett (1979) on the topology of cartographic lines.

Considering maps as *images* has resulted in considerable application of principles from cognitive and perceptual psychology. The map as a visual image depends on *subjective* human interpretation to convey meaning. Besides the linework, additional aspects of a map (e.g., symbology, shading, coloring, and overall arrangement of elements) also convey information for the formulation of an overall visual impression. As noted by Blades and Spencer (1986), the psychophysical paradigm of cartographic communication has dominated research in cognitive cartography in the past (e.g., the eye–brain interconnection). More recently, the importance of the higher level cognitive processes in the interpretation of map information by the map reader has received significant attention.

It is thus evident that researchers from a variety of fields investigating cognitive questions about how individuals communicate despite the variability of views and the inherent inexactness of the communication process, have gathered substantial evidence that the *processes* in the construction of perceptual world views are the same for everyone. More specifically, this overall viewpoint holds that processes used to organize information are innate and either largely independent of environmental input or dependent on kinds of environmental input that no human being can avoid encountering. This input also implies that a common, fundamental cognitive *structure* of the information contained in the world view results from these common processes. This structure, at a minimum, represents a unifying idealization of the mappings between reality, visual images, and other sensory input, language, graphics, and other forms of communications, even though the content of that structure varies considerably from individual to individual.

COMMON THREADS

Through this brief sketch of the history of philosophical and scientific thought, it is strikingly apparent that a number of fundamental ideas have persisted and have indeed been refined, from antiquity to the present, regarding the nature of space and time and their representation. At the most fundamental level, historical views of time and space can be categorized into what can be termed both *continuous* or *discrete,* and *absolute* or *relative.*

The continuous view focuses on space and time as the subject matter. All objects are contained within space and time. The discrete view, in contrast, focuses on objects as the subject matter. The absolute view assumes an immutable structure that is rigid and purely geometric. The relative view, in contrast is subjective; assuming a flexible structure that is more topological in nature. Relative space and relative time are defined in terms of relationships between and among locations or in terms of relationships between and among objects. On the one hand, the absolute view involves measurement referenced to some constant base, implying nonjudgmental observation. The relative view, on the other hand, involves explicit interpretation of process and the flux of changing pattern and process within specific phenomenological contexts. The relative view, by virtue of being defined via intervals between objects or locations, is also bounded. In contrast, the absolute view is unbounded.

Referring back to the dictionary definitions at the beginning of this chapter, the first definition of space (distance extending without limit in all directions) and the 19th definition of time (unlimited duration extending into the past, present, and future) can clearly be seen as referring to continuous space and continuous time, respectively.[3] Similarly, the second definition of space and the first definition of time clearly refer to space and time that are both discrete and relative.

Discrete space was described by the Greek atomists, who reduced everything to distinct bodies adrift in space, with space itself being the container of these objects: the Void. From this view arose the notion of space and time espoused by Newton, with space composed of points, time composed of instants, and both existing independently of the bodies that occupy this space–time. Within the Newtonian perspective, the movement of a body changes the position of that body, but space and time are viewed as a backdrop with a rigid, unchanging structure that is absolute. The Newtonian view can thus be interpreted as *both* discrete and absolute: objects existing within constant space and constant time.

Leibniz, in contrast, subscribed to the notion of relative space and time, as did Kant, who viewed space and time as a priori functions of intuition, thereby making them contextual. Nevertheless, Newton's absolute view predominated in science until the beginning of the 20th century and the work of Minkowski and Einstein. The relative view of a combined space–time continues to dominate modern physics as well as 20th-century science in general. This view is also the primary perspective of cognitive psychology, linguistics, and related fields.

In everyday terms, there is a common modern presumption of a combined space–time matrix derived from the Minkowskian view used in modern science. But this is more valid from a relative rather than an absolute view. While absolute space and time are highly interdependent and share many characteristics, they are not interchangeable in the sense of a four-dimensional, mathematically defined space–time hypercube.

There are a number of important differences between absolute space and absolute time. In absolute time, everything everywhere progresses inexorably forward in time. Nothing can travel backward in time save, of course, in the sense of a historical retrospective. We accordingly experience absolute time as unidirectional (July 24, 2000, will never happen again). In space, by contrast, we can travel backward as well as forward. Given the forward flow of absolute time, it follows that processes—spatial as well as nonspatial—are evolutionary (continuous and progressive) in nature. For example, everyone continually grows older; the map of political states continually changes. Both space and time are continuous, yet for purposes of objective measurement, they are conventionally broken into discrete units of uniform or variable length. Time is divided into units that are necessarily different than those for space (we cannot measure time in feet or meters). For example, temporal units can be seconds, minutes, days, or season.

Within the relative view, however, we can interpret patterns of occurrences through time. The four main mathematical characterizations of temporal pattern that have been developed are reminiscent of Aristotle's temporal categories: steady-state, oscillating (cycles and rhythms), chaotic, and random. The term "chaos," in its modern meaning, has been defined by Roderick V. Jensen as "the irregular, unpredictable behavior of deterministic, non-linear dynamical systems" (in Gleick 1987, p. 306). Chaotic behavior characteristically amplifies small uncertainties through time, allowing only relatively short-term predictability within an overall random pattern of occurrences. We can also use the term "chaos" to describe spatial distributions that are not completely irregular and unpredictable. The steady-state, oscillating, chaotic, and random characterizations of temporal distributions also have corresponding characterizations of pattern in spatial distributions: regular, clustered, chaotic, and random. Identifying the type of pattern present—such as distinguishing oscillation from chaos—and examining temporal discontinuities are fundamental tasks in the study of temporal processes (Young 1988).

So what is the resolution between absolute and relative views? Is one superior to the other? A fundamental thesis of this book is that absolute and relative views of space and time are complementary and interdependent. The same holds true for continuous and discrete views of space and time.

A graphical schematic of this double duality is shown in Figure 2.1. The two extremes along the horizontal axis are the absolute view at one end and the relative view at the other. The two extremes along the vertical axis are the continuous view at the top and the discrete view at the bottom. The actual axes are not drawn since the surface of this space–time framework is not regular. What would be exactly halfway simultaneously between relative and abstract views, and discrete and continuous views would be hard to identify. Indeed, what this schematic represents is a space–time "space." I therefore do not attempt to locate individual philosophies or viewpoints within the diagram. Nevertheless, it still serves as a convenient graphical device for presenting the framework and showing how various views of space–time are related in a more general sense. A similar conceptual view of space was espoused in a book by Sack (1980), whose purpose was to lay a foundation for a critique of science from a spatial perspective. I offer my own interpretation of this dualism below.

The absolute side of the diagram shown in Figure 2.1 assumes an unchanging structure that is independent of human perception. As such, it is objective (i.e., uninterpreted). The relative view relies upon subjective judgment drawn from social, religious, or other, broader contexts and prior individual experience. As such, relative space and time are humanly internal, contextual, and interpretive. Thus, at one extreme of this continuum, on the side of relative space and time, is the domain of myth and metaphor. On the other extreme of the continuum, on the side of absolute space and time is the domain of external observation and measurement: external "truth."

The space–time schematic in Figure 2.1 can also be seen in a generalized sense to coincide with the division between the physical and the social sciences, between those sciences concerned with direct observation of physical space in the discovery and study of natural laws, and those concerned with the study of the way people "see" the world and of the humanly created environment. Given geography's position as both a physical and a social science, how space should be viewed has long been an issue within the geographic litera-

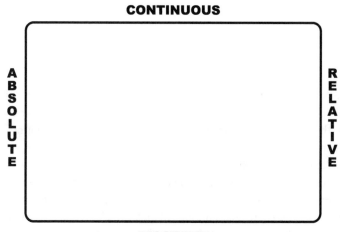

FIGURE 2.1. Varying views of reality.

ture. Some starkly geometrical ideas were proposed by Schaefer (1953) and then extended by Bunge (1962), Nystuen (1963), Berry (1964), and others. Objections were raised concerning the rigidity and narrowness of this approach in light of the goal of a universal way of viewing space. In the 1970s, the absolute, geometric approach gave way to the perceptual approach emphasizing the subjective interpretations of individuals. This in turn led to interest in geography, as well as in other social sciences, in alternative approaches to viewing human behavior in space–time, including structuralism, idealism, Marxism, and postmodernism.

Among the sciences, physical sciences such as physics and chemistry can be seen as closest to the absolute side of this absolute–relative continuum in their emphasis on the understanding of external reality; progressing toward the relative view, somewhere in the middle, is geography, as both a physical and a social science, then the social sciences, including sociology and economics, with their emphasis on human-created environments and institutions. Psychology would be closest to the relative side, with an emphasis on pure human interpretation and perceptions of reality. Art would certainly occupy the relative side of the framework to about the midpoint, representing art forms and styles, from those with the intent of faithfully representing reality, to highly abstract forms.

Mathematics becomes a unique case with regard to the proposed framework in that it does not fall within any single area within this framework but rather seems to have aspects or subfields throughout. For example, Euclidean geometry is absolute and very continuous in nature, but topology is relative and discrete.

SUMMARY

This chapter has outlined the development of concepts concerning the basic nature of space and time. Although these notions have varied among cultures and over the course of human history, the most fundamental notions have a striking consistency. That there is a difference between reality as it exists and as it is understood (i.e., external and internal reality), and that these two realities are interrelated has been acknowledged in Western cultures since the time of the ancient Greeks. How these interrelate has been debated among philosophers as the basic epistemological question. The existence of an internal reality that differs from external reality also means that knowledge is highly individualistic, yet there are marked commonalities in "world views" among individuals which make communication and organized social systems possible. The key to these commonalities is that the process of gaining knowledge is the same, and that knowledge itself is a construction based on the interaction of the biological organism with its environment. What is the nature of the process involved in gaining knowledge of the external world and construction of a "world view"? What are these commonalities in internal world views, and how do these relate to commonalities in representations of geographic space? How do these internal representations relate to external representations used to communicate that knowledge? As we have already seen, these questions have gone beyond the realm of philosophy relatively recently and have become basic issues in a wide range of scientific disciplines, including psychology, linguistics, computer science, and geography. I explore these questions from the human point of view in the chapters that follow in the remainder of Part I.

NOTES

1. As noted by MacEachren (1995), semiotics has two parallel roots from different sides of the Atlantic, with very different points of view. One started in the United States with Peirce (1839–1914), who was trained as a chemist. His perspective was the "science" of symbols as a logical system. The other started in France with Saussure (1857–1914), who was trained as a linguist. He focused on the interpretation of signs within a cultural milieu.

2. For quotations from Peirce (Hartshorne and Weiss 1931–1958) in this and subsequent chapters, the first numeral after the year in the reference refers to the volume number of this multivolume work; the second numeral indicates the paragraph number within the specified volume.

3. The fact that continuous space appears first in the dictionary definition of space, but absolute time does not appear until much farther down the list of definitions of time is also intriguing in light of the current discussion.

CHAPTER 3

ACQUIRING
WORLD KNOWLEDGE
The Overall Process

UNDERSTANDING THE WORLD AROUND US

From Descartes onward, cognitive theory has considered ideas as objects of knowledge that are distinct from sensory perceptions. Kant also argued that since everything we know is filtered through our senses, it is impossible to separate *what* we know from the *process* of acquiring that knowledge. Beyond these basic concepts, the specific scheme used for classifying types of world knowledge varies with different disciplinary perspectives on the subject. Even though the details of the theory are open to criticism, the overall types of knowledge and the nature of the process of development of cognitive world models in children espoused by Piaget and his associates have had a major impact on cognitive psychology (Piaget and Inhelder 1956). Behavioral geographers' categorization of knowledge types derived from Piaget focuses on how adults and relatively older children learn about their environments. Those involved in artificial intelligence and robotics emphasize a geometric and deterministic approach that has focused on how to use the computer and computing techniques to imitate outward human behavior, without necessarily modeling human thought. In other words, developments in these latter fields have often in the past been computer-driven. Since Part II of this book examines geographic representations from the computing perspective, the discussion of how these fields represent geographic knowledge is mostly reserved until then.

In the current chapter, I discuss the process of learning about one's environment and the accumulation of environmental knowledge in the context of everyday human activity.

THE NATURE OF THE PROCESS

As previously discussed in Chapter 2, Kant asserted that space and time are the only a priori concepts. The presence of these concepts is a necessary precondition for perceiving anything. We also perceive phenomena in a spatiotemporal order; we perceive physical objects next to each other in space, and in sequence through time. Time and space thus also become the way in which the human mind inevitably arranges phenomena. In this, Kant revives the Aristotelian notion of the category; however, he greatly extended this notion with the idea of basic "categories of thought." These basic categories or concepts include quality, quantity, relation, and modality, just as a tennis ball has color, size, location, and velocity. (The structure of categories is discussed in detail in Chapter 4.) These categories enable us to make *judgments* concerning sensory input and constitute the basic mechanism for synthesizing and acquiring knowledge.

The basic categories, in themselves, however, are not sufficient. A *mechanism* for synthesis is also necessary. According to Kant, there are three separate mechanisms through which we acquire knowledge:

1. *Sense*—the perception and recognition of specific sensory stimuli.
2. *Apperception*—the recognition and categorization of combinations of remembered and current sensory inputs as a unified and sensible experience.
3. *Imagination*—the association and synthesis of ideas without direct sensory input.

For example, we may sense redness within our visual field. However, we may also instantly recognize this as an apple in our association with redness, as well as a shape and surface texture given our concept of an apple and its expected characteristics. We may also pick up the apple and heft it—using an additional known characteristic—to verify that it is indeed an apple and not a wax imitation. This association of sensory characteristics with a particular conceptual classification (apple) is apperception. We may also imagine that the apple has been sprayed with pesticide, although there would be no visual clues. This last level of experience is the realm of imagination and "pure reason."

All of these mechanisms are utilizing judgment. The acquisition of knowledge is a process of judging the true and the false, the useful and the nonuseful. According to Kant, this is not a simple absorption of facts that we then group into categories. Rather, the real power comes from the derivation of new knowledge. In the preface to the *Critique of Pure Reason* (1950), Kant states that instead of supposing that all knowledge must conform to all objects, he was going to experiment by supposing that all objects conform to our modes of cognition. This summarizes Kant's view of the importance of pure thought, regardless of sensory input.

Hegel (1817/1990) recast this into a more systematized structure, discussing at length that the way the mind achieves an understanding of nature (i.e., the external world) is through logic (i.e., thought). Being and becoming gradually advance the mind to the notion of the absolute idea, derived from the dialectic process of thesis, antithesis, and synthesis. Thus, a person may already possess the notion of "chair" (thesis), yet a newly encountered chair has different characteristics in terms of size, color, shape, and so on (antithesis). In rectifying the two (synthesis), a higher level understanding of "chair" is achieved.

Cassirer also expanded upon Kant's notion of the nature of objects as a construction of the mind: that learning is not a mere looking-at process or mental copying of reality. He called space and time the principal modes of experience and was one of the first to view the human experience of space and time from a developmental perspective (Cassirer 1944). As shown schematically in Figure 3.1, Cassirer posited three progressive modes of experience corresponding to levels of spatiotemporal knowledge: *expressive* or *concrete*, *perceptual*, and *symbolic* or *abstract*. At the first level, understanding is not separated from what is seen, heard, and touched. The character of known elements is judged by their outward appearance through the emotive system. This is a literal and immediate realm of space–time. "Things" experienced at this level were described by Cassirer as encompassing the total sensory experience, being inherently "luring or menacing, the familiar or uncanny, the soothing or frightening" (Cassirer 1973, p. 67). Everything is encountered "as it comes," with little or no interconnection in space–time. Certainly, an infant recognizes its mother at this level. In a cultural context, this is the level of myth and art. At the second level, the world view consists of cognitively enduring "things-with-properties" that are ordered in space–time. What is seen (or otherwise detected through the senses) is distinguished from meaning; sense is interpreted from *signs* detected via the senses. The perceptual level is distinguished by focusing upon specific sensory data and realized as comparatively constant, relevant for action, or otherwise significant. Specific

DEVELOPMENT

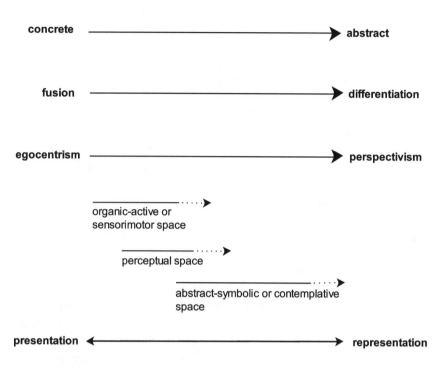

FIGURE 3.1. Levels of experience and corresponding world views recognized by Cassirer, Werner, and Piaget. Adapted from Hart and Moore (1973) by permission of the authors.

sense data thus has become the cue for representation of something else. It is these *fixed centers of orientation* within the perceived environment that become the basis for mental representation of the world. At the third, purely conceptual level, the environment is experienced vicariously through the interpretation of symbolic representations that do not necessarily have any connection with direct sensory experience.

Cassirer thereby also elaborated on an important distinction between *concrete acquaintance with*, and *abstract knowledge of*, space and spatial relationships (Hart and Moore 1973): "Acquaintance means only presentation; knowledge includes and pre-supposes representation. The representation of an object is quite a different act from the mere handling of the object" (Cassirer 1944, p. 46). How, then, do we humans make the connection between the outside world and our constructed view of it? The mediator between the two was called the *transcendental schemata* by Kant. Cassirer, in contrast, interpreted the entire process as one of transforming and integrating experience into *symbolic forms*. For Cassirer, no meaning can be attributed to anything except in reference to pervasive symbols arranged within a framework of space, time, and causality. It is through such symbols that sensory information is translated into meaning.

Each of Cassirer's three levels are *qualitatively* different from the others. The key difference arises out of a new "point of view" with respect to past and present experience. Even though old concepts are retained, they are seen on the new level in a different way, with the incorporation of specific types of knowledge into a new form of representation. According to Cassirer, language is a key to organizing and communicating knowledge in that it provides an entire, shared symbolic system.

Kant's mechanisms of sense and apperception are clearly functioning within Cassirer's three levels, in a sequence of decreasing dominance. Although Kant also emphasized that the use of all of these mechanisms also requires judgment, the basis for this judgment was developed completely from within the individual, seemingly in isolation. In contrast, Cassirer emphasized that the interpretation of sensory information is always influenced to some degree by cultural context. Space itself is physically modified by culture through the construction of buildings, and so on. From the very beginning, then, culture influences what we "see." All previous knowledge is also brought to bear on interpretation of sensory input and can include myth, religion, or knowledge of geometry.

Cassirer asserted that his three levels of knowledge also apply to cultural evolution. This can be readily seen in the broad historical perspective as described in the previous chapter. In the early stages of human history, there was reliance upon myth and magic. Humans considered themselves as one with their environment. Phenomena were explained in human terms, and the world took on human attributes. The world, or elements within it, could see and hear, and therefore could be called upon to act in requested ways. Rituals and ceremonies are full of visual and other forms of sensory symbolism. At the other end of the progression, the field of mathematics provides perhaps the best example of knowledge at the abstract level. Mathematics is symbolism expressed in purely abstract terms, with no reliance whatsoever on sensory experience. The development of culture in general and of science in particular is a continual constructive process: from eras of "expressive" space to "perceptive" space to "abstract" space and modes of spatial experience. Certainly, we can see this progression within Western culture, from primitive man to ancient Greece and Rome, to the Renaissance, to modern times. Nevertheless, no level of knowledge or mode of experience disappears once acquired. The expressive mode of experience verifies and informs purely abstract areas of knowledge. For example, abstract geometries require interpretation at the perceptual level if they are to make sense, particularly for those initially endeavoring to learn such concepts. This is why geometry texts are full of graphical diagrams and exercises set as real-world tasks.

To summarize Cassirer's view, each act of knowing is built upon *all* knowing that has gone on before, on both cultural and individual levels. As such, the collective knowledge of culture, including science, is highly intertwined with individual knowledge.

In contrast to Cassirer's focus, Werner and Piaget focused on individual development of knowledge. They also built upon Kant's and Cassirer's philosophical ideas within the field of psychology through empirical observation. Werner formalized his theory as the *orthogenic principle* to describe shifts in knowledge: "Wherever development occurs, it proceeds from a state of relative globality and lack of differentiation to a state of increasing differentiation, articulation, and hierarchic integration"(Werner 1957, p. 126). He studied how spatial knowledge is acquired in children in the initial building of a spatial knowledge structure. Werner also recognized three progressive levels of development, which he called sensorimotor, perceptual, and contemplative (Hart and Moore 1973; Werner 1948).

The most comprehensive and influential theory of spatiotemporal cognition was developed by Piaget. He studied the process of thought in children, as did Werner, through different stages of development. Piaget asserted that logical concepts about space, time, and objects are acquired through an invariant sequence of stages determined by biological constraints. According to Piagetian theory, children progress sequentially through higher stages of development from an initial, practical, sensory–motor stage, in which knowing is tied to the child's own actions, to a mature stage, which includes what he referred to as logicomathematical thinking. This progression is characterized by an increasing importance of abstract, reflective thought relative to direct sensory stimuli from the external world. The degree of organization in the cognitive world model reflects this process. Piaget also emphasized that the essence of this process involves a change in the viewpoint of the individual, from one of pure egocentrism, where all space and time are viewed relative to the self and one's own actions, to one of perspectivism. The perspective view can be briefly described as an integrated and objective view of the world.

The details Piaget described for each of these stages, the sharpness of the boundaries between them and the invariability of the sequencing, have been seriously criticized (Boden 1979; Brainerd 1978; Hooker 1995; Siegel and Brainerd 1978). Nevertheless, Piaget's theory, taken broadly, has had a significant impact on cognitive research and views on learning and cognition in a variety of fields.

The empirical work of Werner and Piaget provides a substantial body of evidence that although environmental and cultural influences may vary considerably, both the *process* of knowledge acquisition and an initial biological motivation, upon which all other cognitive constructs are built, are universal. This universal process progresses from concrete acquaintance with, to abstract knowledge of, the world. Each of these stages is also linked with a qualitatively different view of space–time.

THE START OF THE PROCESS: INNATE CONCEPTS

In agreement with the philosophy of Cassirer, and in explicit contrast to the Kantian perspective, Piaget's biological–developmental approach places emphasis on the continuing importance of sensory input. He describes the acquisition of spatial knowledge as an interactive process, in which new sensory input causes adaptation of previous ideas instead of "merely to submit the data of space perception to a process of logical deduction capable of analyzing them indefinitely" (Piaget and Inhelder 1956, p. 3). In addition, organization within space of sensory–motor behavior results in mental constructs, complete with their own functional rules.

Another critical difference between Piaget's theory and that of Kant is that space–time concepts are *not* totally innate. All space–time concepts must be learned via experience, except for a few rudimentary spatial and temporal relationships that enable initial learning. According to Piaget, the only innate spatial or temporal knowledge possessed by the newborn infant is relationships directly involved in the biologically driven behaviors (sucking, touching, etc.), and consist of;

1. *Separation*—distinctness of perceived objects
2. *Proximity*—nearby-ness of elements within the same current perceptual field
3. *Order*—spatial succession of objects and temporal sequencing of events
4. *Enclosure*—differentiating inside from outside

Thus, Piaget claims that space–time concepts, as known by adults, are not innate. Nevertheless, the inborn relationships from which all learning begins deal with space and time in a very rudimentary fashion.

The inherently spatial relationship of proximity begins with the sense of touch (the sense for the infant of touching a blanket and clothing; the mother touching, holding, etc.). The same is also true for enclosure—the sense of being secure, wrapped snugly inside a blanket. Separation for the infant is the ability to recognize a rattle or a person's face as a distinct entity within its visual field, perceived as a spatial image. Space suffices for the simultaneous ordering of the positions of objects within the infant's visual field, but as soon as movement is seen (and thereby a change in spatial position) by the infant, successive ordering of spatial states in time is also introduced. Temporal order, however, also begins with the sense of touch: Soon after birth, even before the infant's eyes can focus, there is the temporal ordering of the infant's crying, and then (frequently) being fed.

Rudimentary perceptual space is egocentric in the sense that it revolves around the infant as the organizational point of reference. Thus, at this initial stage, space is also purely relative and topological. Sensory inputs are initially constructed in terms of separate spaces but are united in a notion of continuous space early on. The notion of a single, continuous space is a prerequisite for developing the concept of objects that exist separate from one's self, and for understanding how these objects behave in space (e.g., an object cannot be in two places at the same time).

The concepts of space and time are constructed gradually through what Piaget called an "elaboration" of a system of space–time relationships (Pulaski 1980). For example, the separation relationship changes as the child grows older and learns to distinguish parts of objects that are also attached to the whole. Touching also changes as the child grows older and other factors of sensory organization develop, such as eye–hand coordination. Corresponding changes also occur for the other innate relationships as the child develops both physically and mentally, and acquires experience with his or her surroundings.

With progressively increasing mobility, the child's geographic realm of experience gradually widens. The child begins to navigate *through* space, thereby actively exploring his or her environment instead of passively experiencing space as elements in the environment that come to him or her. In addition, increasing mobility enables space to be experienced on progressively larger scales. Out of this increasingly complex interaction between the child and his or her environment, the child develops a structure of thought that is not only progressively more complex in itself, but is also qualitatively different at each progressive stage. The cognitive *form* of knowledge is thereby tied to *how* that knowledge is constructed over time. I now discuss theories relating to this process in more detail within Piaget's developmental framework.

TYPES OF KNOWLEDGE

Piaget and his followers suggest that there are two basic types of knowledge: *figurative* or symbolic, and *operative* (Piaget 1963, 1969b; Pulaski 1980). Figurative knowledge, tied to the recollection of experience, is the first to develop. It consists of familiarity with elements within the environment (people, things, places, and events), and grows to include awareness or belief that something exists, knowledge of its characteristics, the ability to recognize it when encountered and, often, to attach a name to it. As previously mentioned, the spatial scale of figurative knowledge increases as an individual's scale of movements increases to include knowledge about one's neighborhood, city, and so on. The temporal scale of figurative knowledge also increases with the length of time actually experienced, and as people's memories of things, places, and events accumulate.

Figurative knowledge begins as knowledge about specific, individual objects and is subsequently generalized to classes, with specific objects as examples of specific class concepts. In other words, specific objects become recognized as belonging to classes that have shared and distinguishing characteristics. Characteristics for "things" can include size, color, spatial location, and temporal duration. For locations, distinguishing characteristics can include the objects and/or events that occur there. Spatial relationships that relate to specific objects, such as inside, and temporal relationships, such as before and after, are also included. Figurative knowledge focuses on the static appearances and configurations of objects and representations, and can include visual images that may or may not be faithful recollections of sensory input; they also may not be the result of any visual input whatsoever.

Operative knowledge develops later, and is related to a specific mode of thought. This type of knowledge includes generalized knowledge about how objects dynamically interrelate with each other in space, as well as how the individual's behavior in space can be modified. Operative knowledge was described by Piaget as an introspective or coordinating process by which we construct reality-as-known, distinct from reality-as-experienced in figurative knowledge. According to Piaget, operative knowledge thus acts as a self-enriching, internal process that ultimately becomes the principal source of knowledge for an adult. Operative knowledge thereby provides the logical framework upon which our mature knowledge of the world is based.

Piaget breaks figurative and operative knowledge into many more finely-divided gradations of knowledge types that are acquired progressively as a child develops. Recognizing that this process is less discrete and more variable than Piaget described, I will outline only the major stages here in order to elucidate his core insight: that learning is a process of construction that is achieved through active interaction with one's environment. Moreover, this process is constrained by one's preexisting knowledge and overall world view.

The first stage of development was called the sensory–motor or intuitive–figurative stage by Piaget (see Figure 3.2). At this stage, the infant has no preexisting knowledge structure and is limited to direct sensory input within its immediate physical proximity. The infant first learns about space on its own constrained, personal spatial scale, and about individual objects that it can directly see, touch, hear, and smell within that space. According to Piaget, no "things" exist as known objects as far as the young infant is concerned. Rather, elements in the environment are simply remembered as consistent sensory sensations. Similarly, time is limited to familiar successions and durations given by direct sensory experience.

Piaget theorized that small children do not develop the notion of object

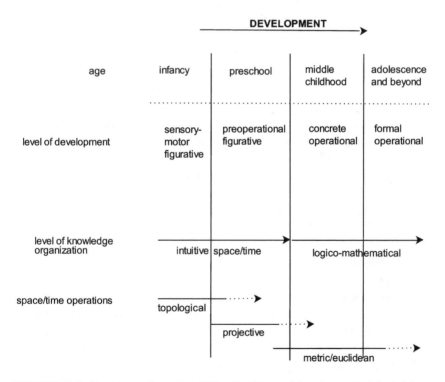

FIGURE 3.2. Summary schematic of Piaget's theory of development. Adapted from Hart and Moore (1973) by permission of the authors.

permanence or a notion of self as distinct from the environment until about the end of the first year—if something isn't being seen, touched, or heard, it doesn't exist. Whether or not small children possess the concept of object permanence is a current topic of debate (Adolph, Eppler, and Gibson 1993; Baillargeon, Spelke, and Wasserman 1985; Bogartz, Shinskey, and Schilling 2000; Bogartz, Shinskey, and Speaker 1997; Meltzoff and Moore 1998). However, there does seem to be consistent evidence that even very young infants are keenly aware of themselves as distinct from their environment, and that they can actively affect things and events within that environment (E. J. Gibson 1993).

At this stage of development, the time and space of that environment is still viewed in a discrete and egocentric way: home, Grandma's house, time for dinner, time for bed, and so on. Relationships remain topological in nature, although these gradually increase in complexity.

During the next major stage of development in Piagetian theory, the pre-operative–figurative stage, the child begins construction of a world representation. With increasing interaction with the world, the child recognizes shared characteristics among individual experienced objects and groups objects into conceptual classes. This increasing conceptual knowledge organization continues, and class groupings extend to other elements of knowledge: places, events, and relations.

At this stage, the child also develops the notion of a conceptually continuous and connected space and time populated with objects, and of relationships between those objects. With the notion of continuous space and time, the child develops the ability to perform mentally rudimentary projective operations on physically experienced objects within that continuous space. From about 4 years of age, the child can navigate a familiar neighborhood between two places that are not simultaneously within the field of view.

Eventually, the child can find shortcuts and detour successfully around obstacles. Nevertheless, the child can only retrace a previously learned route as a sequence, and only in the direction in which it was learned. Similarly, the child can only act out events in the same order in which the events were originally experienced. With the egocentric world view characteristic of the figurative level, the child therefore judges space and time solely based upon this personal experience and fails to grasp the idea of an objective, uniform space, and of an objective, uniform time common to all phenomena that are separate from his or her own experience of the world. It is because of this that a child at this stage cannot conceptually reorder experienced routes or events. For the same reason, space and time also expand and contract according to the child's own experience and memory of it.

At about age 8, the child can cognitively represent space and time utilizing a "decentralized" view, with a coordinated, overall structure. It is this representational ability that Piaget associates with the beginning of operational knowledge, which he called the *concrete–operational* stage. It is only at this stage that the child acquires a grasp of metric relationships, in addition to topological and projective relationships, with the construction of a comprehensive space of coordinates in a Euclidean sense. At this stage, the child also progresses beyond the naive time of the small child to an intuitive understanding of time based on an external succession of events with duration and intervals between them. Thus, the child finally becomes able to perform freely the necessary spatial transformations to modify routes and reverse direction, and to reorder events mentally. Grasping the idea of a nonpersonal, objective space and time is perhaps *the* most important attainment in the development of a mature cognitive representation of the world. As stated by Piaget, this attainment of a "decentralized" view is

> tantamount to freeing oneself from the present, to transcending space by a mobile effort, i.e., by reversible operations. To follow time along the simple and irreversible course of events is simply to live it without taking cognizance of it. To know it, on the other hand, is to retrace it in either direction. (Piaget 1969a, p. 259)

Operational knowledge continues to develop at a progressively higher level of abstraction until, at adolescence and beyond, spatial and temporal operations can be completely disassociated from any real objects, spaces, or actions, and can be dealt with as totally abstract concepts. This is what Piaget called *formal–operational* knowledge. It is finally at this highest level that the possibilities of space and time can be mentally explored as purely mathematical and theoretical entities. A distinguishing characteristic of formal–operational knowledge is the development of a hierarchical organization of elements. Category structures and prototypical effects are integral components of that process and of the resulting cognitive knowledge structure.

Starting from the innate spatiotemporal relationships in the Piagetian theory of cognition, then, acquisition of notions of space and time occurs in parallel. Nevertheless, notions of time are dependent upon our notions of space: "Space is a still of time, while time is space in motion. The two taken together constitute the totality of the ordered relationships characterizing objects and their displacements" (Piaget 1969a, p. 2).

According to the Piagetian view, how humans function within their environment is *dependent* upon the *forms* of knowledge they possess. These forms of knowledge change as that person develops into a mature adult. This also means that increasing knowledge is not simply a process of accumulation. The progression from the infant's rudimentary figurative knowledge to adult's operational knowledge is essentially a process of adaptation, organization, integration, and abstraction of an individual's internal knowledge *representation*.

Although there may be reorganization of knowledge at certain stages, current thinking among psychologists departs from the strict Piagetian view, in that knowledge stored within each type of representation is neither mutually exclusive nor strictly progressive. It has been argued that, at least in some cases, abstract representation may precede concrete knowledge based on experience even in young children (Simons and Keil 1995). It is also currently assumed that we do not abandon one form of knowledge representation (or thinking) for another. Certainly, even though the adult possesses abstract concepts, there is also remembered imagery that can be associated with some of those concepts. Also, the knowledge construction process is never complete. As experience continues, sensory experiences are selectively remembered. Current sensory experiences are continuously being compared to memories of past experiences, and our knowledge is continuously being adapted. What we perceive thus depends in part on interpretation based on our current level of knowledge, just as our current level of knowledge is dependent on current and past experience. Moreover, we are better able to *employ* the functions of higher level representation to process new information as knowledge increases.

Following this latter reasoning, the acquisition of knowledge representation *forms* as a component of child development has also been called into question (Bullinger and Chatillon 1983; Millar 1994). The issue here is whether the capability to encode knowledge in a certain way must be acquired, or whether the representational capability is innate and the mere lack of experience initially inhibits the organization and integration of remembered experiences into more abstract representational forms. For my purposes here this is somewhat beside the point. The important aspects of the Piagetian view are that the acquisition of knowledge is, generally speaking, a progressive, complex, and self-adapting process that relies on interaction with, and reasoning about, the outside world.

TYPES OF KNOWLEDGE AT GEOGRAPHIC SCALES

Geographic-scale space, or space on a scale that is beyond immediate personal space, presents special problems. In the macroscale physical environment, it is possible to perceive directly only a small portion of the total environment at any given moment. Therefore, spatial knowledge at geographic scales serves an additional function: to prevent getting lost.

Work on the representation of geographic-scale space is predominantly found within the fields of geography and urban planning. Much of this work relates to navigation, or way finding. Indeed, Lynch (1960) argued that the *primary* function of a cognitive geographic representation is to facilitate location and movement within one's larger physical environment. A key element of spatial representation, according to Lynch, is "landmarks." These denote dominant features in the landscape that readily impress themselves on the senses and, as a result, tend to be retained as an organizing element of a cognitive representation of geographic space. Thus, city hall is a landmark in most cities and towns, providing a locational reference or "anchor point." The location of many other features in the town are remembered as being, for example, behind city hall. In this context, the building itself may be distinctive (and thus easily recognizable), but the primary feature of importance within an overall cognitive representation is its location, with other features in the town being located in relation to it. A cognitive representation of the town in which a person lives typically contains a number of such landmarks. In finding one's way, landmarks provide locational cues for navigating through geographic space (Golledge 1988). Certainly for the child, "home" and "school" are landmarks providing strategic foci for early navigation. Lynch's landmarks correspond to fixed centers of orientation within Cassirer's theory.

Another key element of representation in theories regarding geographic scale space is routes, which also include landmarks as anchor points. From a logical point of view, it can be argued that landmarks and routes are perhaps the necessary and sufficient elements for "minimal" representations that allow way finding to occur (Siegel and White 1975) Although most way-finding research has viewed landmarks as discrete objects within the landscape, Smith, Pellegrino, and Golledge (1982) empirically verified Lynch's notion that environmentally based navigational cues also can include grouped objects, or scenes. From everyday experience, this type of image cue is certainly something that everyone has used in navigating through an area in which there is perhaps no single "distinctive" feature. In such cases, we must rely on recognizing the overall scene produced by the specific juxtaposition of more common objects within the landscape, such as a row of trees with a house behind it.

Elaborating upon the Piagetian perspective, Siegel and White (1975) provided a theory for the development of geographic knowledge based upon the notion that geographic knowledge arises from the integration of successive perceptual experiences. For children as well as adults, this proceeds from the recollection of specific landmarks, to route learning, and ultimately to a fully integrated map-like or "configurational" view. This sequence of stages, as in preceding theories, is also considered to be invariant. Thus, landmark knowledge and route knowledge are eventually integrated to form what has been variously called configuration knowledge, plan knowledge, or survey knowledge (Shemyakin 1962).

Siegel and White (1975) describe this first stage of remembering landmarks as a special kind of figurative knowledge, because the spatial and temporal context is soon remembered in association with the landmark itself as experience accumulates: "I remember we passed that same church yesterday." Gradually, this recognition in context leads to development of a choice-point decision system between landmarks, which leads to the formation of individual routes. Once a number of routes are learned, these become interrelated into a network-like assembly. With continuing experience, the gaps are filled in and geographic knowledge for the given area becomes an overall, "configurational" view. Although young children are able to develop configurational representations, it takes them longer to do so than adults.

Siegel and White (1975) attribute these learning differences to the lack of both a preexisting spatial reference framework and information *capacity* in children. Millar (1994) has argued, however, that these differences can be explained entirely by the former. In other words, without a preexisting knowledge base, children require a greater amount of concrete, sensory reinforcement to compensate for a relative lack of higher level, more abstract, and general-purpose knowledge that can be brought to bear. Siegel (1981) explicitly rejects the Piagetian notion that young children are limited to topological spatial relations, because a configurational representation requires metric relationships.

Golledge built upon the work of Piaget, Lynch, Hart and Moore, and others. His concern, as well as that of other researchers in behavioral geography, has focused upon how different types of geographic knowledge are represented and used among adults and older children (beyond about age 6) rather than in childhood development per se (Golledge 1988; Golledge, Dougherty, and Bell 1995).

Golledge (1988) described route knowledge as a specific sequence of procedural descriptions involving the starting place, or anchor point (e.g., one's home), and a subsequent sequence of landmarks and/or distances. More detailed route knowledge may also include intermediate landmarks along the way. Knowledge of specific paths through complex environments connecting objects via specific spatial relationships provides the ability not only to navigate from place to place but also to plan the route in advance, evaluating known alternatives.

Golledge (1988), however, equated landmark knowledge with declarative knowledge, and route knowledge with procedural knowledge. Declarative and procedural types of knowledge are described in the cognitive psychology literature as "knowing what" (e.g., "Paris is the capital city of France" and "Three plus five are eight") and "knowing how" (methods for performing tasks and solving problems), respectively (Way 1991). Although landmark and route knowledge thus do not map directly onto declarative and procedural knowledge, this linkage perhaps helps in understanding how various types of knowledge are used in navigational tasks.

Golledge's survey knowledge is described as knowledge that provides an understanding of how the various elements of the environment are interrelated. This level of abstract, associational knowledge goes beyond simply "filling in the gaps" about what is where, as knowledge of objects within the environment accumulates. This is truly an integrative type of knowledge, the type of geographic knowledge commonly described as "the bird's-eye view" or the cognitive map.

In accordance with Cassirer's view, the progression from landmark to configurational, or survey, knowledge also seems to occur in adults acquiring knowledge in a new domain. As a simple example, learning to navigate in a new city often begins with being driven by others (in a private car, in a taxi, on a bus) or otherwise assisted. During this phase, people tend to acquire memories of individual locations and other geographic elements of significance (landmarks, street names, etc.). Associations among these are gradually accumulated, including both spatial connectivity and sequencing for navigation, and spatial groupings (the restaurant district, etc.). Gradually, people are able to navigate between specific locations (home and work, home and shopping, etc.) along specific routes, and finally, with the accumulation of broadening spatial exploration, are able to devise new routes in an ad hoc fashion, as needed. These two modes of navigation reflect a progression from figurative to operational thinking in Piagetian terms (i.e., progressing from a focus on static and remembered configurations—a concrete "reality-as-experienced"—to a more generalized knowledge about how objects dynamically interrelate—an abstract "reality-as-known." This corresponds to the shift from what Golledge called *route knowledge* to *plan knowledge* within a geographic context (Freundschuh 1992; Golledge 1988, 1992).

Thus, the process of learning about one's environment can in very general terms be characterized as beginning with the accumulation of individual, discrete, and disconnected locations of "significance" to the individual. Then, the individual gradually accumulates connections and relationships between these locations with increasing familiarity and established groupings, until, finally, he or she obtains an overall view of how everything interrelates. This includes purely conceptual groupings, such as "places to shop," as well as spatial relationships, such as distance, direction, and adjacency. It is important to note that this form of knowledge is information-driven. In other words, it is influenced by the amount of information available rather than the form of knowing available to the individual. Bullinger and Chatillon (1983) provide a detailed discussion of this issue and its historical evolution. Experimental evidence of this information-driven view is provided by Millar (1994), Siegel (1981), and Pick and Lockman (1981), among others.

Because of a focus on learning in early childhood, the Piagetian perspective emphasizes learning through real-world, sensory–motor experience. It must be remembered that learning is also accomplished through many other means, although not as much so for very small children. The acquisition and subsequent use of language represents a significant advantage in learning about, and dealing with, the real world. Knowledge about people, places, and things also includes knowledge acquired by means other than direct sensory experience and introspection. Knowledge is often acquired from verbal descriptions, photographs, maps, and even watching television. Certainly, such indirect means are the only way to learn about things on a geographic or temporal scale too large to be experienced directly under usual circumstances, such as the shapes of the continents (or the shape of the earth) and continental drift. Research on spatial cognition within behavioral geography has utilized learning from a variety of sources, including verbal descriptions and reading maps, as well as from actual experience in navigating geographic space. Recent research using experimental means other than purely sensory–motor experience (e.g., through map drawings) has shown that the sophistication of the child's world view has been underestimated, although the basic process still holds (Siegel 1981).

Relating to this, the modes of thought are cumulative—just as experience is cumulative. Instead of replacing one mode of thought with another, people retain the ability simply to remember visual scenes, as well as to perform abstract logical reasoning. As such, all of these modes of thinking and types of knowledge are elements of an adult's repertoire for learning about, and dealing with, the world. This means that although the dominant mode of thinking within a given domain about which the individual knows little may be limited to specific remembered examples, higher level knowledge, represented in logicomathematical form from another domain, may be brought to bear to assist learning and to solve problems in new situations. Similarly, the progression of thought that Piaget observed in young children is due to a general progression of experience, starting from an overall level of inexperience, rather than from any nonavailability of certain knowledge structures or modes of thought, through a continuing (and lifelong) process of adaptation, organization, integration, and abstraction.

HOW TYPES OF KNOWLEDGE RELATE TO COGNITIVE REPRESENTATION AND LEVELS OF KNOWING

Knowledge, then, can be seen as self-enriching. Piaget equated knowing with structuring, which has aspects that derive from the intrinsic structure of the organism itself. Although the role of biology in Piaget's theory falls far short of biological determinism, it is a fundamental element in his theory, in that biology plays a large part not only in how we learn but also in the commonalities of knowledge among individuals in similar environments. It is this biological factor that allows different individuals to react in similar ways when exposed to the same sensory stimuli. Without such a physically determined normative compass, individual perceptions and subsequent "world models" would be totally subject to individual invention. This would also mean that communication between individuals in any true sense would be impossible. Communication of knowledge or information about the world requires shared representational models with agreed-upon rules.

There are also limitations of purely empirical observation, whether in scientific or everyday contexts. For example, Western civilization held that the Earth was flat and that the sun, moon, and stars were affixed to a series of concentric crystal spheres revolving around the Earth at their center. The use of indirect observation and the accumulation of related knowledge over time were required to deduce the truth. The limitation of direct observation needs to be balanced by the use of *deduction* based upon previous knowledge, whenever possible. In other words, the key to gaining new knowledge is the use of prior knowledge.

Learning and the acquisition of new knowledge may include interpretation of current sensory perceptions, reinterpretation of remembered perceptions, and discovery of connections and similarities with other knowledge and observations. The use of analogy and metaphor is thereby an important component of the deductive process. It is also impossible to use direct observation to learn about things that we may never be able to see. For example, I have general ideas about such things as the structure of DNA or daily life in Medieval Europe. My knowledge of them is derived largely from communication and the use of external models. Direct observation, then, is the starting point, but there are many ways of acquiring knowledge. This means that the learning process in general can be characterized as a complex interplay between *percepts* and *concepts*. A number of researchers have also associated different types of learning with different types of knowledge. Certainly, Piaget's theory associates progressively more sophisticated ways of learning with the use of progressively more sophisticated forms of representation as the child develops.

Moreover, new knowledge acquired throughout life does not always result in merely the storage of that knowledge in memory. People do not and cannot remember everything they have ever seen or heard, nor do they remember everything they ever learned. Rather, new experiences and new knowledge act to modify our store of knowledge, adding new connections and associations among elements, and generally refining our knowledge (i.e., increasing our understanding of the world). Given that what we know is a cumulative and selective construction through the sum of our experiences, the knowledge of any individual is also unique, although the process and the modes of representation are universal.

The term "knowledge," commonly used by cognitive psychologists and behavioral geographers, and therefore used also in the preceding discussion, is a generic term meaning "anything that is known." Although the terms "data," "information," and "knowledge" are frequently used interchangeably, particularly in everyday speech, they have very meaningful distinctions within the scientific context. *Data* are characteristically raw observations that have been remembered or recorded in some way, whereas *information* is data that are ordered and contextualized in ways that give them meaning. Information is thus selective with regard to data, separating the important from the relatively unimportant. *Knowledge* is a cumulative understanding of information based upon a "world model" (i.e., an overall representative structure and a set of generalized rules that pertain to the relevant phenomenon).

In science, deriving information from raw data is considered a process of separating signal from noise, with noise taking the form of information that is simply not relevant within a predetermined context as well as error or anomalies. Again, this distinction can only be made within the context of a preexisting model (Willmott and Gaile 1992). The progression from (1) recording observational data as objective, uninterpreted "facts," to (2) interpreting, organizing, and filtering sets of observational data in order to derive information, to (3) synthesizing the eventual accumulation of knowledge as a generalized phenomenological model is the basic paradigm of science in its most fundamental terms, regardless of discipline, and is known as "the scientific method."

SUMMARY

Expressive/perceptual/symbolic, figurative/operational, declarative/procedural/ survey, and data/information/knowledge are different terms that are appropriate to varying descriptive contexts for progressive types of knowledge. Each term within a given sequence relates to a very different type of knowledge in a universal progression that is integrally tied with the learning process, whether development in the child, continued learning in the mature adult, or scientific inquiry. The processes involved in this sequence of attaining progressively higher levels of knowing consist of observation, selection, integration, and abstraction. Because this sequence is progressive, deductive logic plays an increasingly important role in further learning, and direct observation plays a progressively less central (although certainly ever-present) role. This is also consistent with the three forms of development earlier identified by both Cassirer and Werner, progressing from a predominance of action-in-space as direct sensory input, to perception-of-space (interpretations of sensory experience), to conceptions-of-space (abstract and contemplative) (Hart and Moore 1973).

Figure 3.3 schematically relates the learning process and the progressive types of knowledge associated with learning to the conceptual framework of varying views of space–time in Chapter 2. The learning process, from a very generalized perspective, corresponds to a diagonal traversal from the discrete–absolute corner of the diagram in Figure 3.1 to the relative–continuous corner. Along this continuum, we can think of the process of learning as gaining a more continuous and relative world view. Increasing levels of abstraction that accompany this traversal proceed from individual and disjoint perceptions to a generalized and coherent understanding of space-as-experienced, and ultimately to the incorporation of a purely abstract view and the ability to make metaphorical and symbolic associations. Figurative knowledge, in Piagetian terms, is on the left, occupying the absolute side of the framework. This corresponds to knowledge that is dependent upon direct interaction with the environment—perception, imagery, and memory of external reality. Operative knowledge is on the right, occupying the relative side of the framework, and corresponds more to internalized thought and abstract understanding of reality. This is a continuum, and there is no sharp dividing line.

Human cognitive development and learning can be mapped (see Figure 3.3) as a progression beginning with very absolute and discrete and moving toward higher level knowledge in the form of generalizations and abstractions, and more comprehensive and spatially continuous knowledge, through

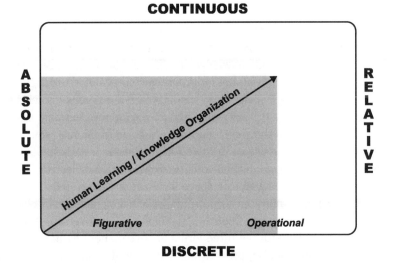

FIGURE 3.3. A generalized view of the progression of human learning and knowledge organization.

increasing experience of a world with an existence independent of individuals' perception of it. As knowledge accumulates, the area of the schematic encompassed by that knowledge also grows (shaded area). It must also be remembered that the process of learning is cumulative in every way. The very absolute and discrete "world view" used for learning in early childhood never disappears. Rather, these world views become enriched; space can be seen cognitively through varying world views, and relationships can be combined indefinitely. What we have discussed in this chapter does indeed confirm the notion introduced at the end of the previous chapter: The edges of the absolute–relative and discrete–continuous continuum shown graphically in Figures 2.1 and 3.3 are connected; the highly relative and discrete perspective of space—the domain of metaphor—helps inform the domain of measured and absolute space, and vice versa. Remember the earlier mention of the importance of graphics in geometry textbooks to help convey purely abstract concepts.

In this chapter, I have examined the nature of the process involved in gaining knowledge of the external world and construction of a "world view." This process, and the resulting cumulative *types* of knowledge we cognitively store, are universal, although our individual life experiences and cultural contexts may differ considerably. This process is also cumulative, utilizing varying world views and relationships in increasingly complex ways. At its most fundamental level, the process of acquiring knowledge, whether as children or adults, is essentially one of abstracting and structuring that knowledge. Our understanding of the world and our ability to acquire new knowledge is due in large part not to the amount of information about places, people, and things we can remember, but rather to the way in which we *structure* that information.

What, then, can be said about the form in which our world knowledge is stored? Are the different types of knowledge stored separately in different forms? Is the cognitive structure of knowledge the same for everyone? These are some of the questions that I address in the remainder of Part I. In Chapter 4, I discuss the basic elements that comprise the form of our world knowledge. In Chapter 5, I look at how sensory experience functions as an important component of a cooperative knowledge acquisition and storage system. In Chapter 6, I discuss how we go beyond experience in the construction of abstract concepts and an integrated world view. In Chapter 7, I examine how we gain information and knowledge through indirect experience—being told through the use of language and maps. Then, in Chapter 8, the last chapter of Part I, I draw some conclusions about the structure of our world view and how different types of world knowledge work together.

CHAPTER 4

STORING WORLD
KNOWLEDGE
Some Elements of Conceptual Structure

EMBODIMENT OF SPATIAL KNOWLEDGE

Structuring knowledge is a key adaptive mechanism that we use to cope with the immense complexity of the environment in which we live. We accumulate knowledge and our overall world view from our experiences by integrating, organizing, and encoding the essential information from these experiences into complex and interconnected structures. As discussed in Chapter 3, the process of structuring knowledge is an integral part of learning and the ability to use abstract thought. Because we cannot remember every experience from birth, or even from the time we got up this morning, we rely more on these knowledge structures than on exact recall of individual experiences. We can therefore say that, overall, the *form* in which our spatial knowledge is stored is more important than *what* is stored.

According to the behavioral viewpoint, the structure of human knowledge, as well as the capacity for meaningful thought, is abstract and not necessarily embodied in any physical sense. In Kantian terms, concepts and rationality are *transcendental*, in the sense that they transcend, or go beyond, the physical form of any organism. Meaningful concepts and abstract reason may happen to be embodied in human beings, in other organisms, or in machines, but they exist abstractly, independent of any particular embodiment. In other words, our knowledge and the means to learn are not "hardwired." Rather, certain physical capabilities, are just that—*capable*-ities, physical attributes that *allow* learning and the storing of knowledge.

56

A number of biologically based models of spatial cognition have been developed as an alternative to the behavioral approach to understanding cognition and mental representation. Rumelhart and McClelland (1986) provide an extensive historical review of this approach. Because the brain represents the physical embodiment of the mind, most of these models rely on theories of the behavior of neurons in the cerebral cortex as parallel and cooperative processing units. This gave rise to models referred to collectively as parallel distributive process (PDP) models. The biologically based line of investigation as a means for understanding human cognition has been criticized in part as an overly literal acceptance of the biological argument used as a foundation for Piaget's theory (Lakoff 1987). Although the brain is an amazing combinatorial device and worthy of study for many reasons, increasing evidence indicates that trying to find specific groupings of neurons and synapses that represent specific stored knowledge of things, places, or relationships is a futile endeavor. Nevertheless, PDP emphasizes the parallel nature of the processing in the brain, the use of distributed representations and distributed control. Its use as an approach to gain insight into these specific mechanisms may still prove useful.

One functional characteristic of the brain, similar to the multistability behavior of a fluid dynamic system, may be seen as an enabling characteristic for learning, storing, and recalling knowledge in a very flexible and powerful manner. As a result of this characteristic, a fluid dynamic system is capable of realizing many flow patterns out of a potential set of patterns. Which specific flow pattern is realized at any given time depends on the initiating conditions that in turn set off interactions among elements within the volume. Thus, a set of patterns is not stored in any static fashion, but is dynamically generated anew each time. Such a behavior also applies to the human brain, which it is now believed does not permanently store patterns. Rather, the brain generates patterns on-the-fly, by means of learned synaptic strengths that govern the cooperation between neurons (Portugali 1992). This reinforces the idea that the process of acquiring and using knowledge is the key to understanding how knowledge is encoded.

Utilization of the basic idea of PDP models—the idea of a highly distributed, parallel processing mechanism where connections are made on-the-fly has become an important approach in deriving computer-based models of human cognition. PDP models had been most successful in simulating lower level functions (e.g., visual recognition and simple learning) (Eliasmith 1997), but have also recently gained wider credence as providing a very useful perspective in considering higher level cognitive functions, particularly in deriving higher level abstractions.

Experimental research in support of a more behavioral, in contrast to a biological, perspective indicates that the Piagetian theory of how knowledge is acquired may also extend to other species in addition to humans (Gallistel 1990). At a fundamental level, key differences among various species may simply lie in the extent of the progression achieved in using logical thinking and abstract reasoning, with humans progressing further than the rest of the animal kingdom. Such a behavioral approach also allows for research findings indicating that, aside from persons with severe mental impairment, both the learning process and the resulting types of knowledge are the same for all humans. This includes those who, for example, have been blind from birth (Millar 1994).

In Chapter 3, a sharp distinction was made between percepts and concepts. The former are based directly on sensory information, while the latter are our abstracted beliefs about the world (Neisser 1987b). The two are highly interdependent within the context of how mental representations are formed. As Kant wrote: "Without sensibility [perception] no object would be given to us, without understanding no object would be thought. Thoughts without content are empty, intuitions without concepts are blind" (1950, §§A51, B75). Nevertheless, it is the examination of understanding that leads us directly to the problem of mental *representation*. What, then, is the form of knowledge in the mind? This is certainly another of those most fundamental of all questions, and again, literature on the topic abounds. It therefore helps to return to the focus on the representation of geographic knowledge specifically.

OUR WORLD VIEW AS A COGNITIVE MAP

The most frequently used metaphor for describing the form of our geographic knowledge is the *cognitive map*. This term within the context of human cognition, usually refers to the cognitive representation of geographic-scale space and includes the immediate space of one's own neighborhood, as well as very large and complex spatial entities such as towns, cities, and whole environments. Because of their size, such entities normally cannot be seen in their entirety. It is the total collection of these mental maps, each representing a specific domain (my home town, the way to work, known countries of the world, etc.), that together comprise our world view. The cognitive map is a convenient and obvious metaphor for the representation of geographic space. The map as a graphic product is, after all, one of the most familiar means of storing and communicating knowledge about geographic space. As such, the metaphor of the cognitive map has been often used without further elaboration or explanation. Some implications of this metaphor can be very misleading with regard to conceptual structure, and care must be taken to avoid other than the very broadest interpretations.

The original use of the term *cognitive map* is usually attributed to Tolman (1948). One of the first attempts within cognitive psychology to approach the notion of spatial knowledge as being hardwired in the brain was O'Keefe and Nadel's book *The Hippocampus as a Cognitive Map* (1978), in which, as reflected in the title, they suggested that the hippocampus functions as a cognitive map. The idea of a cognitive map was also inherent in the work of Hägerstrand (1967, 1970) on the spatial diffusion of innovation. Tobler (1963, 1961) explicitly raised the question of cognitive maps that people have of their environment, but his focus was on distance distortions in spatial judgments, not on cognitive representations per se. Kevin Lynch, in another classic work (1960), investigated the different images people have of the urban landscape.

The term "mental map" was widely adopted within behavioral geography subsequent to a number of groundbreaking works on human decisionmaking. Gould (1963, 1966) argued that movement activity could be examined with reference to people's mental maps of space preferences. Burton and Kates (1964), as well as a number of other geographers, focused on the perception of environmental hazards and the spatial implications that such images have for spatial decisions. Downs (1970), in his study of people's cognitive maps of shopping centers, suggested that such spaces have definable sets of cognitive dimensions that are equally, if not more, important than those same attributes when objectively measured.

The idea of mental maps has wide intuitive appeal, as demonstrated by Saul Steinberg's famous tongue-in-cheek depiction of a New Yorker's view of the world. This is a bird's-eye view westward from Manhattan, with skyscrapers and streets dominating most of the drawing. Beyond a prominent Hudson River, the space is rapidly compressed. Asia appears as small islands in the distance, beyond an essentially empty Midwest. Steinberg's composition struck such a chord that it has been copied by cities, states, and countries throughout the world to portray people's geographic biases. This intuitive appeal in some cases also led researchers astray, as mentioned by Golledge and Timmermans (1990) in their thorough review of behavioral research in geography. They discussed problems arising when space preferences of individuals are graphically rendered and subsequently used in attempts to explain or predict behavior.

The mental map metaphor conjures up the notion of a unified representation of geographic knowledge as a graphic artifact inside each individual's head, namely, that the cognitive representation of geographic knowledge is isomorphic with the graphic map, and that we retrieve information by reading this "master map" with "the mind's eye." Our knowledge, like the graphic map, is highly interrelated. Piaget's work, as well as later work in cognitive psychology and behavioral geography, certainly supports this notion (Deregowski 1990; Hayward and Tarr 1995; Kosslyn et al. 1989; Montello 1992; Piaget and Inhelder 1956; Shepard 1984). Nevertheless, most known aspects of our cognitive representation of geographic space do not fit the mental map metaphor. A number of cautionary articles have been published about its use (Downs 1981a, 1985; Foley and Cohen 1984; Kuipers 1982; Lloyd 1989).

On the one hand, then, the use of metaphor can guide our thinking and allow insight into something that is very complex or unlike anything we have previously known. On the other hand, the danger is that the metaphor—in this case, the map—can easily be taken too far, with the characteristics of the map becoming the *only* model of knowledge representation, thereby coloring and restricting our own expectations. In the case of the map as a metaphor for human knowledge representation, as discussed by Downs (1981a) there is an additional layer of difficulty in that the map is itself a representation of the world created from our own knowledge of it.

Examining the characteristics of a graphical map display relative to cognitive structure can nevertheless be illuminating with regard to the characteristics of the mental representation of geographic space. It seems intuitive that we do not literally store geographic knowledge as maps (using cartographic symbology, with dashed lines for dirt roads, etc.) (Downs 1981b). Even though we often talk about "mental imagery," mental images are also not exact copies of reality, analogous to a photograph (Anderson 1978; Kosslyn and Pomerantz 1977). They are perhaps best described as representations possibly, but not necessarily, derived from some visual stimulus. Most adults are familiar with imagery conjured up by reading a novel or a poem, or by listening to music. According to Piagetian development theory, there are image-like representations early on in learning, but as the active component of memory, they play an increasingly subordinate role as thought processes mature and become capable of abstract, logicomathematical operations, and knowledge representation itself becomes more abstract.

Although Piagetian theory asserts that our cognitive spatial representation progresses to include Euclidean relations in adolescence and beyond, it has also been shown experimentally by subsequent researchers that cognitive spatial representation is not always Euclidean, as is the usual map representation or graphical image (Montello 1992; Tversky 1992, McNamara, 1992). Indeed, examination of humans' mistakes in judgment, such as in relative distance and direction measurements, have proven useful in providing insights into the structure of our cognitive-geographic representations. Some critical components of our cognitive world view are not definable in any Euclidean geometry, but rather in a hierarchy of concepts.

A number of researchers, some from theoretical extrapolation of Piagetian theory and many based on experimental observation, have asserted that knowledge is encoded as a combination of image and nonimage forms including pictorial, schematic, mathematical, textual, and auditory (Anderson 1978; Hayward and Tarr 1995; Ioerger 1994). But does this not take us back to the metaphor problem again? Do we really have knowledge—particularly most of our "abstract" or "higher level" knowledge—encoded into words, mathematical equations, or schematic (graphic?!) diagrams? Downs has asserted that knowledge within the mind has no form at all—that it is pure relation (Downs 1981a, 1985). In his view, knowledge takes on an explicit form only for the purposes of communication, to convey information and/or facts to another. person. This, by definition, is external (as opposed to internal) knowledge representation. External representation can take many forms, including words, images, diagrams, and maps (Liben 1981). We must therefore maintain a careful and conscious distinction between external and internal forms. To (externally) describe the form of internal knowledge, using the normal, external forms we use to communicate, seems almost automatic, whether or not they are really appropriate.

Some further examination of external–internal and explicit–implicit representations may prove helpful. Undeniably, we can indeed conjure up in our minds visual imagery, that may be the result of a remembered visual experience or a mental creation of a real-world scene, drawing, or map. We have similar "imagery" with respect to specific words and mathematical equations—even music. All of these can be recalled as sensory sensations, whether they are completely imagined or based upon actual sensory experience.

From the Piagetian perspective, sensory sensation is the beginning of learning. We also recall that the first type of knowledge within this framework is variously called figurative, declarative, or (on a geographic scale) landmark knowledge. This consists of what might be termed experiential or observational knowledge (i.e., stored sensory sensation and not abstract or derived knowledge). On this basis, it would seem reasonable to say that such knowledge could be cognitively encoded as stored sensory sensations, which indeed could include a visual recollection of words on a printed page. Can we say that all *other* types of knowledge (i.e., the more "abstract" or "higher level" knowledge) is pure relation, as Downs has suggested?

UNIVERSITY OF HERTFORDSHIRE LRC

Learning has been portrayed as a process of grouping and abstracting. Would this not at least include the modification of remembered sensory sensations over time, to fit accumulating knowledge and evolving points of view (in addition to simple forgetting)? Another argument potentially supporting the idea that at least some "abstract" or "higher-level" knowledge is encoded into visual, auditory, or other external forms is, again, that such external forms can and do channel how we think.

Tulving (1972) made a distinction between episodic and semantic memory. He described episodic memory as the kind of memory that receives and stores information about specific events and groups of events (i.e., episodes), and the space–time relationships among those events. This is the type of memory that deals with remembered experience. Semantic memory is the organized knowledge a person possesses about concepts and their interrelationships. Unlike episodic memory, semantic memory does not refer to unique episodes or events, but rather to universal principles. Information in episodic memory is recorded directly from perception and is susceptible to forgetting. Semantic memory, although it can be recorded directly (by, say, reading a textbook), is often derived through a combination of perception and thought. Through the learning process, certain events and episodes become associated with concepts in semantic memory as examples. It therefore seems reasonable to view our cognitive representation of geographic space as having *both* semantic and episodic elements (Garling 1985). Reading or talking about a neighborhood in our hometown, or about some other familiar city, may prompt visual memories of a restaurant we visited there. Thus, events and episodes remembered as sensory sensations are also a component of "higher-level" knowledge.

Another problematic aspect of the cognitive map metaphor is that it easily leads to the implicit assumption that the cognitive structure of our spatial knowledge is dynamic only in the sense of its original construction and subsequent modification as we learn. Nevertheless, a significant amount of empirical evidence shows that we have multiple cognitive representations for the same geographic environment (Bryant, Tversky, and Franklin 1992; N. Franklin 1992; W. Franklin 1990; Garling, Book and Lindberg 1984; Kosslyn 1980; Kuipers 1978). Not only can we have representations in point-oriented (landmark), linear (route) or survey form of a given domain because of our level of knowledge, but also our representation for knowledge belonging to the same domain changes depending on the task at hand. Thus, people have a route-oriented representation for the path they drive to and from work every day and may generate a route-oriented representation when asked to give directions. In a specific experimental example, Taylor and Tversky (1992a) have shown that people who read about an extended environment from either a route- or a survey-perspective description could answer questions by utilizing apparently flexible perspectives. The perspective used in the text for the learning task did not seem to affect people's ability to answer such questions, further supporting the notion that our brains structure representations on-the-fly.

In contrast to the properties of a map as a graphic artifact, then, the structure of our knowledge representation is not static and monolithic but dynamic and multifaceted (Montello 1992). Our cognitive spatial representation can be discontinuous or linear (route-oriented), segmented, and incomplete. Accommodation of these properties has led Tversky (1993) to describe our world view as a cognitive collage rather than a cognitive map. This, however, derives from the map metaphor and retains some of the same problems, such as not reflecting the dynamic nature of our cognitive spatial representation. Perhaps a better way to describe it is as a multistable system. As such, the various representational forms of the environment are dynamically generated, depending on the initializing circumstances, and can assume any of a number of forms and modes (visual, verbal, etc.). This multistable characteristic, then, also allows us to form conceptualizations of space at many levels, including the following (Lakoff 1987):

- Our immediate reality, as perceived
- Past situations, as remembered
- Future situations, as imagined
- Fictional situations, as related in books, paintings, movies, and so on
- Hypothetical situations, such as those theorized by scientists, economists, or the insurance industry
- Totally abstract domains (mathematical, etc.)

We quickly see that how humans cognitively represent spatial knowledge is an extremely complex process. This is also one of the reasons why research on this topic has been so fragmented. Tversky (1992) asserted that it may be possible to reduce the various properties of cognitive representation to a few simple, general properties. That this view is acquiring an increasing level of support is partly reflected in an edited volume by Portugali (1996). It is this same view that I use for the remainder of this chapter and, indeed, for the remainder of Part I. In the three sections that immediately follow, I explore what might be described as the basic, encompassing properties of cognitive representation: sets of elements and relational rules by which these elements are combined.

GEOGRAPHIC KNOWLEDGE AS A
MULTIREPRESENTATIONAL AND DYNAMIC SYSTEM

To summarize the discussion in this chapter so far, the picture of our knowledge of geographic space is that of a highly dynamic, multirepresentational system. What, then, can be said about its primary components? Certainly, the distinction between the external world and our internal view of it is key, and it is helpful to explore this relationship further, from a process-oriented perspective.

The classical approach assumes a complex internal representation in the mind that is constructed through a sequence of perceived stimuli that generate specific internal responses. In very general terms, this reflects the Kantian view that people are observers of an already existing reality. This is also very much the Piagetian perspective, with specific stimuli triggering specific types of responses and a gradual acquisition of an increasingly complex internal knowledge structure through interaction with the external environment.

Research dealing specifically with geographic scale space has worked from the perspective that the macroscale physical environment is extremely complex and essentially beyond the control of the individual. Research such as that of Lynch, and of Golledge and his colleagues, has shown that a complex of behavioral responses generated from correspondingly complex external stimuli are themselves interrelated. Moreover, the results of this research offer a view of our geographic knowledge as a highly interrelated, external–internal system. Use of landmarks encountered within the external landscape as navigational cues is the clearest example of this interrelationship.

Portugali (1996) extended this view and has explicitly acknowledged a complex interrelationship between our internal representation and the external environment. Furthermore, he asserted, as did Peirce and Cassirer previously on a purely philosophical level (Cassirer 1966, 1973), that elements in the external environment act as an interrelated knowledge representation, external to ourselves, that functions in concert with our internal knowledge representation.

The rationale is as follows: We gain information about our external environment from different kinds of perceptual experience by navigating through and interacting directly with geographic space, as well as through maps, language, photographs, and other communication media. With all of these different types of experience, we encounter elements within the external world that act as symbols. These symbols—whether a landmark within the real landscape, a word or phrase, a line on a map, or a building in a photograph—

trigger our internal knowledge representation and generate appropriate responses. In other words, elements that we encounter within our environment act as knowledge stores *external* to ourselves.

Each external symbol has meaning that is acquired through the sum of the individual perceiver's previous experience and imparted by both the individual's specific cultural context and the specific meaning intended by the generator of that symbol. Of course, many elements within the natural environment are not "generated" by anyone but are nevertheless imparted with very powerful meaning by cultures (e.g., the sun, moon and stars). Man-made elements within the environment, such as buildings, are often specifically designed to function as at least in part as symbols. The sheer size of downtown office buildings, the pillars of a bank facade, and church spires pointing skyward are designed, respectively, to evoke an impression of power, stability, or connection to heaven.

These external symbols are themselves interrelated, and specific groupings of symbols may constitute self-contained external models of geographic space. Maps and landscape photographs are certainly clear examples of this. Elements that differ in form (e.g., maps and text) can also be interrelated. These external models of geographic space correspond to what Donald referred to as external memory, which resides in "a number of external states, including visual and electronic storage systems" (1991, pp. 308–309).

Thus, "mind and environment are seen as two entities existing one inside the other or enfolding each other in implicate relations" (Portugali 1996, p. 14). From this perspective, the sum total of any individual's knowledge is seen as being contained in a multiplicity of internal and external representations that function as a single, interactive whole. The representation as a whole can therefore be characterized as a synergistic, self-organizing, highly dynamic network. Portugali introduced the term interrepresentation networks (IRN) and bases his interpretation on J. J. Gibson's ecological approach, George Lakoff's experiential realism, and Gerald Edelman's ideas within neurobiology.

ONTOLOGY: ELEMENTS OF KNOWING

What, then, can be said about the elements of our *internal* representations? Our world, as conceptualized in our minds, is full of "things." Even very young children can identify recurrences of discrete elements within their environment. We quickly begin to generalize observed "things" into types of things, or categories, and more abstract concepts, building a semantic memory. The entire inventory of things, categories, and concepts is called an *ontol-*

ogy. From the time of the earliest philosophers, the definition of what "things" exist in our minds as memories or concepts has been a central component of how we structure our understanding about the things in our world.

Ontology is the study of the elements in our world as we see it, whether real or imaginary—the basic elements of the conceptualized world and their relationships. These elements can be concrete or abstract, divisible or not. This is in contrast to metaphysics, which attempts to understand the nature of those elements (Heylighen, Joslyn, and Turchin 1993). Thus, my discussion of representation versus reality in Chapter 2 involved metaphysics, examining various views as to the nature of space and time from the ancient Greeks and other cultures through the centuries. Any metaphysics—for example, whether we consider space to be the container of objects or a property of objects, finite or infinite, continuous or discrete, and time cyclical or linear and infinite—also implies an ontology. In the current context, the basic ontology includes concepts we call "space," "time," and "objects." In discussing how knowledge is structured, whether cognitively, on a map, or in a database, some additional ontological elements must also be defined as the basic elements represented.

Ontology has always been tied to language and logic, studying the way in which words, mathematical symbols, and logical propositions function, as a means of gaining insight into how the knowledge represented by the elements of these external representations are internally (i.e., cognitively) structured. In one of his earlier works, Aristotle (1984a) pondered the structure of everyday language and presented 10 ways to describe or differentiate things. He distinguished the following 10 different predicate types, which he called categories: substance, quality, quantity, relation, activity, passivity, having, situatedness, spatiality, and temporality. Although this entire list was not accepted by Aristotle's contemporaries, the first alternative did not appear until Kant's *Critique of Pure Reason* (1950) developed a revised list of categories. His definition of a category differed from Aristotle's. Instead of linguistic predicate types, Kant examined the different ways in which logical propositions function to indicate different forms of human judgment. He listed the following four groups of three as differing logical procedures:

Quantity	*Quality*	*Relation*	*Modality*
Unity	Reality	Inherence and substance	Possibility
Plurality	Negation	Causality and dependence	Existence
Totality	Limitation	Community (reciprocity)	Necessity

He further grouped the first two lists as "mathematic," or belonging to the things themselves, and the second two as "dynamic." Kant explicitly proposed these procedures as fundamental to the *form* of knowledge, regardless of the specific knowledge represented.

Hegel took a very systematic approach concerning the structure of knowledge and attempted to derive a broad ontology from an elemental starting point. Hegel's system is composed of a hierarchy of triads, as reflected in the table of contents of his *Encyklopädie der Philosophischen Wissenschaften im Grundrisse* (1817/1990).

C. S. Peirce was influenced by Hegel's insistence on triads, in that there must be only three types of predicates, or elements, of concepts. He defined categories as follows: "The List of categories ... is a table of conceptions drawn from the logical analysis of thought and regarded as applicable to being" (Hartshorne and Weiss 1931–1958, 1.300).

Peirce (1868) reduced Kant's four groups into a new list, totaling only three categories: quality, relation, and representation. He later termed these, respectively, "firstness" (possibility), "secondness" (actual existence), and "thirdness" (relations), in support of his theory of signs, or semiotics (Morris 1938). Peirce's notion of thirdness was more than simple relation between things and events. He stressed that no relation could be completely defined in a strictly didactic manner; rather, it could only be completely understood within the larger framework of general truths and principles, the structural framework that governs the relation. Peirce's categories are fundamental not only to his theory of signs but also to his views on logic, the working of normative science, and all of human thought.

It is important to note that the term "category" in the ontological sense discussed earlier is very different from our normal use of the term as conceptual groupings of "things" or types of objects. The fundamental ontological categories, as viewed by Aristotle, Kant, Peirce, and other philosophers, were attempts to place modes of description into categories at the highest and most abstract level. These various categorizations all address the basic *modes* of description, or ways of distinguishing one object from another—how we initially distinguish different "things." (How we in turn group these things into categories in an everyday sense is described in the next section.) This philosophical heritage thus provides insight into the *types* of basic and elemental structures from which a representation is made.

FORMAL ONTOLOGIES

The term "formal ontology" has been coined within the artificial intelligence and database communities in order to distinguish the design and construction of knowledge models and database models from the study of ontology in a purely philosophical context. While philosophers usually take a top-down approach to determining what types of things exist, taking an abstract and human-focused perspective, computer scientists and programmers tend to work from the bottom-up, representing limited information within a specific task or problem context (Sowa 2000). Even current artificial intelligence systems are limited to microworlds with limited ontologies. The blocks world has become a standard context for machine vision, robotics, and spatial problem solving, with its ontology consisting of regular geometric solid shapes and a flat surface (Winston 1992). Development of robots that can autonomously navigate in geographic space or solve other types of geographic scale problems are still active research areas. Dealing with geographic-scale concepts is thus viewed as a problem of "scaling up" in an ontological sense.

The term "semantic modeling" has been used to denote a focus on capturing more meaning, in the sense of extending database models to capture the meaning of more than a specified range of observational values, and perhaps beyond immediate constraints (Date 2000). While this also remains an active area of research within the database management systems community, the field early on in its history had devised a context-free ontological framework for database design. In the paper that originally introduced what he called the Entity-Relationship model, Chen (1976) introduced the following three elements (1) entities, (2) attributes, and (3) relationships—as the basic elements of database design. The term "entity," as used, is synonymous with what I have been calling *object*. An entity is a distinguishable *thing*. Thus, a chair, a person, a mountain, a street, and a county are each individual entities that can have attributes associated with them. Attributes of an entity describe or characterize that entity. Attributes of specific entities might include size, shape, orientation, color, and height. Location may also be an attribute associated with an entity, such as the street address of your child's school. There are also relationships between and among entities. Relationships between entities might include larger–smaller than, older–younger than, distance, direction, inside–outside, and so on.

In a geographic representation, locations can also be considered entities, since we can associate properties relative to a specific location (e.g., the corner two blocks down is usually where I catch the bus). Specific locations also have interrelationships, including distance and direction. Geographic locations also have properties associated with them (cold climate, forested). Similarly, events can be considered entities that we remember with specific properties (e.g., Christmas). Events have relationships as well, such as before, during, or after (Peuquet 1984, 1994).

As noted by Date (2000), the notion of entities with attributes and relationships among them, was subsequently extended by Codd (1979), who introduced the notion, within a formal database model sense, that entities can be grouped together by various types of relationships into entity sets, resulting in a model that includes *entity types* and *subtypes*. This is a critical notion within computer databases, for it means that more than individual, observed entities can be represented within a database, and it introduces a conceptual hierarchy with derived, more abstract entities.

Recalling the distinction of factual knowledge of things from events and knowledge of universal principles, as discussed in the philosophical literature, it is fairly evident how a representation of factual knowledge can be stored using just the three organizational elements of entities, properties (or attributes), and relationships (e.g., the grocery store is *next to* the gas station). These elements also apply to the storage of universal principles or concepts. Concepts, however, also include classes in the sense of sets or groupings of entities. As such, class membership becomes a primary relationship (Notre Dame *is a* church), and common or characteristic properties of classes (churches *have* steeples) are of key importance.

Chen (1976) put the notion of entities, attributes, and relationships forward in his original proposal of the Entity-Relationship model without citing any of the philosophical literature dealing with ontology. Rather, these notions were derived via a natural but separate progression within the field of computer science in the development of a theory of database model design. That progression had its origins, like ontology, at least in part in logic and natural language. In the computing context, however, these were applied to the development of computer programming languages and database query languages.

This basic ontological framework still serves the computer science community in a number of subfields, although the names used for actual implementation may vary. Entities are classically called records in primitive files, tuples in relational databases, and frames in artificial intelligence. Where use of this ontological framework can go awry is in assuming that (1) the structure of stored knowledge is inherent in the reality being represented, independent of any human understanding, and (2) the content of that structure is also externally determined and therefore also constant—in other words, that there is a passive and objective reality, which is certainly not the case. This is a trap resulting from the bottom-up approach. If there is too much focus on implementation issues, it is easy and expedient to make simple assumptions about higher level abstractions and lose sight of the larger context regarding the nature of what is being represented.

From a more top-down perspective, if human knowledge can be broken down into these three elements, devising a model of how that knowledge is stored and used could be viewed as a mapping into a set-theory structure. This would fit Tversky's assertions mentioned earlier that the various properties of cognitive representation may eventually be reduced to a few simple and general properties. While entities, attributes, and relationships seem to provide a useful classification of stored empirical facts about the perceived world, it is obvious that the *form* of human knowledge is far more complex. Certainly, a categorization of basic elements relating to the form of knowledge, like knowledge itself, is subjective. Thus, philosophical discussion regarding the definition of elements within Kant's (or any other) scheme, or how one scheme maps onto another, may serve to derive insights into human thought, but there can be no "true" result. These schemes change to suit, depending upon context. And at least within the context of this book, the simplest scheme of entities, attributes, and relationships will suffice.

CATEGORIES

Categories, as conceptual groupings of known entities, provide an organizing structure for our knowledge of places, times, and things that is rooted directly in sensory experience. We group things into categories based on our perceptions of similarities and differences of elements in our environment. This gives us a means to reduce the vast amount of information we store into manageable proportions and allows faster recall. The inherent grouping and ordering of this knowledge is also how we, in turn, interpret the structure of our world and the things in it—in the *construction* of our world view.

As Lakoff (1987) stated, categories are the repositories of our beliefs about the world. They are where our available synthesized information is stored, including known entities, the known properties of those entities, and relationships between them. Of course, we can form groupings of more than just *things*. We can categorize places, events, actions, spatial relationships, social relationships, and many other types of entities, both concrete and abstract, over an enormous range. We also group categories into higher level categories that are related in complex ways.

A category is a grouping of things that somehow are considered similar or treated in a similar way. They are assumed to share common properties that are inherent to the thing or abstract concept. Thus, how entities are grouped into categories is determined by these properties. Given that all members of a category share these common properties, all must be equally good representatives of that category. This is a commonsense view, and until fairly recently, categories were thought to be well understood, essentially as containers with clearly defined boundaries. Things were either in or not in a category. This view has become known as classical category theory. It was in the work of Eleanor Rosch (1973a, 1973b, 1978) that it was empirically demonstrated that although this view is not entirely wrong, it is far from the whole picture. Neisser (1976) attributes the long acceptance of the classical view to an artificialist approach to research within the field of psychology. Laboratory experiments were conducted using accessible subjects (e.g., mice and college students) and contrived tasks (running mazes and responding to flash cards) in an artificial, laboratory environment. Little attention was given to the categories people use in dealing with the everyday world. In the 1970s, research shifted from outward observation of the classification process to seeking the cognitive form of category representation (Estes 1993).

The first person acknowledged to have questioned this view was the philosopher Wittgenstein (1953). He pointed out that a category such as *game* does not fit the notion of a classical category, because there is no clear dividing line between what is or is not a game that can be determined on the basis of shared properties. Rather, Wittgenstein based categories on what he called *family resemblances*—just as members of a family resemble each other in various ways, such as hair color, eye color, or musical talent, but do not necessarily share any single, common property. Similarly, bridge and Old Maid share the property of using cards and are based on competition; solitaire is also a card game, but without any interpersonal competition. Golf is a game, but it does not involve cards. All involve amusement, but when golf is played professionally, does it cease to be a game? In short, games—like families—share a wide but variable set of properties. As such, there are recognizable resemblances, with no single property or collection of properties that determines membership or nonmembership.

Wittgenstein observed that inexactness of category boundaries also implies that those boundaries can be extended or constrained. Certainly, in the case of the game category, we have recently seen a significant extension of this concept, with the introduction of computer games. Wittgenstein cites the example of the category *number*. Numbers originally included only integers but were later extended to rational number, real numbers, complex numbers, and others. One can also, for some purpose, limit the members within a category, such as teaching a small child elementary arithmetic, or define very clear boundaries. The category *number* is very precisely defined within the field of mathematics. Another implication of a group of shared but not prescribed properties is that some members are more central than others within a category. Thus, integers would be intuitively more central to the category *number* than, say, complex numbers. In other words, for any definition of the category *number*, it would be very difficult to exclude integers, but easier at least to ignore complex numbers.

Others in philosophy, linguistics, anthropology, and other fields continued to reveal the inadequacies of classical category theory through the 1950s and 1960s. These included Austin (1961), who asked, "Why do we call different things by the same name?", Zadeh's (1965) fuzzy set theory, Lounsbury's (1964) studies of Indian kinship systems, and Berlin and Kay's (1969) study of basic color terms. Lakoff (1987) provides a review of these developments. It was, however, the work of Eleanor Rosch (1973a, 1973b, 1978) that provided an overall perspective into the issues raised and firmly established category theory as an area of scientific research.

As Lakoff also accounts, this change in the scientific community was traumatic, because to change the basic idea of what a category is means also to change our basic understanding of the world. On a scientific level, there is comfort and appeal in the simplicity and deterministic nature of classical category theory. On an individual level, we understand the world in terms of our own stored knowledge, which is organized into categories. We have categories for everything we think about. Therefore, we also tend to attribute a real existence to these categories. The research of Rosch and others has since demonstrated that categories, as known in the mind, are a function of properties of the elements within the environment, but as *interpreted* by the perceiver.

Rosch's work was based on two basic assertions; first: that the perceived world comes as structured information; and second, that the function of category systems is to store and use maximum information with the least amount of effort (Rosch 1978). In her assertion of a perceived world structure, she stated that the perceived world is not an unstructured set of equiprobable occurrences of elements with unpredictable or arbitrary attributes. Rather, entities in the world are perceived to possess a high correlational structure (e.g., that feathers, and not fur or scales, are more often seen on wings). This is, as she put it, "simply an empirical fact provided by the perceived world" (Rosch 1978, p. 29). Furthermore, this perceived structure is strongly influenced by (J. J.) Gibsonian affordances. Thus, given the physical ability of humans to sit in a particular posture, chairs are more likely to be perceived as having "sit-on-able-ness" than objects with the appearance, say, of a bird, related to specific physical attributes of chairs that allow "sit-on-able-ness." Thus, the relevant notion of a "property" is not something in the world independent of any perceiver. It is, rather, the result of our interactions as part of our own physical characteristics and cultural contexts.

In her assertion that the form of category systems in memory allows the retention and use of a maximum amount of information with minimum effort, Rosch (1978) posited that categories are arranged in hierarchies. The vertical dimension of any particular hierarchical system of categories—such as animal, mammal, dog, collie—in a particular category concerns the amount of inclusiveness of the category. Categories at the same horizontal level of the same hierarchical system—such as dog, cat, horse—would have the same amount of inclusiveness. This, then, forms a taxonomy wherein categories are related to one another by means of class inclusion. The greater the level of abstraction, the greater the inclusion, and the higher the location in the hierarchy. Information common to groups is stored with minimum repetition, and information relating to a particular object or object class can be retrieved by means of direct association or inference.

The implication of this is that some levels of the hierarchy are more useful than others, in that they relate more directly to how elements of the world are perceived. These levels were termed by Rosch to be *basic-level categories*. It is at this level that properties attributed to the specific categories mirror those perceived directly in the real world. These are characterized by overall shape and motor interaction, and are the most general level in the hierarchy at which a mental image can be formed. Basic-level categories are located in the middle of any given taxonomy and are the first to be learned. Rosch determined experimentally that basic-level categorization is mastered by the age of 3. Later, we acquire the superordinate and subordinate categories established by our particular culture. Because basic-level categories tend to represent types of things with which we physically interact directly, Rosch asserted that they tend to occur on a human scale—as do chairs, dogs, and birds.

What, then, are superordinate and subordinate categories good for? Tanaka and Taylor (1991) hypothesized that as a person becomes expert in a particular knowledge domain, the additional knowledge is added as subordinate categories. This provides increasingly specific information about specific instances. Thus, a child can distinguish a bird from a dog but may not know what kind of bird. A bird expert, however, would be able to distinguish a swallow from a thrush, and then to identify specific species of birds within each of these categories based on their physical and behavioral characteristics. From a functional perspective, we become increasingly accurate in our judgments as we add subcategories and the increasingly fine distinctions between them, up to a certain limit, as our experience and stored knowledge accumulates.

Murphy and Wisniewski (1989) hypothesized that superordinates may be more useful when referring to groups of objects. Observation seems to indicate that superordinate names are often used as plurals in everyday speech to refer to groups of objects rather than to individual objects. This also relates to the research of Markman and Callanan (1984), who found that children understand superordinate terms as collections rather than classes of things. In a series of experiments, Murphy and Wisniewski found that superordinate categories are useful in accessing information about scenes and groups of objects within them. Superordinate categories also provide spatial, taxonomic, and other types of linkages between entities as explicitly stored relationships.

The research of Tversky and Hemenway (1984) demonstrated that our knowledge at the basic level is mostly concerned with *parts*. This agrees with Rosch's theory, in that parts usually correspond to functions and how we physically interact with objects. For example, chairs have *seats*. This is a critical part for all things we categorize as chairs because of the nature of how we physically interact with this category of object. Parts also determine shape and, hence, the way that an object is perceived and imaged. This also means that members of the same basic-level category tend to *look* alike. Rosch and her associates have shown that no such result can be obtained for categories higher in the hierarchy (i.e., superordinate categories), because different items of furniture, for example, simply do not look alike. Because of this direct association with the experiential world, basic-level categories also contain more information in terms of attributes and functional characteristics of objects.

Tversky and Hemenway also suggest that we impose this part–whole structure on events, and that our knowledge of event categories is structured very much like our knowledge of physical object categories. A similar point was made by Lakoff and Johnson (1980), who suggested that event categories and other categories that deal with abstract entities are structured metaphorically on the basis of structures from the realm of physical experience. This means that our stored knowledge in general is organized by *kinds* (taxonomies) vertically within the hierarchy, and by *parts* (partonomies) within categories. According to Tversky and Hemenway,

> Taxonomies serve to organize numerous classes of entities and to allow inference from larger sets to sets included in them. Partonomies serve to separate entities into their structural components and to organize knowledge of function by components of structure. The informativeness of the basic level may originate from the availability of inference from structure to function at that level. (1984, p. 169)

According to Rosch's (1978) theory of categorization, properties of entities have varying *cue validity*. This is the idea that the presence of some single properties provides a level of probability for membership within a specific category. Thus, if some living thing is seen to have feathers, there is a very strong probability that it is a bird. The probability would be essentially zero that it would be a fish. On the other hand, a living thing that is blue could equally be a bird or a flower—given just this property alone.

To increase the distinctiveness and flexibility of categories at all levels vertically in any hierarchy, categories tend to become defined in terms of *prototypes* that contain the attributes most representative of items within that category. Following the ideas of Wittgenstein about centrality within a category as it would apply to prototypes, Rosch devised the theory of *graded* internal structure of categories. Thus, beanbags and barstools would be less members of the category "chair" than the straightbacked, four-legged type usually found in kitchens and dining rooms. This gave rise to the theory that the similarity of a given example to a category prototype would correspond to the centrality of the example's membership in that category. This would also imply that category membership could be simply measured on the basis of similarity to a prototype.

Armstrong, Gleitman, and Gleitman (1983) eventually showed that there is no way simply to reduce category membership to similarity among category members. Rosch also abandoned graded categories as a universal principle in her later work. Some categories, such as birds and chairs, were found to have strict boundaries, whereas others, such as color (red vs. blue) and number, seem to have inexact boundaries and graded internal structure. Categories with strict boundaries are not graded but still have some sort of internal structure that produces goodness-of-example ratings. This is important, because it means that categorization is not based on the characteristics of the categorized objects themselves. The claim that something is a bird does more than assert some degree of similarity to a prototype. Because it results in asymmetrical inferences, there must nevertheless be some internal structure that is part of our concept of what a bird is and plays a role in our reasoning (Lakoff 1987; Rips 1975).

Lakoff extended Rosch's theory of prototypes to the notion that many categories have a *radial* structure, with their most central members being predictable through typical properties of the category. Other, noncentral members are "motivated" by central members in that they have family resemblances to the more central members. Less typical members are related to central members of a category through various mechanisms, including family resemblances and image schemata (Lakoff 1987).

The nature of cognitive categories as often not having sharp boundaries was a key finding that suggested categories are not defined in terms of stored lists of characteristics that are compared with perceptual input, nor on an everyday basis do we need to compare visual input with stored visual proto-types. The way categorization works has been described by McClelland, Rumelhart, and Hinton (1986) as more like a neural network, with patterns of weights, and with some characteristics and combinations of characteristics generalizing better than others. McClelland and his associates' implementation of a neural network model that simulates how memory abstracts categories over the course of experience was interpreted as a pattern activation process.

There also seems to be a major shift that occurs in the basis of categorization of physical objects for small children as they develop, as well as for people of any age as they progress from naiveté to expertise within a specific knowledge domain. Keil and Batterman (1984) described this as the "characteristic-to-defining shift." Younger children and novices in a given area tend to rely on perceptible attributes. Older children and experts tend to employ more theoretical assumptions and models that may also include explicit definitions, such as for an *island*. Keil and Batterman's experiments demonstrated that most fourth graders believe that no amount of tinkering with physical characteristics can change a skunk into a raccoon. Building on their observations, they also speculated about how social norms and the social functions of specific categories determine the definitions that people form. It is interesting to note that the characteristic-to-defining shift is in agreement with Piagetian development theory, as described in the preceding chapter.

Barsalou (1983) studied "ad hoc" categories. In contrast to "natural" categories (e.g., birds, mammals, chairs), ad hoc categories are formed on-the-fly to handle special situations and the nonroutine aspects of daily living. Examples of ad hoc categories would be "things to take on a camping trip" and "places to look for antique desks." These categories seem to arise spontaneously for use in specialized contexts and violate a number of characteristics of natural categories in that they (1) are not well established in memory, (2) are not part of a hierarchy, (3) tend to violate the correlational structure, and (4) are not categories commonly held by most people. Some ad hoc categories may be used frequently enough by individuals that they become ingrained in their memory. Even though they may still violate the other characteristics of common categories, they take on a similar function in a person's knowledge structure. For example, after going on repeated camping trips, the category "things to take on a camping trip" may become well established in memory. This seems to parallel the initial development of a category structure in children.

In summary, what is now known about human cognitive categories goes beyond the classic view. Classic categories do not serve in most cases of conceptual categorization. There is no unambiguous mapping between the world and our knowledge of it. The arrangement of categories, both between and within categories, is strongly influenced by physical characteristics, cultural context, and the previous experience of the perceiver. Although physical attributes and cultural context dictate a significant amount of commonality in the category systems between individuals, categories vary from individual to individual, and they are always changing for any individual, depending upon the situation at hand.

GEOGRAPHIC CATEGORIES

How does category theory apply to geographic categories? For objects that have a physical manifestation in the real world (i.e., they are not totally abstract), shape is still among the most important criteria for identification, even though such objects cannot be interacted with in the same ways as human-scale objects. The shape identified with a particular type of geographic object can include an overall shape, as well as the shapes of key component parts. As an example of geographic-scale objects, we distinguish churches from other kinds of large structures primarily by their distinctive overall shape, provided mainly by the distinctive shape of steeples, a key element characteristic of churches. Characteristic elements also commonly have important connotations within specific cultural contexts. These connotations, and the deliberate use of the symbolism attached to elements, particularly in the man-made landscape, have been explored in the geographic literature (Duncan 1993; Knox 1991; Meinig 1979; Tuan 1977). For example, steeples were originally designed in Medieval Europe to draw people's eyes heavenward, and their size, as well as the overall size of the building, was intended to convey an impression through the senses of permanence and power. Western church architecture thus conveys a powerful symbolism connected with the teachings of Christianity.

Geographic-scale entities, like other entities encountered in the environment, tend to be arranged in nested hierarchies. For example, country, region, state, city, and neighborhood are conceptually arranged in a hierarchical order. There is also a container relationship built into this hierarchy. In other words, for any set of specific examples (e.g., the United States, New England, Massachusetts, Boston, and Beacon Hill), each entity is spatially contained within the entity directly above it in the hierarchy. A nested hierarchical structure of cognitive geographic knowledge has been experimentally confirmed by a number of researchers (Eastman 1985; Stevens and Coupe 1978;

for a review of this work, see also McNamara 1992). Besides the obvious spatial containment that can be seen in cartographic portrayals of such geographic entities, cartographers also select symbolization to represent spatial categories on maps, so that their hierarchical relationships are clearly portrayed.

Tversky and Hemenway (1983) investigated whether Rosch's basic-level categories could be extended to environmental scenes as well as geographic-scale objects. In one experiment, preselected geographic terms were arranged in a skeletal hierarchy, with one level each of superordinate, basic-level, and subordinate categories. They used *indoors* and *outdoors* as their superordinate categories. Basic-level categories under *indoors* were *home, school, store,* and *restaurant.* Under *outdoors,* basic-level categories were *park, city, beach,* and *mountains.* The subordinate categories included terms such as *lake beach* and *ocean beach* under *beach,* and *industrial city* under *city.* Subjects in the experiment were provided with both pictures and textual descriptions, and were asked to provide lists of attributes for specific elements at all three levels of abstraction.

They found that the preferred level of description of objects embedded in spatial contexts was the basic level that provided the greatest amount of information about the scenes and objects involved. Thus, *beach* and *city* provided much more information than the superordinate category *outdoors,* and subordinate categories *industrial city* and *Midwestern city* provided only marginally more information. Most information that respondents provided for basic-category scenes and objects was observable, perceptible properties rather than more abstract types of properties, as would be expected given Rosch's description of basic-level categories. Indeed, the mention of the term *city* to most people will evoke images of tall buildings, dense residential development, and lots of traffic, among other attributes (which today might include high crime, cultural amenities, etc.). The grouping of information in basic-level categories, whether objects or scenes, seemed to correspond to a naturally occurring correlational structure of properties. They also found that scenes form categories in and of themselves (with a hierarchical structure of kinds and parts within categories). Moreover, they are the spatial contexts in which basic-level objects appear. The implication is that scenes allow us to determine the situational and spatial context of objects across a spectrum of possible contexts. This applied to the temporal dimension as well. Event structures form the temporal context in which objects are elements.

Lloyd, Patton, and Cammack (1996) also performed an empirical investigation of geographic basic-level categories using a skeletal, three-level hierarchy. The category *place* was the single superordinate category; a series of common geographic terms hypothesized as basic-level were in the middle, and specific examples of those categories comprised the subordinate level (cf. Figure 4.1).

Their focus was based on the observation that basic-level categories can occupy many different geographic scales. For example, the question "Where is your home?" can evoke a number of basic-level responses. I can say that my home is in College Heights, State College, Pennsylvania, the Northeast, or the United States, depending upon the context. If I am in State College, I would answer College Heights, but in Amsterdam, I would answer that I am from the United States. These answers correspond to specific instances of the basic-level geographic categories of neighborhood, city, state, region and country, respectively, as shown in Figure 4.1. The notion that key information contained within basic-level categories includes parts also translates into a container relationship for spatial categories: Countries contain regions, which contain states (or provinces, departments, etc.), which contain cities, which contain neighborhoods. The hierarchical structure of spatial knowledge expressing containment as well as other spatial relationships has also been suggested by others (Eastman 1985; Stevens and Coupe 1978). Specific instances of categories are cognitively associated with basic-level categories as subordinate categories. In this way, information learned about any specific city (e.g., New York) can be generalized to provide information about all cities.

According to Rosch's theory, basic-level categories have more information stored within them as a result of their typically human scale and involvement with direct human perception and interaction. Certainly, people cannot directly perceive and interact with countries on the same level as chairs or

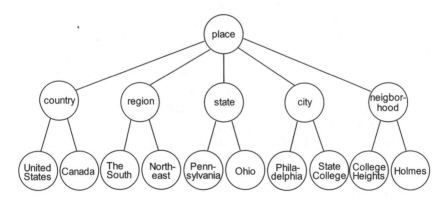

FIGURE 4.1. A three-tiered hierarchy of geographic entities. Adapted from Lloyd, Patton, and Cammack (1996). Copyright 1996 by Blackwell Publishers. Reprinted by permission.

birds. Human interaction with anything at geographic scales is fundamentally different. It can also be argued that interaction with things at widely differing geographic scales (neighborhoods vs. countries) is also fundamentally different. This raises two questions concerning the geographic basic-level categories tested by Lloyd and his associates (1996): First, do geographic basic-level categories at widely differing geographic scales contain approximately equal amounts of information? Second, would the amount of information stored in geographic basic-level categories be significantly more than that stored in superordinate or subordinate categories?

Their results confirmed that geographic basic-level categories do contain about the same amount of information, regardless of geographic scale, and that this amount of information is significantly greater than that in superordinate and subordinate categories. They offered several possible explanations for why this characteristic of basic-level categories should *not* hold true for geographic basic-level categories. However, they offered no explanation for the results they found. It would have been illuminating if they had tested for specific types of information stored at differing scales. How can this characteristic for basic-level categories continue to hold true, even without the possibility of direct body-level interaction?

Another issue not addressed as yet in the literature is geographic categories with inexact boundaries. As can be seen in the work just described, research in geographic categorization has so far dealt with categories that have clearly defined boundaries. One exception is the work of Suchan (1997), who investigated the structure of the categories "rural" and "urban."

The hierarchical structure of stored geographic knowledge and how we use that knowledge, is also evidenced by systematic distortions in judging relative geographic locations. The first experimental example was documented by Stevens and Coupe (1978), who asked subjects to indicate the direction from a given city to another by drawing a line in the proper orientation from the center of a circumscribed circle, the center of which indicated the "from" city. North was noted on the diagrams, and the "to" city was only named. They found in subjects a strong tendency to mistakenly attribute the relative direction of superordinate geographic units to those cities. One example was the direction from San Diego to Reno. Most people indicated that Reno is east of San Diego, although it is actually north. Stevens and Coupe argued that this mistake in judgment occurs because of the hierarchical structure of the stored spatial information. If the actual relative direction for two cities is not known directly, the direction is inferred on the basis of the states containing them. Because Nevada is generally east of California, the wrong inference is made. Stevens and Coupe also hypothesized that spatial relationships between two locations are only stored explicitly (i.e., retaining Euclidean measurement) if they are within the same, containing, superordinate unit. Similar experiments were conducted by Wilton and File (1975) and Maki (1981), as well as Hirtle and Jonides (1985) and Lloyd and Heivly (1987) on the intraurban scale.

Another question of particular relevance for the encoding of geographic-scale spatial knowledge is whether there is any difference in how knowledge that is *not* acquired through direct sensory interaction with the environment is stored. Knowledge acquired directly has been called *primary knowledge*, whereas knowledge acquired indirectly (being told by another, or by reading books and maps) is *secondary knowledge*. Certainly, all geographic knowledge at national and global scales is secondary knowledge, except perhaps that acquired by the occasional astronaut. Studies have shown that use of secondary sources allows knowledge to be acquired faster, in what seems to be the same representational form (i.e., hierarchical categories). As already mentioned, Lakoff and Johnson (1980) have suggested that abstract knowledge is structured metaphorically on the basis of experiential structures.

A number of researchers who have investigated the effects of map reading versus real-world experience on orientation biases, including Evans and Pezdek (1980), Presson and Hazelrigg (1984), and Presson, DeLange, and Hazelrigg (1989), as well as others, have found that learning spatial information, such as paths, in an orientation-specific manner biases later judgments. Maps, of necessity, are more prone to presenting geographic information in an orientation-specific manner. Geographic information learned from maps is therefore also more prone to orientation biases. Although the cartoon in Figure 4.2 overstates the problem, the point it makes has some validity. MacEachren (1992) tried to determine whether this problem could be overcome by showing the same information on multiple maps with various orientations. The curious result was that, when asked, people took longer to determine direction between two specific entities. The effects of cartographic representation on learning and communicating geographic knowledge is discussed in more detail in Chapter 7.

SCHEMA: THE LINK BETWEEN PERCEPTS AND CONCEPTS

Regardless of the form that a knowledge representation may take or the basic components we use for modeling that representation, rules or patterns are needed to interpret what we perceive, and to relate those perceptions to what we know. This follows from the observation that there must be something in between the sensory appearance of, say, a church on one hand, with the concept of church on the other, that makes this association possible when a

BEETLE BAILEY

FIGURE 4.2. Orientation bias in maps. Reprinted with special permission of King Features Syndicate.

church is actually seen, heard, felt, and so on. This mediating form was called the *transcendental schema* by Kant (1950) and the *symbolic form* by Cassirer (1973). These forms, however, have been viewed in very different ways by many authors, leading to much confusion.

Kant's *schemata* are the pure form in which the matter of knowledge is organized, and are as such value-free. In Kant's view, knowledge is an affair of judging the true and the false, the good and the bad, the relevant and the irrelevant. He reasoned that there must be some mechanism, some guiding principles for applying pure concepts to appearances in order to make these judgements. Kant's *schemata*, then, consist of the rules for judgement, rules for categorization/inclusion, association, ordering, and so on.

Cassirer's *symbolic forms* are imbued with meaning and have value as defined within the specific cultural context. Cassirer became intrigued with this notion of a mediator between the realm of sensory data and the realm of the intellect. Unlike Kant, who saw these two as separate and distinct, and understanding as a spontaneous leap, Cassirer saw them as a continuum. He explored the implications of this in his multivolume work *The Philosophy of Symbolic Forms* (1953–1996). He mentions three distinct levels of understanding, which have a distinct similarity to Piaget's developmental stages:

1. The expressive level.
2. The intuitive level.
3. The conceptual level.

The expressive level is the first level along the sensory–conceptual continuum. Cassirer described this as the level of pure immediacy, with the world understood from an emotive–affective perspective (frightening, cheerful, exciting, soothing, etc.). In this mode, all of reality is interpreted purely on the basis of what is seen, heard, touched, and so on. Therefore, there is a total congruence between meaning and sense data. Cassirer also described this level as involving a form of primitive thinking. From a cultural perspective, this is also the domain of myth and art.

The distinction between meaning and sensory data becomes more apparent at the intuitive level. On this next level, fixed centers of orientation are distilled out of the flow of sensory data. These become the basis for the representation of something else—"things"—with properties and interrelationships: "Thus it is clear also . . . that the "thing" is by no means grounded in the sensory character of perception in the mere impression, but is of a reflective character" (Cassirer 1973, p. 121).

At the conceptual level, on the opposite end of the sensory conceptual continuum, there is a recognition of fundamental order inherent in sensory data. From this, the principles that govern various aspects of perceived reality are understood as such. These principles cease to resemble physical things directly, but it is precisely this separation that provides the transition to pure knowledge as abstract and generalized truths. At this level, to recognize a perceived object means to subject the complex of sensory data to a rule, or set of rules, that enables the sensory data to be placed within this universal, known ordering. It is this highest level that corresponds to Kant's notion of *schema*.

The field of psychology adopted Kant's notion of schema, along with the term. The psychologist Otto Seltz (1927) interpreted schemata as networks of concepts and relations that guide the thinking process. Bartlett (1932) asserted that a schema is an active and dynamic organization, and showed experimentally that people organize known elements into larger patterns, and that recalled material tends to assume conventional forms defined within the culture. Seltz used Kant's schemata as the basis for his notion of *anticipatory schemata*, where, given one concept, another concept can be found by setting up expectations. Anticipatory schemata represent overall patterns that guide thought in the most promising directions toward a resolution, utilizing the linkages between concepts.

A number of researchers have found evidence of this mechanism at work in people playing chess (Chase and Simon 1973; de Groot 1965). de Groot was interested in what differentiates master chess players from less skilled players. He showed chess players a logically coherent chessboard arrangement and asked them to "think aloud" and talk through their process of finding the best move from that point. One finding of his analysis was that instead of analyzing all possible moves, a master chess player simply "sees" which moves are worth considering. He also noticed that master-level players were able to reconstruct a coherent chessboard arrangement almost perfectly after about 5 seconds, but that this ability decreased rapidly for players below this level. If the chess pieces were placed randomly on the board, the players performed about equally, regardless of their chess-playing skill.

Chase and Simon (1973) asked the question "What is it that the master is perceiving during a brief exposure to a logically coherent chessboard configuration?" They studied three chess players—a beginner, a class-A player, and a master—and asked their subjects to reconstruct a number of specified chess positions, sometimes from memory and other times with the original board configuration in plain view. They found that the master-level player recognized familiar or meaningful groupings of pieces that were already structured in memory, including logical configurations, such as pawn chains with blockading and defending pieces, castled king positions, and even clusters of pieces of the same color. Chase and Simon called these groupings "chunks," each consisting of about four to five pieces, which were combined by the player into complete board configurations. Many of these groupings had names (e.g., "castled king") that acted as labels to help in the recall process. These groupings were interlinked in the sense that various groupings were associated with each other to characterize specific situations.

A similar anticipatory mechanism has been found at work in the task of sight-reading music. Wolf (1976) interviewed four pianists noted for their sight reading ability. All the pianists interviewed agreed that musical sight-reading is essentially a pattern recognition task, and not one specifically of playing individual notes. This work confirmed an early study by Bean (1938), who used a greater number of subjects with various degrees of expertise. The sight-reading musician scans the page, looking for familiar patterns, including groups of notes as chords and sequences, as well as rhythmic and phrasing patterns. A good sight reader is aware of the melodic patterns, ornamentation, and rhythms characteristic of specific periods of music, such as the baroque, and even specific composers.

A movement from a Handel sonata was used to illustrate how a musician scans for familiar patterns. The following example looks quite complicated:

But, as Wolf noted, it reduces to a simple scale (circled notes), with each note of the scale separated by "changing notes." Changing notes, three-note embellishments, are quite common in baroque music:

So the visual pattern of the notes on the page are associated with the concepts of musical scale and changing notes, which in turn set up expectations regarding the overall pattern of the piece. Such cues can also be obtained by listening to a short passage of a particular piece. A musician anticipates much about a piece, even before playing it for the first time, using a variety of visual and/or audio cues. Skilled musicians are so familiar with such cues that they do not consciously attend to them while playing. Musicians process the cues automatically as conceptual chunks, relying on conditioned "motor memory" associated with those chunks to get the fingers to the right notes at the right times. It was also revealed that much guesswork is involved in playing a piece for the first time. The musician sees a few clues and fills in what seems appropriate to complete the recognized patterns. A similar anticipatory process is apparent in tennis. One of the essential elements of becoming a good tennis player is to be able to anticipate where and how the next stroke needs to be executed, based on visual cues of ball trajectory and body language of the opponent, even before the opponent hits the ball back.

Wolf suggested a close correspondence between musical sight reading and reading conventional text, where groups of letters form words and phrases that—particularly in the case of poetry—also have their own rhythm. Letter-by-letter processing is resorted to only when an unfamiliar word is encountered, once one is beyond the initial learning stage of a small child. Patterns soon become recognized, familiar, and, eventually, anticipated. In music, there is also simply not enough time to read and play note-by-note, if the appropriate tempo is to be maintained while playing a musical composition. Thus, both letters and notes are in similar ways used as cognitive building blocks of a larger symbolic unity. Just as the chessboard looks like a jumble of pieces to someone who does not know how to play chess, a page of music and a page of text are only meaningless squiggles on paper to those who have not learned to read them. Similarly, a spoken language sounds like confused babble to someone who does not speak that particular language. Schemata operate on various levels to give structure to games, music, and language; to aid in recognition of individual elements (game pieces, notes, and letters) and meaningful groupings; and ultimately to attain a more global meaning.

Kant understood the *schema* as the mediating element between pure concepts and appearances, percepts and categories, allowing what we gain through the senses to be interpreted and thus to be given meaning. He stated that the schemata are "void of all empirical content" (Kant 1950, §§A138, B177); that is, they are the means by which we *construct* order. As such, they are not receptacles of stored knowledge in any direct sense; rather, they contain patterns for organizing and gaining information from interaction with the world, as well as from purely abstract thought. In a very general sense, then, the schemata are central to the *process* of thought, whereas *categories* are where the synthesized information derived from experience is stored. In a certain sense, schemata can be viewed as templates for ontological categories. The overall mechanism can be viewed schematically in Figure 4.3. Drawing upon available information, the schemata direct our actions, for example, by focusing on specific elements in our visual field, interpreting them as important or interesting, and allowing us to disregard other elements. Any elements, or characteristics of known elements, that are judged to be new or different would then result in a modification of our stored categories relative to those elements, as well as potential modification of the schemata (i.e., the rules for interpreting) themselves. Such modifications involve the primary mechanisms of imagery and metaphor, as well as induction, deduction, and abduction.

As discussed in Chapter 2, Kant viewed *all* schemata as being based upon space and time. These are also the only two forms of sensory experience. We may be able to think abstractly of things that have no extent or position in space or time, but we cannot perceive them. It is through knowledge of space and time that we can also manipulate elements on a purely abstract level, completely disconnected from sensory experience, as in mathematics.

Whereas Kant asserted that the schemata are unchanging, Neisser (1976) built upon Bartlett's notion, emphasizing that schemata are never

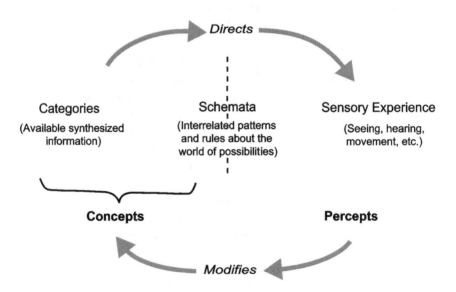

FIGURE 4.3. The overall process of spatial knowledge acquisition.

static or rigid, but are very dynamic. These mediating structures, in Neisser's view, not only direct all perceptual activity of all modes (seeing, hearing, etc.), but also are modified as this activity occurs, as a result of increasing world experience. This process never ceases, regardless of the individual's level of development or expertise within a particular knowledge domain. Neisser's schemata are also interpreted differently in various situations. In Neisser's view, applications of these fundamental cognitive patterns are the best way to understand all forms of cognition.

In cognitive psychology and cognitive linguistics, the terms "image schema" and "embodied schema" are used interchangeably to distinguish them from the propositional interpretations developed by Rummelhart and others (Gibbs and Colston 1995; Johnson 1987; Lakoff 1990; Talmy 1988). These two terms are intended as explicit reminders of the notion of schematic patterns grounded in and determined by our bodily interactions with, and sensory perceptions of, our environment in space and time. This includes input from all sensory modalities, not just vision. They differ in meaning from Kant's notion of schemata in that image schemata or embodied schemata do not include schematic patterns for purely abstract concepts, including mathematical concepts such as a schema for "number" (i.e., the quantity of something). Image schemata are also not images of objects in a literal sense. For example, no image could ever adequately define the concept of a triangle. It would never attain that universality of pattern that would adequately characterize all triangles, regardless of the relative lengths of the sides or the magnitudes of the interior angles, or whether flat or draped over a sphere or other surface shape.

Studies in cognitive linguistics suggest that over two dozen different image schemata appear regularly in people's everyday thinking, reasoning, and imagination (Gibbs and Colston 1995; Johnson 1987; Lakoff 1990), including schemata for *container, balance, path, source–path–goal, cycle, center–periphery,* and *link*. Although these are called image schemata, they are fundamentally spatial in nature. Our schemata for spatial and temporal orientation are so fundamental and pervasive in our experience that they are usually taken for granted. They have also been extended into many seemingly unrelated contexts. Cognitive linguistic research has shown how these schemata can be extended into many contexts through the mechanisms of metaphor and metonymy, as well as other transformations (Gibbs and Colston 1995; Johnson 1991). Consider the word "stand" in the following sentences:

Stand at attention.
She won't stand for such treatment.

The law still stands.
The president stands firm on his policy.

Certainly, the first sentence relates to the physical act of standing. The others use the word in a more figurative sense, seemingly related to multiple, basic image schemata—perhaps, at least in part, the image schema of *balance*, with its implied pattern elements of the relation of subject to ground and nonmovement, providing the single core-sense usage over the range of these diverse examples.

Lakoff (1987) has generalized this context in his *spatialization of form* hypothesis, which maintains that *all* conceptual structure is understood in terms of image schemata, through the use of metaphorical mappings from physical space into "conceptual space." Within this mapping, image schemata, which structure space, are related to abstract configurations, which structure concepts. Furthermore, the metaphorical mappings needed in order to achieve this are themselves understood in terms of image schemata. Therefore, our basic notions of space are fundamental to learning and understanding in all domains. Specifically, he maintains the following:

- Conceptual domains are distinguished from one another in terms of *container* schemata.
- Mappings from entities in one domain to entities in another domain are understood in terms of *source–path–destination* schemata, although the path is unspecified.

Image schemata in this view are not only concepts that are directly understood and used in everyday dealings with the world around us, but they are also used to structure and relate other complex concepts.

Studies in visual cognition have also demonstrated the importance of time in understanding. In a famous experiment, Johansson (1973) placed lights at the major joints of people dressed completely in black and photographed in the dark. If these people did not move, observers reported seeing no identifiable pattern. If, however, the people moved in performing some ordinary activity, such as walking or dancing, the observers immediately were able to identify the appropriate number of people engaged specifically in that activity. If the people then stopped moving, the observers reported that the lights returned to a random pattern, with the people seemingly disappearing. Johansson concluded that the perception of the gestalt pattern of an event progressing in time is basic to human cognition.

Given this description of schemata, it seems that not even the very youngest individuals are completely without schemata, although they would lack experience-derived information (i.e., categories). It is the *schemata*, then, that constitute the innate concepts that allow the infant to begin learning about the complex world around it—the infant directs actions in order to gain at least some initial sense out of the barrage of sensory data. I have already discussed the idea of innate concepts in the previous chapter. Piaget's innate concepts of separation, proximity, order, and enclosure can be interpreted as simpler versions of the regularly occurring schemata enumerated here, for example, *container* derives from separation (distinctiveness of objects), *cycle* derives from order (space–time succession), and so on.

Neisser (1976) uses this notion of schemata to interpret the development of perception in infancy and to explore the relationship between the cognitive map and the schemata embedded within it. He uses Kevin Lynch's (1960) book to illustrate his assertion that cognitive maps and schemata are related in two important ways: A cognitive map accepts information and guides movement. In this respect, it *is* a schema, albeit on a larger scale, with containers (buildings), center–periphery (e.g., downtown–suburbs), source–path–goal (work–home path), and so on. Cognitive maps thereby constitute a very complex, orienting schema that allows one to navigate autonomously through the environment. Schemata are also used in a completely abstract fashion. For example, one can plan a complex trip to accomplish Saturday-morning errands while sitting at the breakfast table, or imagine the impact of a proposal to build a new highway through town.

The problem with the notion of a *schema* is what *form* does it have? As previously mentioned, Seltz (1927) described schemata as forming networks (or possibly trees) of subschemata. Each subschemata represents the conceptual constituents of the concept being represented. Thus, for example, the schema for a face would be made up of a certain set of subschemata including a subschema for "mouth," another for "eye," and so on. Each of these may also have subschemata ("mouth" would include a subschema for "lip," etc.). Neisser interpreted schemata to be like input formats in a computer programming language, with some schemata also functioning as plans. As formats, they specify the ordering that must occur for information to be interpreted correctly, and they separate the relevant from the irrelevant not only in sensory input but also in subsequent thought. A schema also operates at varying levels of generality. Thus, we can interpret an object we see as a chair or the letter *B*, even though chairs and letters can occur in a wide range of physical forms. As plans, some schemata provide a sequencing of elements and events for interacting with the world (i.e., a pattern of action). In both forms, however, the distinction between form and content is blurred, because information that fills in the format and defines the plan at one moment becomes the plan for the next action and because we continuously accumulate experience with the world and adapt to it. In a survey of schema theory in cognitive research, the mainstream view of a schema was expressed as

> a cluster of knowledge representing a particular generic procedure, object, percept, event, sequence of events, or social situation. This cluster provides a skeleton structure for a concept that can be "instantiated," or filled out, with the detailed properties of the particular instance being represented. (Thorndyke 1984, p. 167)

This view still holds. Many associated forms have been developed, particularly frames, scripts, and semantic networks. Frames, originally developed by Minsky (1975), retain the notion of schemata as modularized yet interlinked knowledge patterns, which can be populated with specific, known instances. Frames provide an explicit organizational scheme for knowledge representation but they require something else to put them in motion. Semantic networks are similar to frames, with particular emphasis on the inter-relationships among knowledge elements, in contrast to frames, which concentrate on the knowledge elements themselves. Scripts, as proposed by Schank (1976), in contrast, represent a sequence of activities, such as cooking dinner or hosting a party. Nevertheless, a script is just what the normal connotation of the term implies: a predetermined sequence of actions that define a well-known situation. Scripts do not provide a mechanism for dealing with variations or combined situations, such as different kinds of parties. Rumelhart (1975) proposed a representation intended specifically to mimic the dynamic nature of human conceptual schemata, building upon the rewrite rules from generative linguistics. Fillmore's theory (1982) of frame semantics is similar in many ways to Rumelhart's schema theory.

All of the models mentioned here, except for scripts, are structures for encoding propositional information related to specific elements in the environment and specific situations. All are attempts to provide a means to represent human knowledge in the form of a computational model and have been used for computer implementations within the realms of language understanding, computer vision, and artificial intelligence—within both spatial and nonspatial contexts. Nevertheless, these computer implementations exhibit limitations that derive directly from the representational approach used. These limitations often stem from assuming that such models of basic knowledge patterns provide a more precise, more discrete, and more complete form than is actually appropriate. A *proposition* is a function that tests for a condition specified by its arguments, and is either true or false. A script is simply a description of a sequence of events, with a starting state and an ending state. Thus, neither form of representation can accommodate the mechanisms of metaphor or metonymy. Furthermore, no known representational technique can yet truly imitate or otherwise model that triumph of the human mind— creativity, the generation of what seem to be completely new ideas and concepts.

To summarize, schemata can be said to be the knowledge of pure and abstract relation. The problem with determining the form of schemata may be that they have no permanent form at all, but rather are instantiated as needed for solving the problem at hand. This relates directly to Downs's view of a cognitive map (Downs 1981a, 1985). From the previous discussion, we can see that there are nevertheless a number of fundamental and generally acknowledged characteristics of schemata:

1. They are central to the *process* of interpreting sensory data and acquiring new knowledge.
2. They constitute the rules, or patterns by which things can be evaluated and that guide action.
3. They are rooted in the dimensions of space and time.
4. They are highly interrelated and dynamic, operating at many spatial scales, contexts, and levels of complexity.
5. They are always present in the cognizant mind.

SUMMARY

As we saw in Chapter 3, the process of learning is universal. Children, starting from infancy, and before any language or cultural context is learned, proceed through the same stages of spatial and temporal understanding. Learning begins primarily with perceptions of the real world that are based on our innate physical capabilities of seeing, feeling, hearing, and so on. The essence of the learning process, the construction of a knowledge *structure*, involves our continued organization of knowledge by acquiring an understanding of generalized patterns and rules that relate various elements in the world (schemata) and knowledge regarding specific objects and types of objects (our ontology and categorical structure). As knowledge grows, more commonalities, distinctions, and connections are discovered, and groupings are further categorized, abstracted, and related as higher level concepts.

A category is always defined in reference to a cognitive model. This cognitive model begins as mostly physically based schema that we use to interpret our experiences. Categorization begins at the basic level and is closely tied to direct perception of physical attributes and affordances. This process becomes less direct as acquired knowledge and cultural context increasingly filter and direct perceptions through the schemata. The process of categorization, therefore, also begins with directly perceptible objects and scenes, and includes progressively larger scale and more abstract objects and scenes—objects and scenes that cannot be directly perceived.

Through direct (sensory) as well as indirect experience (e.g., reading), a hierarchy of categories gradually evolves, with the addition of superordinate categories. These represent progressive groupings of elements and increasing levels of abstraction, as well as the addition of subordinate categories, which represent progressively specific instances within a particular knowledge domain.

Acquisition of knowledge in a previously unfamiliar domain for which there is no preexisting knowledge structure is the same for both adults and children. One important difference between learning in adults and in younger children, however, is that adults already have accumulated a significant store of knowledge (i.e., a more extensive ontology). This knowledge is represented in separate but related mental models, each relating to a different knowledge domain and stored as a distinct knowledge structure. All of these taken together form our world view.

The higher level concepts, in the form of both generalized schemata and very abstract category structures, are used to facilitate dealing with new situations within an unfamiliar domain. The mechanisms of metaphor and metonymy are important in making connections with these preexisting knowledge structures. The presence of highly specific schemata and highly detailed category structures that experts acquire through long experience allows such individuals to recognize very fine distinctions and variations. Although the hierarchical nature of category structures allows for a highly efficient means of encoding and storing information, it has also been shown that this structure can lead to errors in spatial judgments.

In Chapter 5, I focus on sensory perception and its role in the larger system of spatial knowledge acquisition and use, particularly in the recognition and identification of objects in geographic space. Spatial navigation is part of that discussion, because it is normally highly dependent on sight, as well as all of the other senses. Spatial navigation is also an essential means of directly experiencing geographic space.

CHAPTER 5

ACQUIRING WORLD KNOWLEDGE THROUGH DIRECT EXPERIENCE

SENSORY INFORMATION AND COGNITION

Gathering information through the senses is so effortless that most of the time we are not consciously aware of this process. For example, if you are walking down the street, you might turn the corner and immediately recognize a familiar restaurant. Walking a little farther, you hear a noise behind you and instantly recognize the footsteps of a friend. You turn and see the person's face, verifying that it is indeed the friend you are meeting for lunch. The ease with which even young children can perform coordinated tasks based on sensory information is deceptive, because seeing, hearing, touching, and smelling are truly complex processes. Even the most sophisticated computer vision or speech recognition programs currently do not come close to duplicating human perceptual capabilities, even after several decades of research attempting to do so.

One of the fundamental questions that has concerned philosophers and cognitive psychologists historically is the extent to which knowledge develops as a result of perceptions from a particular sensory modality (Kennedy, Gabias, and Heller 1992). Vision has certainly always been considered central to spatial understanding, since the time of the ancient Greeks. Berkeley (1965), however, believed that touch and bodily movement are the fundamental paths. He stated that each sense provides different information, and that we learn through experience how to combine and interpret the differing sensory data. Descartes and Kant had more integrative views. Kant (1950) conjectured that all sensory sensation, taken together, is the medium through which objects in the external world are presented to us from birth. Descartes also held that the window onto the world through which we gain empirical knowledge is the sum total of all sensory input:

> light, sounds, odors, tastes, heat, and all the other qualities of external objects can imprint various ideas through the medium of the senses; how hunger, thirst, and the other internal affections can likewise impress upon it divers ideas; what must be understood by the common sense (*sensus communis*) in which these ideas are received, by the memory which retains them, by the fantasy which can change them in various ways, and make new ideas out of them. (1980, pp. 29–30)

Among modern researchers, Jackendoff (1987) noted that the information we gain through our sensory perceptions about our world is interpreted as an integrated whole. Barring neurological damage, the ease with which we differentiate sensations of vision, sound, and touch is automatic and obvious. The one exception to this is that taste and smell tend to blend together, as is well-known by both chefs and psychologists. This seems to be a case in which the two senses are intermingled on a physical level.

Besides the personal awareness of differentiation, there is a long-established tradition in cognitive psychology to investigate one sense at a time. Neisser (1976) speculated over 20 years ago that perhaps 99% of perception experiments focus on only one sensory flow. Use of a single-mode approach makes sense from the practical standpoint of experimental design, because this approach provides the simplest case of information pickup from sensory stimuli as separate input.

In cognitive psychology, vision has historically been the primary focus of study and discussion with regard to the role of the senses in our environmental knowledge. Haptics (i.e., the sense of touch), after vision, has been considered the sense that provides the most direct perceptual information about our environment, the things in it, and their spatiotemporal arrangement.

In this chapter, I focus on the process of direct world experience through sensory perception. Because most of the modern theories and experimental evidence concerning sensory perception in general, as well as how it relates to geographic-scale phenomena, are concerned with vision, this is where I begin.

ORDINARY SEEING: VISION AND VISUAL PERCEPTION

The long-dominant view concerning vision, and indeed all perception, is that it is an information-processing task. Visual information processing in the cognitive literature is commonly divided into low-level and high-level processing (Ballard and Brown 1982; Kosslyn and Koenig 1992). Low-level vision is defined as the initial processing of the "raw" retinal image—an array of varying color hues and light intensities that changes as things move within the visual field and as the viewer turns his or her head and changes the direction of his or her gaze. Low-level vision involves figure–ground discrimination, identifying coherent space–time groupings, and other preparatory tasks. This preparatory processing is performed at a fairly "hardwired" neurological level. High-level vision takes the results of this preliminary processing and uses previously stored knowledge to identify individual objects, to associate additional (unseen) properties with those objects, to attribute meaning to the scene as a whole, and to direct subsequent seeing. As such, high-level vision is very much a mental process. Low-level vision corresponds to what Neisser (1976) referred to as the *preattentive* stage of cognition.

For the task of identifying objects, low-level vision and at least the beginning of high-level vision are commonly equated with *pattern recognition*, in which the "raw" sensory data are organized and objects are identified by comparing the sensory stimuli with a particular pattern stored in memory. Four pattern recognition approaches that have been suggested are (1) template matching, (2) prototypes, (3) distinctive features, and (4) structural descriptions.

Template-matching theories begin with a specific pattern already detected within the scene. This pattern in the stimulus is compared to a set of templates already stored in memory. These templates are not descriptions. Rather, they are holistic entities. The amount of overlap between the stimulus and the stored template determines whether there is a match, and an object is thereby identified. The example often used to illustrate how this might work is reading text. In the several examples of the letter B in Figure 5.1, there are obviously significant variations in form, but we can still immediately recognize each of these characters as the uppercase form of the letter B. This letter can be distinguished from the letters D and P, which are similar, but B has two loops on the right side, not one. This approach is in fact used for automated character recognition with standardized character sets, such as the

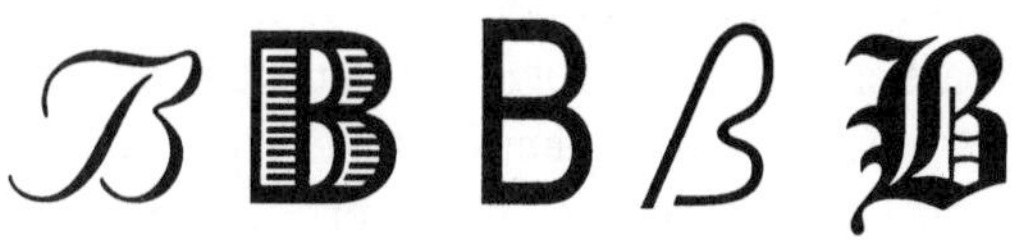

FIGURE 5.1. A wide range of letter forms are easy for people to read.

numbers at the bottom of checks that identify the account and the number of the check. Template models, however, only work for such standardized shapes, and in their complete form (Matlin 1998).

Prototype models are more flexible in that they match more idealized patterns. This is how the text-scanning software now available with many page-size scanners works. However, it still has many shortcomings, particularly in trying to explain human perception. Although it has more leeway than the template approach, the workable variation from a given prototype is quite narrow. This is why standard scanner software only works for text set in standard typeface fonts. Thus, the script and Old English style of the first and last letters, respectively, in Figure 5.1 would not be recognized by such software without the appropriate extensions. Rotation and any degradation in the image (e.g., ragged or fuzzy edges) would also cause problems with this strategy. This also means that any partially occluded objects could not be recognized.

Distinctive feature models work from the assumption that people distinguish among varying stimuli on the basis of a small number of characteristics. This approach has support from biological research, which has shown that the visual system includes physical components that function as "special feature detectors"(Hubel and Wiesel 1965; Marr 1982). Some specialized receptors in the brains of animals respond to horizontal lines, others to vertical lines, and still others to curves, and so on. Perhaps the best-known proponent of this approach was E. J. Gibson (1969), who described the principles of this approach from a developmental perspective. She asserted that children learn to distinguish one object from another visually by learning the objects' specific, distinguishing physical features.

Structural theories are an elaboration of distinctive feature models. The emphasis here is that the relations among the features are as important as identifying the individual features themselves, in order to produce adequate distinctions (Clowes 1969). The problem, however, is similar to that with the distinctive features approach, in the assumption that somehow we are individually matching such primitive elements to elements in our visual array. Whether innate or learned, it makes little logical sense to expect that somehow we match one-to-one such low-level elements for normal, everyday scenes. There would simply be too many elements to perform image cognition as quickly as we do (Sutherland 1968).

Related to this is an issue common to all of the aforementioned pattern recognition approaches: do they provide sufficient explanation for how we perform visual perception? Application contexts and empirical testing of all these approaches have been restricted to very narrowly defined and controlled cases, with the visual field limited to isolated objects such as letters of the alphabet and man-made objects (teacups, scissors, etc.) at tabletop scale. This leads to serious questions about the validity of these approaches for how we see in the much more complex, everyday environment as we navigate within it. How would a person pick out a house, a car, or a street from the very complex and dynamic visual array? Certainly, the process cannot be as complex as these approaches, described in most textbooks on cognition, would infer them to be. Rather, it is immediate, effortless, and veridical (Neisser 1987b).

The critically important question to ask is how do we manage to pick out and identify specific objects with consistency? We are able to pick out and identify specific objects effortlessly out of complex scenes in the environment, regardless of the relative perspective and widely varying sizes (i.e., distance from us), and even with considerable variability in visual characteristics. Noting the variability in natural scenes, J. J. Gibson (1966) addressed this question by first asking what is it about the elements in the real world that allows us to perceive them consistently. If these characteristics can be specified, then and only then can the process by which we perceive be accurately derived.

J. J. Gibson extended the principles of similarity, proximity, direction, and good gestalt (i.e., good form), specified by the Gestalt psychologists in the early 20th century as qualities found in the formal organization of pictures. Gibson asserted that these are *invariant* properties of the real world. In this context, he asserted that these invariants also hold through movement and changes in viewer perspective. One object may also be partially hidden by another object. Johansson's studies on movement (1973) were cited by E. J. Gibson and Spelke (1983) as an example, illuminating how objects and events are identified in a dynamic visual scene through their invariant properties and relationships. Shipley (1991) provided more recent empirical evidence as to how optical invariants are used to detect unified objects in the presence of discontinuities. It is these objectively present properties to which our perceptual systems are particularly attuned—things we cannot avoid. Such properties include light and dark, sequence and proportion. The act of moving through one's environment therefore provides additional information about perceived objects and their interrelationships. For example, when we walk in one direction, we see a given object as progressively occluding another. The rate at which the object becomes increasingly occluded provides information about relative distances and sizes of the objects.

J. J. Gibson claimed that the whole visual system (head, eyes, visual cortex, etc.), should be thought of as "resonating" to the optical invariants. In doing so, the visual system picks up structure directly from the optical array with no thought involved, conscious or otherwise (Neisser 1987b). This direct, physically determined process means that anyone with the same visual system would see these things in the same way. This is also consistent with experimental findings that our ability to pick out objects from a scene begins very early in life. Although a young infant has barely begun the process of attaching meaning to objects and categorizing them, it can discern distinct entities within its visual field as well as any adult, as soon as it can focus its eyes and move its head.

As previously discussed in Chapter 2, J. J. Gibson's overall approach stressed the importance of the constant interaction between people and their environment. In what has been called the ecological theory of perception (Neisser 1987b), two differing aspects of perception, at least on a logical level, can be seen to result from direct, physical interaction. On the one hand, ease and consistency are gained when no form of thinking is involved, including not only consistency for any given individual from one instance of "seeing" to the next, but also consistency from one person to the next. Barring any physical impairment, we all see the same "things" (i.e., we all isolate the same spatially coherent elements from the same visual scene). How we interpret those "things," however, can vary considerably, depending upon the sum of our prior experiences and the specific task at hand. On the other hand, this same direct, physical interaction provides a set of constraints as to which stimuli receive attention and which are ignored. Within this interactive dynamic, we as individuals are constrained by what our particular environment presents to us.

Gibson's ecological approach led to what Marr (1982) called the *primal sketch*. Although Marr was primarily interested in machine vision, his was the first unified theory of vision, and with one stroke, it integrated much of existing theory. He developed a model that derives a completely invariant shape description and carries the process through to object identification. Elements are first derived from the raw image. These are stored as "tokens" that can be assembled and assigned specific values for attributes, such as size or orientation. He divided the process into three representational stages.

According to Marr, primitive features, as regions or lines, are extracted from the raw image, or a series of raw images, to obtain (using a graphic metaphor literally) a sketch-like representation of the image. This primal sketch consists of image elements (lines, curves, areas, etc.) that carry attribute or property values along with them. These are derived by looking for contiguous areas of relatively uniform color and light intensity characteristics, and the discontinuities between them. Elements of the primal sketch are then grouped to assemble more complex elements to derive a *2½-D sketch*. This is a viewer-centered representation of the visible surfaces, with orientation, and some depth information. Finally, a *3-D model* representation is derived that describes individual, three-dimensional shapes, as well as their orientation and interrelationships with respect to each other, in an object-centered coordinate frame.

The shapes of individual objects are built up through a combination of three methods, matching the entire shape and/or its components against shapes of objects stored in memory. First, the entire shape can be matched to a generalized shape that then relates to a potential match to successively more specific shapes, taking orientation, size, and other perceived characteristics into account until the shape is identified. A second method is first to identify a characteristic component that can in turn lead to identification of the entire shape. For example, if one perceives a sinuous, upright tail as attached to a somewhat small, generalized quadruped shape one level lower in the hierarchy, one's descriptions become more sensitive to angles and lengths so that distinctions can be made between dogs, cats, skunks, and so on. A third method is to match components of the object against stored component shapes in successive hierarchical detail until a match is found. For example, if the object is a horse and the person has no "horse" shape that matches the perceived object, then he or she may consider the leg. If there is still no match, then the end of the leg might be identified as a hoof. When this is determined, more specific leg models—those with hooves—can be examined, and so on.

Marr's model thus progresses from deriving tokens from the image—lines, curves, and areas—that are then assembled into surfaces, and finally into volumetric models. Notably, the first two phases of Marr's model are strictly data-driven and in that way correspond to the (J. J.) Gibsonian perspective and also fit the definition of low-level vision. The third component of the process, however, is the beginning of high-level vision, in that it relies upon previously stored knowledge and identifies objects. Elements of the standard pattern recognition approaches can also be seen in this third component. There is a prototype procedure involved in matching generalized shapes to derive the 3-D model, as well as the central ideas of distinctive features and structural theories to determine which components to focus on if a generalized shape is not sufficient to identify the object. What this seems to show is that the classic pattern recognition approaches developed within the field of cognition first assume that the process is much simpler than it really is. Second, they introduce the use of a priori knowledge at the very beginning of the process. What they are, for the most part, focusing on is not low-level vision at all, but rather the beginning of high-level vision. They are dealing with what is the last of the three tasks in Marr's model. From the pattern recognition approaches described earlier, perhaps the one exception to this is the idea of hardwired "special feature detectors" that perceive primitive elements such as lines and curves. This is much more like the process involved in developing Marr's primal sketch.

The overall mechanism proposed by Marr is that descriptive primitives are built up from the undifferentiated array of light and dark through successive groupings, producing objects, hierarchies of objects and spatial patterns. In this abstraction process, the tokens refer to increasingly abstract properties of the image at higher levels of the hierarchy. How to determine "meaningful" groupings can almost never be determined directly from the scene (i.e., the observed data). Some higher level knowledge concerning the nature of the given phenomena must be employed in order to identify the object or objects in the visual field. The higher level knowledge or conceptual view is also organized and used differently from the "raw image" or view.

Further crediting his unifying perspective, Marr drew together the following characteristics of the overall spatial arrangement of "tokens" as a universal and integral set of primitive image elements:

1. *Existence of surfaces*—the visible world can be regarded as composed of continuous, smooth surfaces whose individual spatial structure may be elaborate.
2. *Hierarchical organization*—the spatial organization of entities is often generated by a number of different processes, each operating at a different scale.
3. *Similarity*—the items on a given surface, responding to a process at a given scale, tend to be more similar to one another in spatial organization, size, and other attributes than to other items on that surface.
4. *Spatial continuity*—spatial distributions of items generated on a surface by a single process tend to exhibit some sort of organized pattern.
5. *Continuity of discontinuities*—spatial cohesiveness of individual items and spatial patterns results in a tendency toward smooth-shaped and nonabrupt boundaries between them.

These characteristics are also known individually within other fields, including geography, as fundamental characteristics of spatial distributions. In various combinations, they have been implicit assumptions for a broad range of models for various phenomena unrelated to vision but have never been explicitly enumerated and functionally related.

For example, continuous surfaces are the basic assumption in incorporating a distance decay effect within the gravity model. (The interaction between two places varies in part inversely with the distance between them.) In addition to surfaces, a spatial cohesiveness is also assumed in the absence of barriers, such that there is a smooth pattern. Spatial cohesiveness over a continuous surface is also a basic assumption in the classical Hägerstrand (1967) model of contagious diffusion. Geographers have also been very conscious of the importance of scale in the application of these models to describe various spatial processes accurately. As in the case of interaction between places, the factors influencing intracity trip decisions of individuals vary significantly from those influencing interstate or intercity trips.

Christaller's (1933) central place theory assumes both continuous surfaces and hierarchical organization, although, in this model, surfaces are explicitly subdivided into discrete units for the purpose of description and mathematical expression. In the Christaller and other, related central place models, continuity of discontinuities is also assumed, in the sense that the boundaries between areas themselves form regular patterns, as well as similarity in the functions of central places at specific levels of the hierarchy. Spatial autocorrelation is based on the principle of smooth surfaces and spatial continuity. The application of factor analysis to a geographic matrix, such as an origin–destination or other transactions matrix, assumes similarity of items in space that are responding to a given process.

Putting these similarities together—Marr's principles for the arrangement of image elements, and geographic principles for the distribution of phenomena in space—makes sense from a Gibsonian perspective. It lends additional credence to the principle that our vision is "picking up" on innate characteristics of the real world. Conversely, it is these particular characteristics of the real world to which humans are, at least from a visual perspective, particularly attuned. The fact that these theories were developed by such diverse communities for very different contexts lends mutual support to their validity.

Although Marr's overall scheme must be recognized as a simplified, high-level concept of a complex process, it lacks one key aspect that is of concern here: It assumes a one-way process in which we start with the image and successively assemble features, and then associate them with stored concepts in a bottom-up process. The opposite, Gestalt approach would be first to match large units and then fill-in smaller details in a top-down manner. More realistically, this process works both ways. Knowledge about what is expected in the scene (from stored concepts) is also used to speed interpretation of future images. This can function in allowing inferential "leaps" in feature identification, as well as in selecting and ordering elements in the scene for subsequent interpretation (Neisser 1967).

A basic issue has generated a substantial amount of debate among psychologists since Marr's original theory: How does the human visual system compensate for varying visual points of view in recognizing objects? The two best known theories are Biederman's (1987, 1990) *recognition by components* (RBC) and Tarr and Pinker's (1989, 1990) *multiple views plus transformation* (MVPT).

The basic assumption of Biederman's theory is that any 3-D object can be recognized from a given perspective on the basis of simple 3-D shapes that represent various "natural" components of the object and their relative positions. Also, only a fairly small number (about 36) of these shapes, which Biederman called *gaeons* is needed. A cup, for example, would be recognized by the cylindrical shape, with a loop-shape attached to its side. This is essentially a distinctive features theory for the recognition of 3-D objects. Tarr and Pinker (1989) proposed that the visual system distinguishes 3-D objects by learning to recognize objects from specific viewpoints and then mentally transforming unfamiliar viewpoints until they match one of the familiar viewpoints.

Both of these theories are still being debated (Hayward and Tarr 2000; Tarr in press). Empirical evidence, however, does seem to weigh slightly in favor of viewpoint familiarity dependence (Bülthoff and Edelman 1992, Hayward and Williams 2000).

THE PERCEPTUAL CYCLE

All of the aforementioned visual theories are completely data-driven and do not account for the role of higher level knowledge and its influence in guiding the whole process. Once we have distinguished the object as a coherent entity within the visual field, the identification of the entity and subsequent attribution of properties to the entity, and to the scene as a whole, are in large part guided by our preexisting knowledge. There is also an important connection between schemata, categories, and vision as a process. It is essential to remember that vision, as well as all of the senses, function in a temporally sequential manner. As we move around in our environment and perceive the things around us, we are also reacting to them and actively and deliberately determining our subsequent actions, based on our knowledge and a given purpose. We use our *schemata* to link percepts with concepts. More specifically, it is the *anticipatory schemata*, described in the previous chapter, that not only prepare the perceiver to accept certain kinds of information but also control in part the activity of looking by determining what elements are important within a scene, and what subsequently to look for on the basis of an overall pattern directing further stimuli. We also direct our visual and other sensory processes on a more conscious level by using categories to actually identify objects within any given scene, then adjusting our gaze based upon our understanding of those identified objects and what we associate with them in space and time. Similarly, knowledge stored in categories also draws our attention to inspect further anything unusual in the scene that we identify as either incongruous or unknown.

Thus, we use our knowledge to determine where to move our eyes and focus our attention, and where to turn our heads and move our bodies. What we expect to see sets up one or more anticipatory schemata. The outcome of our explorations and the resultant new information we may pick up modify our categories. Once modified, the cycle continues, as shown in Figure 5.2. The perceptual cycle has been described in some detail by Neisser (1976). The schemata also help in two ways to ensure continuity in perception over time. First, all perceptions are filtered through the schemata and categories. In essence, we see what our knowledge *tells* us we are seeing. Second, our knowledge directs what information will be picked up next. This, however, does not mean that we cannot pick up unanticipated information. In directing our attention, we often may notice something we do not recognize (i.e., something that we cannot match with our existing knowledge through schemata), which itself draws our attention for further investigation. We may in turn shift our position relative to the object to get a view from a different angle. If we

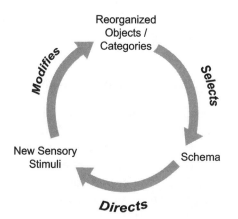

FIGURE 5.2. A simplified view of the perceptual cycle.

still cannot identify it (i.e., it is an unfamiliar object), we often manage a generalized association on the basis of a partial match of the object's visible properties with those of known object classes. Certainly, there are constant instances in which our knowledge of what the image is *supposed* to contain "fills in" missing or noisy sensory data.

In contrast to low-level vision, high-level vision, as well as high-level perception in general, is highly flexible and therefore also highly variable. Chalmers and his associates summarized a number of important characteristics of high-level perception in this regard (Chalmers, French, and Hofstadter 1992):

1. *Perception may be influenced by belief.* Expectations play a large part in the attributes and meaning we place on what we see. On the one hand, a child looking out the window at a significant amount of new snow that has fallen overnight usually sees this as a beautiful sight and anticipates a day off from school to play in the snow. An adult, on the other hand, may shiver at the sight, in anticipation of the need to go outside in the cold to shovel snow.

2. *Perception may be influenced by goals.* A forested landscape that may look like the opportunity for significant profit to the owner of a sawmill would look like a serene escape to a hiker.

3. *Perception may be influenced by context.* A deer in a zoo would be perceived very differently than a wild deer foraging in a farmer's field.

4. *Perception may be radically reshaped, when necessary.* Tropical rain forests were looked upon as dark and unfriendly places by people in more northern climates, until the role of this type of environment in the global climate became known.

The entire role of high-level vision, and high-level perception in general, is to draw meaning out of a given context.

TOUCHING AND HEARING

Although much less attention has been paid to the other senses in the psychological literature, the perceptual process as just described within the context of vision is comparable for all of the other senses. There is a significant overlap between the characteristics of the experiential world and the things within it that can be picked up by the various senses. Vision conveys information about size, distance, shape, and texture, as well as uniquely visual characteristics, such as color and brightness. Hearing conveys information about size and distance, as well as uniquely auditory characteristics. All rely on the recognition of pattern.

Nevertheless, the sense of touch has several unique characteristics. First, the haptic "organ" is not isolated locationally in the body. Rather, the flow of haptic information to the nervous system is through a set of nerve endings dispersed throughout the body. Perhaps a more important difference in a geographic context is that haptic sensation is restricted to objects on a personal scale—things that can be grasped or traversed. Haptic sensation is also more limited to what is detected at any single moment. Visual and auditory sensations are spatially extended in that they consist of an array of simultaneous sites or sounds derived from a range of distances from the individual that can extend for miles at any given instant. Tactile sensation can be extended somewhat by the use of tools, such as a stick or cane, to detect objects blocking the path. Because the sense of touch is more spatially restricted, it has an obvious disadvantage in providing direct experiential information about geographic-scale space. Nevertheless, the study of how individuals deprived of sight at various stages of life provides valuable insight about how we learn about our environment, and how our world view is conceptually stored.

All of the senses are temporally extended to one degree or another. Listening is perhaps a more temporally extended activity than the other senses. There is typically no single moment in which one hears anything, because sound waves themselves are a space–time phenomenon. An initial burst of sound may get our attention, but then we need to listen to that sound for a while to identify it. Was that a roll of thunder or a truck passing by? All sensory sensation must persist for some minimal duration for it to mentally "register," regardless of modality. Even looking at a static image requires some time. Sensory sensation is also temporally dependent in ways beyond mere persistence. Hobbes claimed that motion, through time, is the basis of all sensory sensation.

Certainly, all of our senses perceive motion as change, and it is through change that we identify pattern. We need to be familiar with the patterns of sound that an object makes through time, for example, in order to identify it and to distinguish that particular sound from all of the other sounds being heard at the same time. Without being able to identify such patterns from among the sounds we hear, all we would hear is a muddled cacophony from various sources. Certainly, language is the best example of identifying patterns in sound. If we do not know the language someone else is speaking, we hear it as a continuous stream of sounds without meaning. If, however, we are familiar with that particular language, we can easily distinguish individual words and sentences, and attach meaning to them. The listener also anticipates, based on sensory data already received, what he or she will hear next. People listening to music do this all the time.

Vision, hearing, taste, and touch are all directed by the individual in a continuing and temporally sequential process of exploration, information pickup, anticipation, and further exploration. For touch, we often must rely on exploratory movement in order to detect pattern; a spherical, detached object that gives way to pressure is a ball. The round, hard, cool object we find in the dark at about waist height is a doorknob.

Kennedy et al. (1992) have pointed out that perception and cognition through touch do not require movement, although as already mentioned, they still require time. For example, touch involves a medley of vibrations from the handlebars of a bicycle, providing information about the roughness of the road surface and stability of the bicycle upon it. This brings up the notion of spatial extension through the use of implements, such as the long cane used by the blind. Through active exploration via a cane or rod, a person gains information about distance, hardness, as well as the size of objects (Chan and Turvey 1991). Projection is involved in the use of implements to acquire information through haptic sensations. In a sense, this provides some (limited) spatial extension to the sense of touch that is available intrinsically through sight and hearing; we can potentially see and hear objects that are miles away from us.

THE SENSES AS A UNITARY SOURCE OF SPATIAL INFORMATION FOR COGNITION

As I mentioned in the discussion at the beginning of this chapter, regarding the extent to which environmental knowledge arises from a particular sensory modality, theories that have developed historically can be grouped into three general positions:

1. One sensory modality is central and essential to gaining spatial understanding: vision or touch. Without that key sense, developing spatial concepts and understanding any kind of space—environmental, mathematical, and so on—are impossible.

2. One sensory modality is key at the beginning of the learning process. As infants, we learn through experience to infer or associate information gained from one sensory modality (e.g., touch) with information gained from another—the empiricist view.

3. No single modality is key to our spatial understanding. All sensory modalities work together from birth as a unitary system in providing a suite of interrelated and cross-reinforcing information.

Millar (1994) provides an extensive review of this progression and the evidence to support its validity or nonvalidity. This is a variation of an earlier grouping derived by Fletcher (1980), specifically within the context of vision, called the *deficiency theory*, the *inefficiency theory*, and the *difference theory*. The *deficiency theory* is the same as item 1 in the previous list. It is included here mostly for historical context, because it is no longer credible. The *inefficiency theory* holds that people who are totally blind from birth do learn to associate information from other sensory modalities with that most easily gained from vision, but the spatial concepts they develop, and their overall knowledge of space, is inferior to that of sighted people. The *difference theory* asserts that blind persons can build up a complete spatial understanding and mental representation of space that is equivalent to that of sighted persons, but they do so more slowly and by different means.

Two approaches have been used in the investigation of how and to what extent specific sensory modalities impact our spatial knowledge. One is a developmental approach and involves the study of infants and children. The other focuses on the effects of the lack of vision or other specific sensory input. I briefly review each, and the development of theory based upon them, here.

Most empirical work investigating a vision-based approach for acquiring and interpreting spatial information concerns the blind. Much early evidence that seemed to support the deficiency theory (Schlaegel 1953; von Senden 1932; Warren 1974; Worchel 1951) has since been discounted. Currently, useful conclusions regarding the role of vision in spatial understanding through the study of visual deprivation can only be gained relative to people who are totally blind from birth—a condition not often differentiated in early studies (Millar 1994). There is also experimental evidence to discount touch as necessary for spatial understanding. Literature documenting this evidence was reviewed by E. J. Gibson and her colleagues (Gibson 1969; Gibson and Spelke 1983). The evidence she cites shows that very young infants can discriminate distance (within limits) as well as the shape and size of objects with vision alone.

Revesz (1950) went against predominant thinking of the time and took a position aligned with the inefficiency theory. He asserted that spatial organization is different, rather than lacking, in people who have never been able to see. Revesz theorized that the blind rely on haptic coordinates; a personally centered (i.e., egocentric) coordinate space based on touch and movement relative to the body. In contrast, the spatial knowledge of sighted adults is externally referenced (i.e., allocentric) and tied to landmarks, paths, and planimetric views. Byrne and Salter (1983) provided empirical evidence for this with adults in large-scale space.

The basic assumption of both the empiricist view and what Fletcher called the inefficiency theory is that if sensory associations are learned, then cross-modal performance should improve with age. Similarly, young children should be better at tasks involving only one sensory mode than at tasks requiring intermodal associations. A significant amount of empirical evidence indicates that neither age differences of children performing the same task nor differing complexity of the task for children of the same age has a significant impact on performance (Millar 1972, 1975a; Rudel and Teuber 1964). Millar (1975) has shown that there can actually be less difference between single- and cross-mode performance in younger children than in older children. A potential explanation she offered is that younger children use simpler, more general criteria for making judgments.

The current body of evidence supports the view that our senses provide a unified and interrelated suite of sensations (the third theory in the list), and that we understand how these sensations are related very early in life. Research has shown that young infants can associate sight with sound or touch (Bushnell 1986; Gibson and Spelke 1983; Starkey, Spelke, and Gelman 1983, 1990; Streri 1987). Although vision is particularly attuned to picking up spatial information, it is not essential to spatial learning and understanding. There is also empirical evidence that vision and touch are equally capable of picking up certain types of information, such as texture, orientation, and size (Jones and O'Neill 1985; Lederman and Abbott 1981; Warren and Rossano 1991).

Millar has argued along the lines of the difference theory, that blind people can understand space in much the same way as sighted people (Millar 1988, 1994), and she provides significant empirical evidence to support this view (1972, 1975a, 1975b, 1985). For example, she demonstrated that children blinded congenitally or very early use the same symbolic scheme in drawing people (e.g., a circle for the head) as do sighted children, and are able to draw the body parts in their correct relative positions translated onto the two-dimensional page from about the age of 10 (Millar 1975b). Kennedy and Gabias (1985; Kennedy et al. 1992), in another example, have shown that blind persons use sophisticated metaphorical simulations in the same manner as sighted people. Additional supporting evidence shows that blind persons can mentally combine and transform information in ways that suggest they use imagery (Hollins 1989; Kerr 1983; Zimler and Keenan 1983).

Further evidence that our senses provide an interrelated and unified suite of sensations is provided by empirical results showing convergence of simultaneous hearing and vision (Massaro and Friedman 1990; McGurk and MacDonald 1976). Besides the various modes working together in selective exploration of the environment, empirical findings on cross-modal matching suggest that having more than one source of concurrent information facilitates processing by functionally providing cross-modal verification. This is particularly helpful in conditions of uncertainty (Dodd 1977; Gollin 1960).

There is also physiological evidence to support integrated and coordinated functioning of our senses for gathering spatial information. Kennedy et al. (1992) reviewed empirical evidence that the parietal cortex of the brain is a primary focus of somatosensory nerves, receiving sensory information from all modalities. It is also highly linked with the motor cortex. The parietal cortex therefore seems to provide a coordinating role in both integrating sensory information and directing motor activity for further exploration and resultant actions.

In a review of additional empirical evidence, Ungar, Spencer, and colleagues argue that visually impaired persons can perform navigational tasks equivalent to those of sighted people—if provided with sufficient aids (e.g., tactile maps) and appropriate experience (Spencer, Morsley, Ungar, Pike, and Blades 1992; Ungar, Blades, and Spencer 1996). A number of researchers (Golledge, Klatzky, and Loomis 1996; Passini, Prouix, and Rainville 1990) provide strong evidence that way finding and other forms of complex geographic problem solving can be performed by the blind. In their experiments, they asked subjects to find new routes, take shortcuts, and perform mental rotations of a route learned from a map. Golledge and his colleagues (1996), however, tempered their results with the comment that although blind persons can indeed attain a level of competence that is comparable to sighted people's performance of many spatial tasks, they require more time and the help of additional devices and strategies in order to do so. Golledge (1993) has also noted a number of commonalities in the learning of small- and geographic-scale spaces, including hierarchical organization and clustering.

Based on this and other evidence, it has been formally posited that compensation takes place when one or more senses are lacking as an explanation for the difference theory (Ashmead et al. 1998; Easton and Bentzen 1999; Rauschecker 1995). According to this view, people who are blind, or who have some other sensory impairment, compensate with the other senses in dealing with and learning about space. Other areas of the brain that deal with sensory perception become more developed. As a result, such people can be at least as capable as people without sensory impairment in performing spatial tasks, such as navigation and general spatial awareness. Some experimental evidence involving way finding and other spatial tasks with blind versus sighted individuals seems to support this view (Klatzky, Golledge, Loomis, Cicinelli, and Pellegrino 1995; Loomis 1993). While the experiments on spatial performance indeed indicate that people with sensory impairment can perform at least as well as those without impairment under the right circumstances, the inferences linking performance on spatial tasks with mental representation and level of environmental knowledge should be viewed with considerable caution. As Liben (1988) warned, "doing" does not in itself prove "knowing." Although experimental evidence supports the difference theory, as previously mentioned, the results of many empirical experiments, particularly dealing with way finding, have been erroneously interpreted as evidence of spatial knowledge. What needs to be distinguished is activities in space that can be accounted for purely on the basis of perceptual feedback from persons that require spatial thought and use of previously stored spatial knowledge.

To summarize predominant, current thinking, the various forms of mentally encoding our spatial knowledge are innate and not keyed to any particular sensory modality. Also, persons deprived of sight or some other sense are still capable of acquiring a complete and equivalent representation of space. Even those who are totally blind from birth can develop integrated world views that are based on external coordinate systems. The factor causing differences in which of these forms we use to encode that knowledge is our level and type of experience. It must be noted here that a representation using an egocentric frame of reference is not intrinsically inferior to one with an external frame of reference. Each is advantageous for specific and different situations. The important point here is that because the body-centered frame of reference, as well as a path-oriented representation, is more advantageous for navigation and other direct interaction with the environment based on touch, development (and thus availability) of an externally referenced representation lags. This also means that development of a complete, integrated world view lags behind in the learning process.

Lack of a particular sensory mode results in decreased information about the environment in two respects: first, the lack of the specific type of information that the sense is specialized to analyze, and second, a reduced level of informational redundancy. Thus, deriving an equivalent world view will take longer and require deliberate compensating strategies for persons with sensory impairment. Different strategies are used to derive spatial information from the senses, depending upon the situation and the senses actually receiving information. Because vision is specifically attuned to gather spatial information beyond one's immediate proximity, and previous visual experience serves to draw attention to simultaneous spatial features, Millar (1994) stressed that blind children tend to use an egocentric frame of reference, because so much spatial information comes from the sense of touch.

In contrast, broad-ranging and nonimmediate spatial information attained through visual experience prompts sighted children to utilize an allocentric coordinate system for noting spatial interrelationships. Blind persons' compensating strategies to augment the information they receive in the sensory stream and restore informational redundancy include sensory aids, such as the use of echolocation devices and specialized mobility training (e.g., the use of a cane). Compensating strategies also include tactile maps, scale models, and verbal descriptions, both for preroute planning and enroute.

The validity of the difference theory is not surprising when we consider geographic-scale spaces. Using vision, a person can process information about his or her immediate environment in a wholistic and continuous manner. When navigating in urban areas, or even inside a building, however, many obstacles along one's path restrict or eliminate the view of the destination (e.g., cars, buildings, etc.). And one cannot see around corners. For geographic scales that we cannot experience directly, we must rely on indirect sources of information.

Regardless of which sense may be most important for providing information in a given situation, all senses contribute continuously to our total experience of the environment. As we walk down the street, we see the visual array before us, hear the mix of sounds, and smell the various aromas as we perhaps pass a bakery or a gas station. We also feel the pavement through the soles of our shoes and perhaps judge how to slow down or balance ourselves on an icy patch as we walk. In meeting a friend along the way, we may stop and carry on a brief conversation, speaking and hearing our own words as well as our friend's, shifting our gaze and other sensory attention while still conscious of the traffic noise, seeing the movement of objects in the background, and feeling the cold.

All of our daily activities, whether walking, driving a car, or simply sitting at a desk and reading, involve the reception and integration of all sensory information from all modalities, subsequently interpreting and thereby better functioning in our surroundings through a complex series of actions and responses. The redundancy in the information obtained from the multisensory array of sensations is useful in confirming the correctness and consistency of that information. An application of this is the well-known technique for improving memory for names and facts through the deliberate use of two or more sensory modes (e.g., reading a name and saying it out loud). The validity of this memory strategy has been verified experimentally (Sanders and Schroots 1969).

We are automatically taking in sensory information in an amodal manner as we actively explore and interact with our environment. In gathering coordinated, unified sensory information, the whole functions in a way that is greater than the sum of each sensory component. Given an inherent overlap in sensory functionality that does not have to be learned, the lack of one sense, either because of a particular situation or as a long-term condition, can be compensated for in the acquisition of spatial information and subsequent understanding. This holds true even for a sense as important for spatial information as sight, although spatial understanding may take longer to acquire.

CONSIDERATION OF DIFFERENT SPACES AND SCALES

The distinction between "doing" and "knowing" brings up a number of fundamental issues of great interest in the context of this book. As discussed in the previous chapters and shown graphically in Figure 3.1, both Piaget and Cassirer emphasized distinctions between perceptual and cognitive space. Piaget clarified the Cartesian distinction of perceptual space as being local and sense-dependent to the individual. This has also been called practical space (Liben 1988). Functioning within it may entail the use of remembered sensory experiences (e.g., knowing one cannot walk through walls), but does not require the use of higher level spatial concepts. Higher level spatial concepts are incorporated in a completely constructed and abstract internal space that corresponds to Cassirer's three "space" progression in the learning process, from a concrete, *presentation* of space through the senses, to an abstract and symbolic *representation* of space. A key difference here is acquaintance-with versus knowledge-of our environment. According to Cassirer, the entire process of this progression is one of cognitively transforming space and the elements within it into symbolic forms. No meaning can be attributed to anything except in reference to these symbolic forms.

From Cassirer's perspective of presented versus represented space, and Portugali's extension of this into an explicitly geographic context, as discussed in the previous chapter, this distinction does not equate to perceptual (external) versus cognitive (internal) space. Based upon one's experience, objects and other elements within the perceived external world take on meaning and are thereby transformed into symbolic form. For example, the sensory image of one's alma mater has much more meaning—beyond the lawns, trees, and buildings—than can actually be seen. Thus, with increasing levels of knowledge derived from both direct and indirect experience, one attains a different point of view. As a result, the sensory stream of information from one's environment is interpreted in a fundamentally different manner. The external world functions as an integral component of our overall integrated, highly, dynamic multirepresentation of our spatial knowledge.

From the standpoint of environmental knowledge, the distinction between small-scale versus large-scale space made by researchers of sensory perception seems to be artificial and somewhat misleading. It is tempting to argue that tabletop-scale space has limited relevance to the experience of the environmental world. The distinction pointed out is that the viewer can experience tabletop-scale space all at once—in a single glance. The individual's perspective is also that of looking down upon the space, separated from it instead of being surrounded by it. The key difference between small- and large-scale environmental space (i.e., immediate and personal-scale space vs. geographic-scale space) is that small-scale space, such as a single room, can still be seen all at once and is within easy reach of the individual. In contrast, large-scale space can only be directly experienced in pieces, sequentially, because everything cannot be seen or otherwise taken in by the senses all at once.

As illustrated by these terms, one basic characteristic of spatial scales becomes apparent: Scales are continuous and relative to the perspective of the viewer. Does a large tabletop full of items cease to be "tabletop space" because we have to walk around it to see all of its elements? At what size does it cease to be "tabletop space"? Does a city become "tabletop space" when it is seen all at once, viewed from an airplane high above it?

No experiential space, regardless of scale, can really be directly experienced "all at once" if we consider the important element of time. This is true even in the very artificial case of a small and static tabletop arrangement. As we experience space from one encounter to the next, regardless of scale, we still must cognitively piece our sensory memories together. We piece together multiple views of the tabletop, moving around it and directing our attention to various components in the same way that we piece together a sequence of visual and other sensations as we walk down the street, turn corners, and so on. Space at any scale is also normally dynamic. In other words, elements within the space commonly change even as the viewer remains stationary. Part of our cognitive piecing together includes noting change.

This consideration also sheds some light on the argument that blind persons are at a disadvantage when operating in and learning about space because they must piece together sequences of tactile sensations as part of compensation for vision. While it is true that vision more readily provides broad-ranging and nonimmediate sensory information about our surroundings than does touch, it is apparent that the processing of all sensory information, regardless of the senses and the spatial scale involved, is one of sequentially piecing together, integrating, and imparting meaning to the sensory data as one goes through life.

J. J. GIBSON'S AND MARR'S THEORIES EXTENDED

If there is no single guiding or dominant sense, then how are objects, places, and events presented to us as unitary things with spatial and temporal cohesiveness? Gibson's view of perception is focused on the specific context of visual perception, as described earlier in this chapter. This view, however, really provides an explanation of perception in general, and how we identify distinct elements from a continuous stream of information from multiple senses so effortlessly, and so early in life. Gibson saw perception as based upon the ecology of the animal: the interrelationships of an organism with its environment. The perceiver extracts the information from the environment that he or she needs to function and survive. Gibson (1979) introduced the term "affordances" as the key to understanding perception: "The *affordances* of the environment are what it *offers* the animal, what it *provides* or *furnishes*, either for good or ill. . . . It implies the complimentarity of the animal to the environment" (p. 127, emphasis in original).

Thus, a floor affords support. A large, solid, horizontal object above our head affords shelter. A large, solid, vertical object can be either a barrier that we must navigate around or a place to hide. All of these interactions are abstract, and, as such, the specific affordances of any object, scene, or event are determined through the continuous flow of sensory information taken as a whole.

Perception does not require memory of past experience, because all of these relationships are also "built in." Gibson (1979) used the term "direct perception" in part to emphasize this. The only thing that changes in an individual's perceptions is that young people gradually become better at *detecting* higher order relations, learning to make their explorations of the environment more focused and systematic over time. Instead of our world passively making impressions on our senses, perception is an active and selective process. We look, explore, and deliberately search out information. We are constantly reaching, moving, and shifting our attention in order to optimize the acquired information to suit a particular situation.

This point of view contrasts with the views of Piaget and Cassirer, which are really concerned with concepts rather than percepts. We do not construct new concepts by combining elementary sensory sensations that somehow are presented to us. Instead of a process of association to build new concepts, the Gibsonian view is one of discovery of familiar, novel, or simply desired percepts through exploration, particularly through the law of common fate.

The various senses seem to provide complementary information that partly overlaps. Intuitively, various types of information about the experienced environment are gained through different senses, and certain types of information leading to cognition are not necessarily restricted to a single sensory path. For example, the shape and size of an object can be determined by touch—if the object is of a size that can be grasped or traversed—as well as by vision. The size of some types of objects can also be determined, at least to a general degree, by the sounds they make. For example, a low-pitched bark would almost certainly belong to a large dog, whereas a very high-pitched bark would identify the dog as very small. Blind persons can judge the distance of objects in their path by the vibration in a cane. They judge the fullness of a cup by dipping one finger slightly into it.

Marr's unified theory of vision and 3-D mental representational model are also compatible with this perspective. The unity and functional overlap of the senses explain how blind persons can still construct 3-D mental representations. Primitive features, such as regions or lines, as well as their properties (e.g., texture) and interrelationships can be extracted from the immediate environment through both touching and seeing. Additional depth information may be detected or confirmed through hearing. Characteristic shape components and characteristics that can in turn be used to infer an entire shape—such as that of a cat, an automobile, or a building—can be subsequently discovered through directed exploration. However, the haptic (nonvisual) route to a 3-D representation differs from the visual route (Jackendoff 1987). The 2½-D sketch of the haptic system would need to be adapted to touchable space rather than to a visual scene. Given this difference in process, a full, 3-D representation can be derived that is conceptually equivalent to that originally described by Marr within the context of vision.

INTERPRETATION AND RECOGNITION: A PROCESS OF COOPERATIVE COGNITION

To summarize, perception is a highly integrated and integral process. It is integrated in that our senses provide a unified informational stream. It is also an integral part of how we gain and verify knowledge. We use our previous knowledge to guide and interpret what we see, hear, feel, and so on. In the previous chapter, I discussed the overall process of how we gain and use knowledge about the world around us, and how specific elements of our knowledge are conceptually structured.

As we have already seen, perception is a crucial component of the learning cycle. Sensory perception, as direct world experience, is particularly important early in the learning cycle of children or newcomers in a previously unknown city or country. Indirect perception (i.e., indirect experience) of our world through language and imagery becomes particularly key when we deal with totally abstract concepts (e.g., mathematics), and when the geographic scale is beyond what we can directly experience. You would not gain much information from this book if you could not already recognize letters and words, for example. You would not be able to read a map if you had no prior experience of what the various squiggly lines and other symbology correspond to in the physical world. Similarly, you would not recognize an object you see on the street as an automobile without prior knowledge of automobiles. More importantly, you would also not truly appreciate the importance of getting out of the way if the automobile were coming directly toward you. Knowledge of "things" in the world includes knowledge of both their physical and behavioral characteristics, and their "affordances" as to how those behaviors and capabilities relate to us personally and interrelate with other things.

How we perceive and interpret our world, and the things in it, depends not so much on our remembered experiences per se as on our preexisting, stored knowledge structures as abstracted and synthesized information, including our schemata for being able to pick out patterns from the continual and complex array of sensory inputs as direct world experience, or the indirect world experience in reading a book or a map, listening to music, and so on. It is this selection function that also helps us to direct subsequent perceptual activity. Our hierarchically arranged category structures allow us to identify specific objects, and to understand their characteristics and interrelationships. Both enable us to interpret within a broader context. All perceptual processes utilize the same fundamental cognitive structures (Neisser 1976). Taken as a whole, the overall process of interpretation and recognition of elements in the world, and the knowledge we gain from the sum of those experiences in guiding and refining subsequent interpretation and recognition, is one of cooperative cognition. Elements in the external environment also act as an interrelated knowledge representation, external to ourselves, that functions in concert with our internal knowledge representation. As such, they function as symbols that carry meaning. These can be everyday elements, such as trees, cars, rivers, and buildings that have acquired their own meaning for us, or they can be elements specifically designed to convey meaning, such as the letters and words on this page, or the symbols on a map.

In the next chapter, I focus on how we link our experiences into a unified world view and go beyond experience in constructing higher level, abstract concepts as part of that world view. The primary mechanisms involved are the imaginative mechanisms of imagery and metaphor, which we use continuously as fundamental and interrelated components of human thought.

CHAPTER 6

FROM OBSERVATION
TO UNDERSTANDING

BEYOND MERE OBSERVATION

Observation and direct experience alone are not sufficient for understanding. Our experiences must be connected with and integrated into our existing knowledge, with similarities confirmed, differences noted and resolved, and links made across domains. This is how we derive a unified world view and build a broader understanding of our environment. This is the realm of the imagination.

Mental imagery and metaphor are the two primary mechanisms involved in deriving understanding. Both focus on pattern and similarity in different yet interrelated ways. Imagery employs our perceptual capabilities, often with purely mental constructions, for quickly perceiving continuities, discontinuities, and pattern. The metaphor establishes linkages and correspondences between different knowledge domains. Imagery and metaphor, as well as facts and relationships, can be conveyed via language and graphical images. Because imagery and metaphor deal with concepts, transcending direct experience, these mechanisms are also very powerful for conveying abstract ideas from person to person.

In this chapter, I examine imagery and metaphor as structures of the imagination. This entails exploring the characteristics of these representations, as well as their complex interrelationships.

IMAGERY AS A MEANS OF UNDERSTANDING

Besides picking up information directly from the environment through sensory perception, people also gain information through the use of mental imagery. Mental imagery is sometimes described as pictures "in the mind's eye." Imagery is often derived reflectively purely through imagination, and also frequently conveyed from person to person through language. Literary works are sometimes praised for their rich imagery (i.e., for the vivid scenes the reader conjures up in his or her mind relative to the story). These are often completely imagined constructions for the receiver of the imagery. Imagery can also consist of remembered visual scenes directly experienced, of compositions and variations from remembered experiences. This means of indirect experience occurs in different sense modalities, and combinations of modalities. Imagine the following:

> The scene out your living room window
> Smelling freshly mown grass
> Standing in a snow storm
> Queen Elizabeth on water skis

Most people would report that they do conjure up these sights, smells, and feelings within their minds when asked to imagine them. Visual images, however, as reported by people, tend to be the most vivid and have been empirically studied more than the others.

Mental imagery is closely related to perception but functions in ways far beyond what we can do with ordinary seeing, feeling, smelling, and so on. It can consist of remembered perceptions from experience or pure constructions. Through the construction of mental imagery, we are able to combine elements in new ways, perform mental transformations on them, and engage in spatial problem solving. For example, if someone asked how many rooms are in your house, you would most likely do a mental walkthrough, counting as you go. We also use mental imagery in reflective thought, as a means of deriving abstractions and commonalities out of existing conceptual knowledge and experiences.

Many mental images are not replays of remembered experiences. Some stored imagery represents a distillation of numerous memories—many of which represent idealized category prototypes, such as that for "church." An image of a church is often "seen" in the mind's eye at the mention of the word, but the image is not necessarily any specific church seen in the past. Also, some visual images are on-the-spot creations of the mind. These are images that we mentally generate and manipulate to solve problems and to deal with novel situations. For example, people use this type of visualization in everyday life to "see" how the living room would look if the furniture were rearranged in a new way, to give directions, or to plan a road trip by, in part, imagining what sights they want to see along the way.

This is also the type of imagery that architects, engineers, and mathematicians use as an essential tool in creative thought. Such visualizations are often of physically impossible scenes in the real world or are composed within some totally abstract or imaginary space. For example, many scientists have described how mental imagery contributed in an essential way to a key discovery or insight. Perhaps the best-known account is Einstein's description of having imagined riding on a beam of light, which led him to the concept of special relativity. Part of his genius came from his ability to use creative imagery productively for solving highly abstract problems and describing the effects, using such things as clocks and rods, and trains traveling near the speed of light, to describe how very high velocities contract space and dilate time.

Physical graphical images such as maps and diagrams are used frequently to gain and to convey new insights. Diagrams, whether drawn on a blackboard or projected from a computer display, are a teacher's basic tool for conveying abstract concepts. Artists convey impressions and feelings via visual images that often have little or no resemblance to real-world scenes. This nonpictorial portrayal was the motivation behind the development of abstract forms of painting such as Cubism—to allow freer expression by lifting the various requirements, such as perspective, needed for pictorial representation. In Western abstract art, what is portrayed in a painting for the viewer is often even different than what the artist portrayed. The intention in such art forms often is to deliberately spark the viewer's own personal impression.

What makes graphical images of all sorts particularly effective (1) for portraying and subsequently retrieving that information and (2) for gathering new insights? Larkin and Simon (1987) investigated this question empirically by comparing individual performance in problem-solving tasks using diagrammatic versus nondiagrammatic representations. They found that, first, information retrieval is facilitated by the specific ways that information can be grouped spatially. Second, spatial representation allows the use of recognition mechanisms that are built into the visual perception system and allow very rapid recognition. Detection of spatial patterns and groupings is hardwired into the human visual system.

Much visual searching is avoided when similar items, or items that are to be used together in a problem-solving task, are located close together. The two dimensions of a spatial display allow this to be done in a physically more compact space. Symbolic labels can also be used to extend spatial association over larger areas, and to use the same space in the display to show multiple associations. In essence, this means that the same display space is used over and over again. Aggregations and associations also become quickly apparent through spatial proximity or common visual properties. The visual system also allows hierarchical search. We can look at the display as a whole, and then effectively zoom in to look more closely at a small area.

But what is it about some maps, diagrams, paintings, and even some realistic visual scenes that sparks the imagination and facilitates the derivation of new ideas, while others do not? What is it about *mental* imagery that does this and even more—as a central and essential means of creative thought and problem solving?

As described in the previous chapter, the process of visual perception relies upon our preexisting knowledge for interpretation and recognition of perceptual images. The more familiar the image or patterns within it, the quicker we can recognize and interpret them. Images that are unfamiliar, or that contain unfamiliar elements, take longer to interpret. For example, the image in Figure 6.1 may take a while to recognize; some people longer than others. This very familiar image is presented in a novel way. What, in effect, we do cognitively is to go "further afield" in exploring our internal knowledge structures in trying to make sense of it. In essence, we are using our imaginations.

Finke asserted that images that spark the imagination seem to possess novelty, incongruity, abstraction, ambiguity, and often some combination of these. He referred to these characteristics as *preinventive* properties (Finke, Ward, and Smith 1992). A novel image is one that contains a combination of components composed in some unique or unusual way. For example, all or

FIGURE 6.1. An unfamiliar image.

some of the elements could be rotated, be out of scale with the surrounding elements (i.e., smaller or larger than expected), or be portrayed in an unusual color. Incongruence in an image is the incorporation of components not normally contained within that image. Ambiguity is the property whereby an image can be interpreted in more than one way because of the interrelationships of its components. There are a number of famous examples of ambiguous images. The image shown in Figure 6.2, in which both a beautiful woman and an ugly hag can be seen, was introduced into the psychological literature by Edwin Boring (1930). Another image, shown in Figure 6.3, can be interpreted as a goblet or as two faces—each face being delineated by what is otherwise one side of the goblet. Abstraction is the property whereby details presented in a normal environmental scene, such as color, shading, and the impression of three-dimensions, may be selectively or in large part absent. Diagrammatic images have this property.

All of these properties promote what has been called *emergence*, which is the process whereby initially unanticipated features, often not explicitly created or anticipated by the creator, are detected. In other words, unexpected properties in the image emerge as visual discoveries. There have been a number of studies of this phenomenon and how it works in images. Pinker and Finke (1980) used a mental rotation task to explore the recognition of unexpected patterns. Presented with a cylinder having four small objects hung at different depths within it, subjects were asked to rotate the cylinder while looking into it. At a specific point, subjects were asked to stop and identify any

FIGURE 6.2. Old woman or young woman?

FIGURE 6.3. A goblet or silhouettes of two faces?

pattern they might detect as being formed by the four rotated objects. The majority of subjects were able to detect a parallelogram, which was not evident from the starting position. Subsequent work on a more elaborate set of rotation tasks by Finke, Pinker, and Farah (1989) produced similar results. Suwa, Tversky, Gero, and Purcell (2001) showed that regrouping parts of a sketch results in detection of new features.

Both physical and mental images can have some combination of novelty, incongruence, abstraction, and ambiguity that in turn results in emergence. What, then, is special about mental imagery, that makes it seem to be even more powerful in sparking imaginative thought? Two additional properties of mental imagery make it a particularly powerful tool in reflective learning and understanding: total manipulability and freedom from physical constraints.

Experimenting with a scene, image, or figure is important in making discoveries and deriving new insight. Adding new elements, changing the relative positions of elements, or changing the perspective from which the object is seen by rotating it are all ways to enhance image properties (novelty, etc.) and the likelihood of emergence. In the field of perception, much empirical evidence shows that certain patterns can be interpreted as something different after some change or manipulation (Attneave 1971; Rock 1973; Shepard 1990; Shepard and Cermak 1973). We, in effect, play around with it to see what emerges. Like mental images, physical images can also be manipulated, rotated, rearranged, modified, and reinterpreted. We do this all the time. The key difference is that the ability instantaneously to create, modify, and manipulate images within the mind as mental imagery allows emergence to arise more easily and more quickly.

The other advantage is that mental imagery is free from physical constraints. Image configurations that would be difficult or even impossible to construct are just as easily created in the mind as any other image. Einstein riding a beam of light is a good example of a physically impossible image. Any amount of detail can be included with equal ease in mental images. Things can also be set in motion, slowed down, or speeded up at will, beyond what may be possible physically, as in time-lapse photography.

THE USE OF METAPHOR IN UNDERSTANDING

We use metaphors constantly in everyday speech, and because of the pervasiveness of metaphors in language, they were historically assumed to be merely linguistic devices. The very long history of the study of metaphor began with Aristotle, but only recently has metaphor been recognized as fundamental to human thought, and that language is only one of a number of potential external means of expressing metaphor. Recent experimental research has also demonstrated the cognitive reality of metaphorical mappings (Gibbs and Colston 1995; Turner 1987).

Metaphor was described by Aristotle as "giving the thing a name that belongs to something else" (Aristotle 1984b, 1457[b]7). Examples include "the saddle of a mountain," "Poverty is a crime," or the use of spatial expressions to denote temporal motions such as "We're getting close to Christmas." In language, the metaphor substitutes one word for another. A closely related form, analogy, is an explicit comparison between two different kinds of entities (e.g., "A system of rivers is to the land as arteries are to the body"). Both metaphor and analogy employ the basic mechanism of mapping across different knowledge domains. The effect is to focus on some attributes brought to our attention through such mappings, and to ignore aspects that are inconsistent with the mapping (Ortony 1979). However, it is claimed that analogy, by virtue of its explicitness, lacks the free-form suggestiveness of metaphor (Black 1993). In trying to better understand the more complex mechanisms involved in cross-domain mappings, most of the literature focuses on linguistic forms of metaphor and, by association, on analogy as well.

One way we use metaphor is to make associations across different sensory or emotional perceptions (Marks 1996). For example, in synesthetic perception, sounds (e.g., music or voices) are perceived to have visual properties, such as shapes, textures, or colors. In another example, physiognomic perception, dark colors or dark-colored objects are perceived as gloomy, and bright colors are perceived as happy.

Reddy (1993) and Lakoff (1987), through lists of many examples and extensive linguistic analyses, show how our use of metaphor is entwined in everyday thought. Many things—including temporal relationships (see Boroditsky 2000)—are, moreover, expressed in terms of spatial elements and relationships. Here are just a few everyday examples:

> "We can't turn back now."
> "His politics are on the far right."
> "Look how far we've come."
> "He's over the hill."

What is particularly striking in the analyses of Lakoff and others regarding metaphor is that such a large proportion of metaphors we use involves a mapping onto the spatial domain. Individual words, such as "tableland" and "heartland," can also have their root in spatial metaphor (Livingstone and Harrison 1981; Tuan 1978).

Common spatial metaphors, such as the specific examples just given, also are so ingrained as frequently used parts of language that they become part of the standard vocabulary and take on the character of idiom. The real power of metaphor, however, comes to bear when dealing with a new domain for which we have no preexisting knowledge. In this situation, metaphor becomes an important mechanism whereby a knowledge structure from another domain is brought to bear, in order to aid learning and understanding by employing observed resemblances.

Among Lakoff's (1993) many examples is the deconstruction of the metaphor "Life is a journey." Here, life is understood in terms of a journey, or the domain of traveling. More formally, metaphor is a mapping from a source, or secondary domain (journeys), to a target or primary, domain (life). In language, this is usually expressed in the following form: *target domain is source domain*, or *target domain as source domain* (Lakoff and Johnson 1980), although they do not necessarily take this explicit form.

To understand a metaphor, a person must recognize the juxtaposition and be familiar with the source domain: the domain being mapped onto. The mapping is tightly structured in that there are ontological correspondences between the two domains on an element-by-element basis. In looking at the example "Life is a journey," the entities and their relationships in life compared to travel (beginning, end, sequence of events, obstacles that need to be overcome) have direct correspondences. Thus, death corresponds to the journey's destination, major life decisions or occurrences (graduation, marriage, etc.) correspond to crossroads, and so on. Thus, saying "life is a journey" is really making a statement that expresses these two domains as having a set of correspondences and implies that this entire set of correspondences is a means of understanding the target domain. The domain mapping allows a rich collection of correspondences and generalizations to be inferred.

Lakoff also provides evidence of specific families of metaphorical relations. Thus, a career, as a major component of life, can be substituted in the life as a journey metaphor (e.g., "As I journey through my career"). Such variations and extensions are based upon how our knowledge is structured. Generalizations that specifically utilize the structure of categories allow novel metaphors (i.e., novel source–domain mappings) to be readily understood throughout or as subordinate and superordinate category hierarchies.

There are many metaphors in which the meaning does not derive from mapping one experiential domain to another (e.g., travel), but rather derive from a more abstract mapping of structural or behavioral similarities. Lakoff uses the example "spinning one's wheels" in reference to some situation being experienced or task to be accomplished. The association represented in the metaphor is both a lack of moving forward and the need for some form of intervention to change from that state.

This metaphor also immediately conjures up an image of car with its wheels stuck (in snow or mud), which is not able to move forward regardless of how fast the wheels are made to spin, or how much the steering wheel is turned. Indeed, many metaphors, regardless of the level of abstraction at which the association is made, conjure up images. Lakoff (1987) asserts that imagistic metaphors tend to refer to basic categories in the source domain. This makes sense, because as discussed in Chapter 4, basic categories are most directly related to sensory experience and are thus rich in sensory imagery. Some recent literature, however, goes even further in asserting that *all* metaphor is imagistic (Ghassemzadeh 1999; Gibbs and Bogdonovich 1999).

Black (1969, 1993) introduced the term "interaction metaphor" to distinguish it from the "translation" or "substitution metaphor" (Black 1993; Livingstone and Harrison 1981). The translation metaphor serves primarily as a linguistic device to remedy a shortcoming in a given language for expressing an idea parsimoniously. In this form, one concept replaces another— by proxy—that could have otherwise been used. For example, the spatial metaphor "The crime rate keeps rising" could also be expressed literally as "The crime rate keeps increasing," although the former may be viewed as a more elegant turn of phrase in this case.

The interaction metaphor, in contrast, allows a collection of characteristics to be inferred. When a generative metaphor is first recognized, there is an identification of some kind of general relationship between two knowledge domains in shared attributes, relationships, or a correspondence in the overall shape of the two knowledge structures. This is typically an entrée into further cognitive exploration of the two and subsequent discovery of something new that is not implied by either of the knowledge structures. The comparative exploration of the two different structures is itself a creative and imaginative process that results in *new* knowledge—learning that is better described as "discovery."

The modern perspective integrates Black's view in his distinction of these two kinds of metaphor, namely, that metaphor is not simply a mechanism of convenience but an inherent and essential mode of human thought. It is both a process and a particular way of looking at things. This distinction between substitution and interaction metaphors has subsequently been examined by a number of researchers (Johnson 1987; Lakoff 1987; Lakoff and Johnson 1980; Ross 1993; Schön 1993).

Metaphor allows the understanding and integration of new knowledge that is, as Petrie (1993) terms it, radically new, in the sense that it allows learning beyond the context of our experience. Petrie emphasizes that because of this, the use of metaphor is important in teaching.[1] The validity of the existence of this form of learning is based on several generally agreed-upon principles. First, how we perceive the world is always filtered through our preexisting knowledge structures. Second, most learning involves adding to those structures that are compatible with our experience and preexisting knowledge structures. Third, learning concepts sometimes involve modification of those structures themselves. This other kind of learning results in what we may experience as an "Aha" reaction, when we gain insight regarding an abstract concept. Although critically important as a heuristic mechanism for learning, it is also a relatively uncommon occurrence. It is through generative metaphor that knowledge structures are modified, with associative links added or rearranged.

Because of the use of knowledge structures in metaphor, particularly interaction metaphor, understanding how they are cognitively created and interpreted can offer valuable clues in learning more about how cognitive knowledge is structured and used in a more general sense. This is particularly intriguing with respect to understanding the structure and use of geographic knowledge, because metaphor so frequently involves a mapping into the spatial domain.

HOW WE MAKE THE CONNECTIONS

There is obvious correspondence between the properties discussed in Lakoff's linguistic analysis and Finke's list of properties that spark the imagination related to *emergence* in his analysis of graphical images: novelty, ambiguity, generalization, and incongruity. Indeed, emergence has been described as "replacing the representation of a structure by another representation—that is, deleting one or many structure variables and replacing them with others" (Gero 1994, p. 279).

This seems to closely relate to the representational redescription hypothesis of Karmiloff-Smith (Clark and Karmiloff-Smith 1986, 1990, 1992, 1993). This Karmiloff-Smith theory of cognitive development attempts to explain creativity in terms of how the human mind acquires the ability for creativity. It holds that the human mind, unlike that of most other animals, is compelled without any external prompting to go beyond simple behavioral mastery of a given problem domain and "redescribe" what it knows in increasingly abstract terms. This ability begins at a certain stage of development and is a process of direct manipulation of knowledge structures that in turn enable humans to derive more functional benefit from the knowledge they possess. In effect, the redescription facilitates the matching and recombination of the knowledge already available, which in turn allows associations and connections to be revealed as a person, consciously or unconsciously, theorizes on the basis of the redescription as a new domain.

Figurative representations, primarily through metaphor and imagery, are used externally to communicate complex ideas and introspectively to make new discoveries and insights. Imagery expressed either graphically or linguistically can include metaphor. The "desktop" metaphor is one good example. The individual elements associated with a desktop, and the interrelationships among them, abound in computer software for personal computers as a mechanism for making the software easy to learn. Actions performed on graphic icons in the display are designed to relate to an actual desktop. Typical office terms such as "file" and "folder," as well as "desktop," have thereby taken on a whole new meaning in a computing context but are immediately understandable even to the novice.

Focusing specifically on how we make new discoveries and insights, Finke calls these figurative representations *preinventive structures*. These nonpermanent knowledge structures are generated for dealing with a specific problem. In addition to being central devices for understanding through abstract thought, preinventive structures are often internal precursors to externalized creative products. In other words, graphical images, mathematical equations, computer programs, poetry, music, and all other externally manifested creative products must start with a preinventive structure within the mind. In the following section, I discuss the processes involved in making these new discoveries and insights.

Recognition of the properties in figurative representations that promote discovery is only half of the equation in understanding abstract thought (i.e., thought and learning that is not directly based on world experience). What are the cognitive *processes* at work when we do this? A number of researchers (e.g., Finke et al. 1992) distinguish between two kinds of processes: exploratory and generative. On the one hand, exploratory processes search existing cognitive knowledge structures by exploiting the inherent properties and associations available in those structures, in order to provide meaningful interpretations. Generative processes, on the other hand, modify the structures themselves, looking for new properties and associations that, on the basis of previous experience and other knowledge, appear to be valid.

Exploratory and generative process are used in an iterative fashion. If there is no satisfactory result from one's explorations of an initially generated preinventive structure, the generative phase is returned to again, in order to modify and manipulate that structure. The exploratory stage subsequently begins again. This cycle continues until a satisfactory result or insight is achieved, or the initial preinventive structure is abandoned. While this often occurs unconsciously (thus, the songwriter who says "The melody just popped into my head"), it is usually a conscious process in professional and scientific problem solving. This kind of iterative creative process is what would occur when locating a new bike path for a given city. The process may typically include the use of physical maps to aid the memory with the details of such a complex structure. Various alternatives can be visualized for a new bike path relative to existing topography, and so on. Then, remembering that an abandoned railroad right-of-way runs through the area of interest, and knowing that abandoned railway rights-of-way have characteristics desirable for bike paths (i.e., they are of necessity flat, cleared of obstructions, are of the appropriate width, and could possibly be acquired inexpensively), the bike path solution presents itself. Creative discoveries also frequently occur quite rapidly, apparently without repeated generation and exploration. This seems to be related to the complexity of the problem or the issue at hand.

In Finke's description of the generative processes, he includes a number of components: memory retrieval, association, mental synthesis, mental transformation, analogical transfer, and categorical reduction. A straightforward retrieval of existing structures from memory, such as a recalled visual scene or phrase, is the most basic generative process. Several scenes or phrases can also be retrieved and then associated in novel ways. A richer variety of preinventive structures results from the mental synthesis of components from varying memories and elements of one's knowledge, and subsequently transforming the structure by rotating it, rearranging various components, and so on. In "analogical transfer," as Finke termed it, the relationship or set of relationships in one context is transferred into another. For example, models of the structure of the veins in the body, and of the bronchial tubes in the lungs, have led to insights into the structure of stream networks (Woldenberg, Cumming, Horsfield, Prowse, and Singhal 1970). Analogical transfer allows properties on one, familiar knowledge domain to be transferred to another, less familiar domain. This, then, is what has already been described as metaphor, although Finke does not differentiate between the explicit and implicit transference of metaphor and analogy, respectively. Categorical reduction is the process of reducing objects into their components or into the basic level. This not only provides a translation of elements and properties into a more concrete level based upon observation, but it also provides an organizational framework within which new associations can be revealed as making sense. This is what is happening in the "Life is a journey" metaphor.

We recall that because of the association with the perceived world, basic-level categories contain much concrete information about the attributes and functional characteristics of objects. Basic categories are mostly concerned with parts; thus, they incorporate a part–whole structure. We also infer characteristics of basic categories, such as shape, function, and so on, onto categories of objects at other levels in the hierarchy. At higher levels, we store more abstract information that consists of generalized knowledge about how whole types of things look and behave. In other words, our knowledge is stored in a distinctive structure that is directly linked with, and dependent on, experience at one level, and becomes progressively abstract and detached from direct experience at progressively higher levels in the conceptual hierarchy. It is categorical reduction that allows abstractions, in a sense, to "be brought down to reality," relating them to something we have experienced.

The distance between the two knowledge domains also determines how many properties can be mapped. If a metaphor is a mapping of two closely related conceptual domains, then the correspondence of properties is high. The greater the cognitive distance, the fewer the number of properties that can be mapped, but the greater the novelty (through novel correspondences) and the potential for new insights. For example, with the metaphorical terms "city-state" and referring to a city as a "concrete jungle," the notion of a state is cognitively much closer than a jungle. The notion of "city-state" imparts the idea of an autonomous city, with its own ability for self-government and self-determination—a slight extension of what the concept of a city normally implies. The notion of the city as a "concrete jungle," however, adds the notion of an entirely different kind of environment and a whole different set of attributes that certainly have negative connotations of danger in an inhospitable wilderness, teeming with life in a bewildering tangle for a person not familiar with it.

Finke's exploratory processes include attribute finding, conceptual interpretation, functional inference, context shifting, hypothesis testing, and searching for limitations. Attribute finding is the process of mentally inspecting an image and looking for new patterns or unexpected features—finding emergent properties. Conceptual interpretation is the formation of an association between some element or relationship in an image to some known concept—giving meaning to it. Functional inference, the process of exploring potential uses for a given object or feature, is involved, for example, when an archaeologist (or even an antiques hunter) uncovers some unfamiliar object.

Context shifting is the consideration of an object in varying contexts, in order to gain insight into possible alternative meanings or uses. Thus, the archaeologist may first think of an object as a kitchen tool and be unable to think of what function it could possibly serve. But when he or she thinks the object as a possible farming implement, it is immediately seen as a planting tool. Hypothesis testing is mentally working through a structure as a potential solution to a problem—mentally thinking it through, or putting the image or feature in motion to see if it might work. In the bike path example, the city planner might imagine young children using it, and in this way notice that it crosses a major road. This may lead to either trying an alternative design or finding a solution to that single obstacle (e.g., a bridge).

Another exploratory process as defined by Finke is searching for limitations. This is essentially the converse of hypothesis testing, in looking for contexts or situations in which the preinventive structure will not work. For novel forms or configurations, this may aid in finding the range of contexts in which the configuration will work by first establishing the boundaries of the possibilities. Certainly, it is not necessary to use all of these processes in deriving intuition or new understanding, but they are typically used in combination.

All of these exploratory as well as generative processes interact with properties of a mental image or other preinventive structure. For example, novelty promotes emergence of new patterns in the process of attribute finding, because our attention is automatically drawn to features we identify as new or different. Abstraction also helps us to focus on important features and minimize distraction from irrelevant features during this process.

With the use of various combinations of generative and exploratory processes, the question arises: What is it that guides and constrains these processes? Certainly, the associations and mental combinations constructed and the patterns found are not random. What sorts of underlying patterns allow some associations and combinations to work readily and others not at all? Are there any constraints on the number and variety of patterns we may use?

As both Finke et al. (1992) and Lakoff (1987, 1993) have theorized, it is our structured knowledge of categories on one level, and our schemata on a more fundamental and abstract level, that help to guide and constrain the complex mappings involved in thought as well as perception, because, as described in Chapter 4, schemata are generalized patterns. In this way, they guide the process of appropriate source domains yet still allow flexibility. Categories consist of objects grouped by type hierarchies that can be either taxonomic or partonomic in nature. Taxonomic hierarchies denote *kinds,* and levels within such a hierarchy are linked by "is-a" relationships. Partonomic hierarchies, denote *parts,* and levels within such a hierarchy are linked by "has-a" relationships. The three-tiered hierarchy shown in Figure 4.1 is a simple example of a taxonomic hierarchy where "Philadelphia" is-a "city" and "country" is-a place. At each lower level, categories, and specific instances of a category, inherit many of the properties or attributes of parent categories. Progressing upward, categories represent more general and abstract concepts at each level. In an example of a partonomic hierarchy, a house has-a roof, a house also has-a door, and a door often has-a window. Obviously, these two types of hierarchies are highly interrelated in that the same category is very often a member of both.

Schemata are abstract patterns or rules of relationships/orderings among various elements that provide a perceived sense of order in our environment, and as such, they are independent of some specific context, the *container* schema being one example. It is in this sense that image schemata, in particular, seem not just to guide but actually to prompt spatial metaphors as a manifestation of these most fundamental knowledge structures. Image schemata provide an inherent sense of ordering within us not only for our world view but also for everything we are aware of or know. Thus, the path schema is the real basis for the "Life is a journey" metaphor, and the center–periphery image schema is the basis for politics being "to the left (or right) of center."

Schemata constrain both generative and exploratory processes such as mental synthesis and mental transformation by providing the overall rules and patterns by which such transformations and associations can be made. For example, the *path* image schema would govern how something would move during hypothesis testing or searching for limitations. What categories are present and memory constrain what can be retrieved as elements in a mental image or metaphor. Certainly, this is the case when the links between categories are navigated.

Lakoff and Turner (1989) originally suggested that what they called the *invariance hypothesis* is the basic principle that both guides and constrains the inventive mappings involved in metaphor and imagery, and more generally, the processes involved in creative thought and gaining understanding. The invariance principle was more recently restated by Lakoff as follows:

> We suggest that conventional mental images are structured by image-schemas and that image metaphors preserve image-schematic structure, mapping parts onto parts and wholes onto wholes, containers onto containers, paths onto paths, and so on. The generalization would be that all metaphors are invariant with respect to their cognitive topology, that is, each metaphorical mapping preserves image-schema structure. (1993, p. 231)

Despite the more specific wording, Lakoff explicitly states that, in addition to metaphor, the invariance principle holds for mental imagery, as well as all types of figurative speech. The invariance hypothesis asserts that correspondences are determined on the basis of a uniformity of *form*, and that a great deal, if not all, of our abstract inferences employ image schemata. As mentioned in Chapter 5, the term "image schema" is used to emphasize that all of our schematic patterns are grounded in, and determined by, our bodily interactions and sensory perception of our environment, and as such, these fundamental patterns are *spatial forms*. This means that all of the basic patterns that we use to guide and constrain how we think at a fundamental level are spatial in nature.

Are there a limited number of these fundamental patterns? As discussed in Chapter 4, all image schemata can be related to a very limited number of fundamental patterns or relations, including the schemata for *container, balance, path, source–path–goal, cycle, center–periphery,* and *link*. To further clarify the connection between metaphor and image schemata, Reddy, in an article originally written in 1977, first asserted that there are general-level metaphors that hold for whole classes of relationships (Reddy 1993). Although he is dealing within the explicit context of metaphors, the issue Reddy addresses is processes of analogical transfer and categorical reduction. With many examples, he demonstrated that all expressions dealing with communication entail a conduit–form metaphor, in which the action involves transfer through space. For example, ideas that can be *brought, sent, received, lost,* or *go right past* someone, all derive from the *path* image schema. Also, the elements involved are containers: Messages can be *empty* and people can be *full* of ideas (*container* image schema).

This process, as very briefly discussed here, can be seen as at least a glimmering of what is currently known about how we gain new understanding and insights. A number of constraints on this process become evident. Certainly, the obvious one is the extent of our current level of knowledge and experience. No matter how abstract the concept, we can derive new understanding and insight only by using our accumulated experience and what we already know. Thus, no matter how gifted and imaginative, the individual will not be able to grasp a concept that is too unrelated to what he or she already knows. Our knowledge is also set within a specific cultural context. Thus, the idea of a city like New York or London would be very difficult for the Australian aborigine of a hundred years ago to grasp. This is also why myths were invented—to explain the unexplainable in familiar terms, such as the ancient creation myths with warring gods, and why time was viewed as purely cyclical by early cultures. There was simply not enough recorded history as shared, cultural memory for the concepts of linear time to have emerged. Time was linked only to the observed, seasonal rhythms of day-to-day life.

Beyond these constraints, there are also the constraints of mental fixation or mental blocking, which simply need to be avoided or overcome to allow new cognitive perspectives. Mental fixation is focusing one's selection of contexts, categories, and elements of preinventive structures to an overly narrow range. This has the effect of significantly diminishing the level of novelty and other attributes of preinventive structures that promote insight and new discovery. This has given rise to the expression "thinking outside of the box" in the corporate world to encourage creative thinking. Mental blocking usually occurs when one solution or interpretation is suggested, effectively blocking or interfering with other, alternative, solutions coming into one's mind. Many empirical experiments on word memory and problem solving have demonstrated this phenomenon (Luchins and Luchins 1959; Reason and Lucas 1984; Smith and Blankenship 1991).

It is also the particular lack of this constraint—in drawing from broadly separated knowledge domains during the generative process and avoiding constraints on thought brought on by habitual thinking—that makes for creative genius. This has the effect of maximizing the utility of the knowledge and experience that one does have in making insights and what are effectively "leaps" of insight. One form of mental blocking is the tip-of-the-tongue phenomenon: when one knows something is there but for whatever reason simply cannot identify the observation or insight.

MYTH, SCIENCE, AND THE IMAGINATION

As discussed in Chapter 2, myth and science parallel each other in that they both function to explain observed world phenomena. In deriving a world view within a culture, both aim to provide a coordinated, epistemological frame-work, and thereby replace chaos and unpredictability with order and predict-ability. However, there is a profound difference between the two, in that myth relies on an intuitive and direct elaboration of experience, whereas science re-lies on the discovery of verifiable rules of the workings of a world already in-tuitively comprehended. This corresponds to the overall learning process de-scribed in Chapter 3, which begins by relating everything in the world to oneself and progressing from a sensory–figurative view to an abstract–opera-tional view that cognitively relates everything in the world. It is because of this differing perspective through an accumulation of learning, as researchers such as Livingstone and Harrison (1981) have pointed out, that myth has served as a precursor to science.

The objectivist/positivist view of science, however, holds that science progresses inexorably toward some final and true description of reality. Fur-thermore, that progress can *only* be made through systematic testing of objec-tive observational data (Johnson 1987). We have already seen that all observa-tion is dependent upon context, so that there can never be truly objective data. Furthermore, the progression of knowledge depends on abstraction and abstract thought: the connection of increasingly disparate knowledge domains at a level that is no longer directly dependent on, or perhaps even directly verifiable by, perceptual experience.

Many scientific breakthroughs have been made through the use of con-ceptual metaphors: relating the knowledge domain at issue to a more familiar knowledge domain (Kuhn 1996; Miller 1996a, 1996b). Even though the use of standardized procedures and models is considered a virtue in science and em-phasized in university graduate programs to aspiring researchers in a wide range of fields, this can inhibit creative thinking. Adherence to procedures in script-like fashion serves to facilitate routine tasks, but unthinking adherence to them can result in a delay in "breakthroughs" and suboptimal solutions to problems, even when better solutions exist (Finke et al. 1992; Schank 1988).

Throughout the history of science, significant shifts in the orthodox view of a phenomenon have tended to take place when the generally accepted model or view is demonstrated to be either wrong or severely lacking in the ability to explain observed behavior—when new observations can no longer be accommodated into the accepted view. Scientists have then been forced to think "outside the box" in finding a new model or solution. One of the most fa-mous and far-reaching metaphors used in the development of scientific theo-ries in the 20th century is Bohr's (1913) description of the structure of an atom as a solar system in extreme miniature. The visualization that this al-lowed—with particles spiraling around a nucleus like planets around a sun—was to view the primary subject (atom) in terms of a secondary subject (solar system), allowing a web of implications to follow. Even though Bohr himself emphasized that certain laws of physics were violated if the comparison was taken too far, it allowed the initiation of new and important insights. The solar system metaphor was abandoned in 1923, in favor of a (nonvisual) metaphor of the atom as a collection of simple, harmonic oscillators. This metaphor led to Heisenberg's formulation of modern quantum mechanics in 1925 (Miller 1996b).

Ley's (1974) description of the black inner city as a frontier outpost was identified by Livingstone and Harrison (1981) as a useful metaphor in the development of geographic theory. This conceptual juxtaposition immediately maps attributes of the black urban community to include facing danger and uncertainty as in a foreign environment. Ley extends this metaphor, putting the geographer into the role of explorer, to chart and document "the nature of unexplored reality" (1974, p. 2). In this case, Ley was also using the central property of a frontier—as an unknown environment and landscape—to stress the need for geographers to suspend their normal assumptions about the urban environment, in order to bring a new outlook into the study of the black urban ghetto.

These examples underscore the important role of metaphor in encouraging imagination and speculation, thinking of phenomena in new ways, and finding parallels—connections—with other phenomena. Models based on metaphor are not testable but often do lead to hypotheses that can be tested (Livingstone and Harrison 1981). The trick is knowing when such speculation will effectively lead to new discovery and, as a result, to better understanding.

So a second, important parallel between myth and science is that both depend on metaphor to explain observed phenomena. Myth, however, is static, with nothing beyond the metaphor itself. Science uses metaphor as an explicit modeling device. Indeed, among the many kinds of scientific models in various typologies is the *analog model*. This can be a physical model, using at least some different materials, or a conceptual model, using a different conceptual domain via metaphor. Black even extended this into a more general context: "Every metaphor is the tip of a submerged model" (Black 1993, p. 30).

Certainly, major scientific breakthroughs on the order of Bohr's insight do not happen every day in any given field. If they did, we would have chaos. But how can we promote creativity and insight and avoid the pitfalls of habitual thinking on a more day-to-day level? First, as Miller (1996b) mentions, the task is thankfully constrained, though the reality is that only a few possible theories explain a given set of observational data. Similarly, there are only a few workable solutions for any given problem. The issue is this: How do we "leave our minds open" (itself a metaphor) to the best solutions?

There are a number of ways to encourage creative thinking, thereby fostering discovery and insight. One straightforward technique, which we indeed use instinctively if we have difficulty seeing patterns in visual images, for example, is to "play with" the image—or whatever form in which the information is given. We can, for example, rotate an image, zoom in, zoom out, change some of the display symbology, or compare it with some other images. What we are really doing is manipulating the properties of the image, which we can do with mental imagery as well. Similarly, metaphor should be consciously encouraged as means of bridging conceptual gaps.

Other ways to encourage creative thinking involve methods of fostering generative and exploratory processes, as described by Finke et al. (1992). One is to simply overcome the fear of creativity. This means occasionally setting aside standard and accepted ways of doing things, which is often perceived as "risky." One can never be blamed for following standard procedures. Another strategy is to explore deliberately the hypothetical possibilities. This helps to invoke the exploratory processes, such as finding functional inferences, hypothesis testing, and searching for limitations. This kind of deliberate exploration within a group is what we commonly call "brainstorming." Another technique that fosters creative thinking is to suspend expertise. This strategy should also be employed when trying to generate creative ideas. Waiting to apply expert knowledge can also frequently reveal an optimal solution that would not have arisen by applying the standard approach and view of the problem. In Finke et al.'s terms, the motivation behind this is to bring a wider range of interpretive categories to bear, beyond those within the "usual" knowledge domain. For example, in testing a new software program, one good thing to do is to let someone unfamiliar with the program use it. Invariably, some mode of use not anticipated will be tried by the unfamiliar person. This is the motivation behind "beta testing" for commercial software products. Of course, the real trick in this strategy is, again, to know when suspending expertise would be productive.

A METAPHORICAL SPECULATION

To summarize, the recognition of connections and similarities between often widely different domains through the use of various figurative forms is one of the triumphs of the human mind. Not only do we depend on it for synthesizing our experiences, going beyond the facts presented, and deriving knowledge, but we also use this mechanism constantly in everyday communication. Our minds interpret everything based on prior knowledge, whether we are dealing with direct, real-world observation, figurative images, or language. All of what we experience is at the same time being continuously compared with, and integrated into, our world view as a continuous whole. Moreover, space and time assume the basic modes of order—the mechanisms by which this comparison and integration is accomplished.

On one level, metaphor allows us to express concepts in ways not possible within literal expression. More importantly, however, it is a thought *heuristic*; a mechanism for exploring our knowledge structures (our existing world view) represented as categorical hierarchies and schemata, within new contexts and juxtapositions that are outside the normal pathways. This exploration and comparison reveals consistencies and differences among categories, and moreover, prompts the creation of *new* categories as a means of resolving these inconsistencies and differences (Lakoff 1987; Tversky 1977).

Both imagery and metaphor serve to facilitate active learning in the integration of new knowledge into the preexisting knowledge structures. In this way, image and metaphor are truly basic tools for imagination and an important part of knowledge integration and abstraction—steps in the Piagetian perspective of how we learn through adaptation, organization, integration, and abstraction. Kant defined imagination as what allows us to make the connections among concepts: "the act of putting different representations together, and grasping what is manifold in them in one act of knowledge" (Kant 1950, §§A77, B103).

As just described, a number of various properties and processes act on the knowledge structures that are involved in creative thought. That these occur in various combinations results in individuals' various creative and problem-solving styles. Some may be more adept at generating preinventive structures, whereas others may be more adept at interpreting them (Finke et al. 1992). This explains why there is frequently more than one way to solve a problem, and indeed, also explains people's *ability* to arrive at varying conclusions or solutions.

In Chapter 2, I discussed the conceptual nature of space and time in terms of a double dualism of absolute–relative and discrete–continuous interrelated continua. Given the discussion in the current chapter, the speculation arises as to whether absolute–relative views of space and time should be viewed not as two ends of a continuum that are disconnected and opposite but rather as *connected*: In other words, is the framework shown in Figure 2.1 perhaps better represented as cylindrical in shape? It does seem that insight occurs across such a connection—the use of highly abstract notions that may have no direct connection with reality whatsoever to gain additional knowledge concerning objective and observational reality, which cannot be gained via direct observation alone. As an example, four-dimensional (4-D) space, as derived by Minkowski, is not an objective "truth" about the real world gained from direct sensory input, nor is it higher level knowledge about the world gained from combined observational experiences. Rather, it is a convenient metaphor by which some (perhaps nonobservable) aspect of reality can be understood.

Given the development of the discussion so far, from linkages among the things we know, to the graded categorical hierarchies and configurational spatial knowledge as key aspects, and how we use that knowledge to interpret new observations, we can take this one step further and speculate (using a visual metaphor) that the connection between experience and knowledge can be schematized graphically as a sphere (shown in Figure 6.4). Relative and continuous space–time converge in the unity of what we know. Moreover, this unity of knowledge—our world view—is essential for interpreting new experiences in the complex interplay of percepts and concepts.

It is the use of metaphor and creative imagery in a highly interrelated knowledge structure constituting the important element of human cognition we commonly call *imagination* that makes simulation of human intelligence within computers so difficult. We think primarily in terms of general patterns and resemblances, not logic. It has therefore also become one of the great unmet challenges over the past 50 years to try to replicate within the computer, at least in part, how the human mind works. The fascination with this is cer-

Experiences

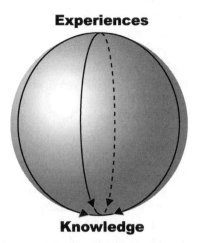

Knowledge

FIGURE 6.4. Absolute–relative and discrete–continuous connect at the extremes of each continuum.

tainly illustrated by the idea of the truly "intelligent" computer having become a cliché in science fiction stories. The paradigm of "mind-as-machine"—both as logical formalism and as actual computer programs—is a useful tool for learning about how we structure and use knowledge. I discuss these issues further in Part II.

In the preceding two chapters, I have examined the basic processes involved in how we acquire information through direct sensory experience, guided and filtered by our schemata and category structures, then how we use our experiences to build an integrated world view. This is the basic cycle shown graphically in Figure 4.3 and more generally in Figure 5.2. In the next chapter, I examine how knowledge is acquired through indirect experience. Indirect experience, through communication of knowledge from one individual, or group of individuals, to others, entails the use of external knowledge representations, including maps and language.

NOTE

1. Petrie also points out in this context that a metaphor may be a substitution and a generative metaphor *at the same time*: a substitution metaphor for the teacher, and a generative metaphor for the student.

CHAPTER 7

ACQUIRING GEOGRAPHIC KNOWLEDGE THROUGH INDIRECT EXPERIENCE

NONSENSORY KNOWLEDGE SOURCES

Much of our knowledge is not based on direct sensory experience of our environment. Some entire knowledge domains, such as ancient history, are acquired through indirect experience. Except for a very few individuals who have traveled into space, indirect experience is also the only way of directly seeing global-scale geographic space. For example, many people have knowledge about the surface of Mars, yet all information was gained via spacecraft and telescope images, drawings, written text, and the word of others.

Graphic images and language are the two primary, essential means of conveying information between individuals, and thereby provide an indirect source of knowledge, described previously as secondary knowledge. Both come in many forms. Graphic images include photographs paintings, drawings, diagrams, and maps. Language can be written or spoken, natural (e.g., Spanish) or formal (i.e., systems of mathematical notation or programming languages).

Besides being able to convey knowledge about places and times we have never seen, both the written word and the image have permanence that can be used to record information and document past events. Both also serve as repositories of information that gradually evolve and accumulate through successive versions, and by different individuals. Thus, besides being essential for providing indirect experience and conveying knowledge, graphics and language serve as key methods of external knowledge representation that provide means of recording and accumulating the collective knowledge and beliefs of a culture, and have an existence separate from any single person.

In this chapter, I examine the nature of these two forms of communication as external knowledge representations, and how each conveys spatial knowledge.

HOW GRAPHIC IMAGES CONVEY SPATIAL KNOWLEDGE

As mentioned earlier, graphic images include paintings, drawings, diagrams, and maps. These are, first of all, visual images, in that they are arrangements of light and dark, color, lines, and texture. The process involved in graphic perception is the same as that in ordinary visual perception, and involves, first, identifying individual elements, and second but more importantly, discovering the interrelationships among these elements. The latter includes their spatial arrangement as well as variation in the visual characteristics of individual elements. It is this second stage of finding interrelationships that is critical in graphic perception, because it is within this arrangement and variation that the message lies.

Graphic images are a very powerful means of conveying information. From a cognitive perspective, the power of graphics arises from a combination of two factors. They first utilize the ability hardwired into the human visual system to derive pattern and coherence instantly. Second, these patterns suggest image schemata (e.g., front–back, center–periphery, part–whole). In this way, higher level associations are made at a very abstract level with the viewer's knowledge. This gives an inherent advantage over textual descriptions and tables in conveying information that goes beyond the individual elements portrayed, and in providing new meaning through a comprehensive view that even the person who generated the image may not have anticipated. Such unanticipated meaning, including the discovery of new patterns in visual display as well as overall comprehension, are known as *emergent properties* (Finke et al. 1992).

Nevertheless, graphic images are fundamentally different than real-world scenes of our environment, in that they are *constructions* designed to portray information or to convey a message: to communicate. In the Renaissance, town portraits became a popular means of embodying and portraying a city's identity and unique character. Certainly, every city had its unique spatial layout, with natural and cultural factors as well as an economic and social context. It became a matter of civic and political pride to generate graphic representations. As described by Nuti (1999) in Italy, these town portraits took the form of an all-encompassing, bird's-eye view taken from some elevated and distant vantage point, and showed the city's overall shape and configuration. In Flanders, Holland, and other areas to the north, the town portrait took the form of a profile view, often from the water or across water toward the city center. Vermeer's portrait of Delft is perhaps the most famous of these representations.

The panorama, a development of 19th-century photography, is an extension of the profile view. These series of photographs taken from a single viewpoint were later joined together to form a 360 degree view. Because this was a photographic process, the viewpoint had to be a physically accessible place, such as the top of a tower or building. The image of the city stretching into the infinite distance in all directions also reflected a feeling of urban optimism in the midst of the industrial revolution and the increasing social and economic importance of the city. The fact that the viewer could not see this representation all at once, but rather had to scan it, also gave the panorama a temporal element that corresponded to the way the viewer would need to scan the real view if standing in the same spot where the photo was taken.

Through selection of viewpoint, level of detail, color pallette, and even decorations placed around the outside, these graphic images were constructed to convey specific attributes of the city portrayed; the 360-degree view stretching into the distance, giving the sense of expansiveness and importance, is just one example. In past centuries, showing many ships in the harbor of the city portrayed economic vitality and importance.

Any graphic is a simplified representation. Particularly in maps and diagrams, specific entities of a phenomenon are selectively portrayed in some schematic form, providing focus upon those while ignoring others. In this way, representation is simplified, focused, and re-formed. Although in most graphics individual objects and the entire visual scene are portrayed to resemble how they would appear in the natural environment as much as possible, the individual scene is selected, as are frequently elements within the scene. Even the most "realistic" graphic images, such as photographs, are simplifications achieved through the selection of the scene, choice of color or black-and-white film, length of exposure, and other devices. Symbolization, such as ships, is also present although not as abstract.

Graphics are external knowledge representations in spatial form. The portrayal of that knowledge depends on the effectiveness of the symbolization, so that individual elements can be easily interpreted, and also on how those elements are ordered and grouped within a meaningful overall arrangement. This is the realm of semiology, which has been discussed in detail by

Bertin (1981, 1983) and subsequent authors (Head 1991; MacEachren 1995) as it applies to maps. Maps must follow some set of symbolization rules or conventions governing their form, order, and arrangement in order for the message to be interpreted as intended by the map designer. As an example of what can happen, the graphic in Figure 7.1 is ambiguous, because it violates normal diagrammatic rules for how lines should be connected.

As for any image, interpretation always occurs within a larger context. Individuals bring their own unique perspective, based on their knowledge, previous experiences, cultural context, and perhaps even the task at hand to the interpretation. Bertin (1981) described two kinds of information contained within a graphic: intrinsic and extrinsic. Intrinsic information is the internal relationships revealed by the graphic, and extrinsic information includes the purpose for which the graphic was intended at the time it was generated, and the nature of the problem at hand when the graphic is interpreted. In the case of art, there is typically not a "problem" to be solved, but the extrinsic information would include, perhaps, an art expert's interpretation.

Bertin also defined three progressive levels of information that can be derived from a graphic image. At the elementary level, a specific piece of information is retrieved, which is the equivalent of reading single entries from a two-variable data table. At the intermediate level, a specific grouping or characteristic of one subset of entities within the graphic is retrieved. Then, at the third, overall, informational level, general patterns and interrelationships of the whole graphic are realized. The level of information retrieved depends upon what the perceiver is looking for and the effectiveness of the graphic in

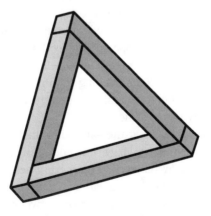

FIGURE 7.1. A well-known ambiguous figure that violates conventional line connectivity.

portraying information at each of these levels. Elementary-level information derives directly from meaningful symbols. Higher levels of information derive not only from simplification but also through ordering and grouping of the various elements portrayed.

THE LEGACY OF MAPS AND MAPPING

Of the many types of graphic images, the map is certainly most closely identified with the geographic context. Maps are both fundamental to how we express and convey spatial knowledge, and ubiquitous (Stea, Blaut, and Stephens, 1996; Tversky 2000). A number of studies have shown that simple maps can be understood and used by children as young as 4 years of age, without any previous experience or training (Blaut and Stea 1971; Landau 1986; Presson and Hazelrigg 1984). Maps also developed independently in cultures as diverse and geographically far-flung as Europe, Africa, Mesoamerica, and the Marshall Islands in the Pacific (Woodward and Lewis 1998). In modern, Western culture, we use them to see what tomorrow's weather conditions will be, where we are in the mall, which subway train to take, and how to find our way to a friend's party or to New York. We use maps to see where things were in the past, where they might be located in the future, and even to understand the spatial layout of imagined worlds, such as Middle Earth in J. R. R. Tolkien's *The Hobbit* and *Lord of the Rings* trilogy.

Maps have been used for millennia to record and convey geographic information. The first known map, identifiable as such, was scratched on a piece of mammoth bone found at an Ice Age campsite on the river Dneper (Harley and Woodward 1987). It depicts a human community that existed 15,000 years ago. Christian Jacob (1999) has discussed the role of the small-scale map, particularly world maps, in the development and evolution of an integrated world view throughout the history of Western culture. It is particularly at this scale that the map serves a key role of "making the invisible visible," providing a means of experiencing geographic reality indirectly on a scale we cannot normally perceive directly through the senses.

In the sixth century B.C.—at the same time that Homer's *Odyssey* provided a unified view of the world, with a notion of continuous, connected space and time within a mythological and literary account—the first Greek portrayals of a unified world also appeared in map form. It was also during this time that there developed an interaction between maps and discourse. The map became a means to help organize and rectify a geography of the known world, with information from diverse sources. Anaximander of

Miletus was one of a group of individuals who decided to write their own accounts of the world as an alternative to the various mythological accounts of the time, such as that of Homer (Jacob 1999). He drew a map of the world as a by-product of his treatise *On Nature*. This work was intended as a factual account of the known world and the interrelationships of the elements within it.

Besides integrating world knowledge of the time in a unified and coherent external representation, this development was also noteworthy in how it was accomplished. These early Greek geographers and cartographers relied upon the written reports of travelers and sailors, and on general knowledge and, indeed, earlier compilations. Eratosthenes probably epitomized the earliest tradition of geographic compiler. As head librarian at Alexandria in the third century B.C., he took it upon himself to rectify the extant small-scale maps available within the library. Besides scholarly geographic texts, reports of travels and sea journeys, and historical maps, the library at Alexandria held the archives of the land surveys done by Alexander's army throughout the Persian empire and eastward, as well as other surveys from around the Mediterranean and Africa. Certainly, this vicarious method was, and still is, the only practical means for compiling an integrated view of the entire world, or any significant portion of it. For a single individual, traveling the known world, particularly then, was certainly an arduous and time-consuming task. There is also the element of interpretation. If our individual recollections of reality are filtered by our personal perspective and tend to be incomplete due to selective and generalized memory processes, then combining multiple accounts provides a means of finding the commonalities and minimizing gaps in an overall picture. So representations of small-scale space, both graphic and textual, are indirect geographic experiences for not only the person or persons generating them but also the person reading them. This also means that the compiler must trust the accounts of others. However, travelers are prone to exaggeration and occasional fabrication. There are well-known cases of mapmaking during the discovery of North and South America in which inaccurate accounts—or inaccurate interpretations of those accounts—have resulted in perpetuation of phantom features or gross locational inaccuracies that persisted for considerable periods of time. A well-known example is the portrayal by a Mexican mapmaker of three major rivers on a map of the American West in 1810: the Timpanogos, the South Buenaventura, and the Los Mongos. These were copied by a succession of mapmakers until the late 1830s (Raisz 1956).

Jacob (1999) discusses how the ancient Greeks were also aware of the problems inherent in relying on the accounts of others, as documented in

Strabo's *Geography* (1917), written at the time of Christ. After reminding the reader of his own extensive travels, Strabo says:

> However, the greater part of our material both they and I receive by hearsay and then form our ideas of shape and size and also other characteristics, qualitative and quantitative, precisely as the mind forms its ideas from actual impressions—for our senses report the shape, color, and size of an apple, and also its smell, feel, and flavour; and from all this the mind forms the concept of apple. So, too, even in the case of large figures, while the senses perceive only the parts, the mind forms a concept of the whole from what the senses have perceived. And men who are eager to learn proceed in just that way: they trust as organs of sense those who have seen or wandered over any region, no matter what, some in this and some in that part of the earth, and they form in one diagram their mental image of the whole inhabited world. (1917, bk. II, ch. 5, sec. 11)

Concerning how humans acquire knowledge, Strabo emphasizes the interaction of direct and indirect experience in gaining knowledge about our environment. This quotation also indicates that Strabo recognized the relationship between the hierarchy of geographic objects and spatial scale, and the need to work from the local to the global scale. He describes a two-step process that has a distinct resemblance both to Golledge's (1988) route versus survey perspectives in geographic knowledge and to Piagetian theory, namely, that we rely on our senses initially, with the areas of geographic knowledge spatially confined and disjointed, but then we shift to a more wholistic and integrated perspective.

Claudius Ptolemy's second century A.D. *Geographia* introduced the use of a general partition of the globe into regular parallel and meridian lines (latitude and longitude). He used this system to inventory the geographic point locations of about 8,000 places. This not only allowed him to locate each place within a universal system of graticules rather than just in relation to other places, but it also provided coherence among separate maps, so that they could be referenced in relation to each other (Jacob 1999).

It is readily apparent that Ptolemy's *Geographica*, with its tables of places and coordinates, was an inventory of places, whereas the works of Eratosthenes, Anaximander, and Hecataeus were intended to render an overall view—a global perspective, symbolized as a structure of abstract points, lines, and shapes. With the use of a universal spatial referencing system, moreover, Ptolemy's map indicates a shift from the use of relative space to absolute space in representation.

Ptolemy's map made it possible to check mistakes in locational calculations and to refine collective geographic knowledge in a systematic way. As such, it served as not only a communication device, as had previous maps, it also served much more effectively as a storage system—an analog database of known geographic facts that could be shared among others to build gradually and progressively, thereby perpetuating and extending a body of knowledge.

The map, like all knowledge representations, is inextricably bound to a cultural context. World maps of the Middle Ages are well-known examples. In Medieval Europe, views of space and time were profoundly influenced by theological beliefs and reliance on biblical accounts of places and events. With Ptolemy's map, as well as other works of the ancient Greeks, forgotten until their rediscovery in the 15th century, mapping again adopted a relative view of space. Time was also included in these medieval maps as an integral component of this theologically based perspective.

Even though the portrayal of the Garden of Eden as a terrestrial feature, and of Heaven as having spatial identity, may be looked upon as cartographically naive today, these maps are quite valid in that they are representations of the dominant world view of the time (Woodward 1985). This same world view found expression not only in maps but in all forms of art and literature (Thrower 1996). It was widely accepted that the Garden of Eden existed somewhere on Earth, although far away and inaccessible. And this is how it was portrayed on European maps from the 6th to the 15th century. It was also viewed as the earthly origin of space and time.

As described in detail by Scafi (1999), this aspect is clear evidence that these world-scale maps were not intended to portray only contemporary space. Rather, they served as visual compilations and syntheses of all spatial knowledge, providing a comprehensive cosmography. Eden never occurred on larger scale maps, nor was this appropriate, because such features were only part of the macroscale world view. The space on world maps in the Middle Ages was never intended to be an accurate representation of observed reality. Rather, they were intended more as space–time narratives of a comprehensive world view (Woodward 1985). Thus, cartography, as viewed through time and across various cultures, also serves to demonstrate that maps are not direct representations of reality but are rather representations of human concepts pertaining to the world (Wood and Fels 1986). This is borne out in the psychological literature (Tversky 2000).

Thus, medieval cartography can be characterized as using space as a *container* for portraying what is known. It was not a mathematical mapping from geographic space to the space on a sheet of vellum. This conceptual rather than perceptual approach to representation has also occurred in other cultures. To Australian aboriginal artists, representing all of the important elements rather than their spatial relationships is paramount. The goal of aboriginal painters is to represent all they know rather than what they see, or might see from some vantage point, as a visual image. A similar approach is used in Aztec maps.

The frontispiece of the Aztec manuscript known as the Fejervary Screenfold, shown in Figure 7.2, was created before 1521, in the Tochtepec area of Mexico. Some aspects of this image represent the geographic layout of the area, although it would not be recognizable as a map to the Western eye. It is highly schematized within Aztec symbology, and although there is correspondence of direction, there seems to be no distance "mapping," and the design simply fills the square layout. There are also intertwined spatial and temporal concepts. The four main segments into which the map is divided in the cardinal directions presumably represent the four provinces into which the Aztecs divided their empire. Each is surrounded by a separate, sacred river and dominated by its own gods. East is situated toward the top. The fire god Xiutecuhtli, who stands in the center, and the other eight gods represent 9 of the 13 months of the Aztec year. The birds represent tribute that each prov-

FIGURE 7.2. Fejervary Screenfold. Reproduced by permission of the Board of Trustees of the National Museums & Galleries on Merseyside, Liverpool.

ince brought to Tenochtitlan at different seasons. The circles above the birds symbolize years. This, then, represents the Aztec cosmology, where space and time seem to be inseparable (Barber and Board 1993).

We must remember that even modern cartographic products are not representations of only a contemporaneous instant in time. These also are compilations of information collected over time that often intentionally portray features that may or may not still remain, and features (such as planned roads) not yet built.

At the beginning of the Renaissance, global exploration and the rediscovery of ancient texts and maps caused a fundamental ontological shift in the dominant world view. With reports of entirely new, previously unknown continents and the rapid expansion of trade, world maps became secular representations of observed space rather than theologically influenced representations of a world view. This also reflected a shift in the use of world maps—moving from primarily the domain of the philosopher into practical use for global-scale trade and exploration. Geographic knowledge became distinguished from theological, cosmographical, and historical knowledge (Scafi 1999). The use again of a uniform spatial referencing system aided in the systematic accumulation of new geographic knowledge, as it had in the time of Ptolemy. The Euclidean perspective of empty, homogeneous, and infinitely extensible space also aided in rationalizing the ambitions of competing nation-states to carve out global empires (Benedict 1991).

Given the "scientific" approach to cartography stressed over the past century, within the cartographic community, maps have been looked upon as representations of an objective reality. Quantitative and thematic mapping techniques of the Renaissance became progressively more refined. Systematic gathering of locational measurements has allowed the world to be mapped and remapped, with great locational precision.

Over the past 30 years, the rendering of cartographic products has become fast, easy, and accessible, with the introduction of remote sensing and satellite imagery, digital mapping, and geographic information systems. Mapping procedures have become highly codified, systematized, and unquestioningly standardized (Corner 1999). Mechanically produced representations of geographic observations made at least in part also through automated means have reinforced the view of maps as neutral mirrors of an objective and true reality. And with this view, maps are used as the justification for many urban planning decisions, as well as policy formulation at regional and national levels. Through the processes of automation and computerization, maps are now tools firmly within the Newtonian paradigm of an absolute space and time as a backdrop upon which the locations and dynamics of physical objects are measured. Nevertheless, the principles involved in mapping remain the same (Wood 1994). What gets forgotten is the fact that human agency remains an integral part of what is measured and how it is measured and interpreted. In other words, even the most "objective" maps remain *constructions* of perceived reality.

HOW MAPS CONVEY KNOWLEDGE

The map is a uniquely complex combination of image, language, and mathematics (Woodward 1992). Much of cartographic representation is also highly conventionalized. Besides being a complex and sophisticated system of symbology, placement of specific features within a professionally derived cartographic product are often based upon precise observational measurement of real-world features. Given some specified scale, map projection, and coordinate system, the corresponding map features can be measured, thus retrieving (to some level of approximation) the original real-world measurements as data.

Regardless of the mathematics involved, there are also other, less-precise, rules of mapping practice that deal with how to portray a given collection of information, in order to convey a specific overall message. Besides the inclusion or exclusion of specific elements, the sizes of the symbols used for specific elements, for example, alter their significance overall, as well as in relation to other elements in the map. The choice of framing—the orientation of the geographic space in relation to the viewer, as well as the selection of the spatial area portrayed—similarly affects the overall message (Cosgrove 1999). Mapping is thereby a creative process.

The power uniquely inherent in the map lies in its ability to provide a portrayal within a spatial context that goes beyond the data and individual elements. In a spatial display of information, local data are given greater meaning via an integration within the broader context, and the entire image may provide insights unanticipated even by the mapmaker. Space becomes both subject and context.

As with graphics in general, the entire mapping process involves both a complex system of signs and "rules of grammar" and an arrangement within which an overall portrayal is shaped through selection, transformation, and arrangement. Thus, as previously mentioned in Chapter 2, the map also has a dual nature (Peuquet 1988a). Its *structure* is algebraic in nature, an assemblage of symbols with prescribed meaning, ordered according to a system of positional rules of interrelationships. It is also a visual *image*, with patterns of light and dark, color, and so on. The entire cartographic "process," including both map compilation and interpretation, is highly complex as a result. Certainly, the context of the map, the meaning of the individual symbols and the basic "rules of grammar," must be known if the map is to convey the intended *meaning* to the person viewing it. Thus, the map reader should be able to say, "Ah, this must be a map of Pennsylvania, and I can see by this particular linear symbol that a railroad runs through here." Mapmakers also realize that there are intangible qualities of perceived reality that can only be conveyed cognitively, utilizing the nature of the map as an image, for example, the use of reds and yellows to convey danger in the intensity of thunderstorms in television weather maps (as opposed to shades of green for milder rain). Given all of these characteristics, mapping inescapably involves both art and science.

Maps, as known in the modern Western world, are unique in that they are also *mappings* in the mathematical sense. Each location in the represented geographic space corresponds to a location in the map space. Because of the diminution in physical size, however, many locations in geographic space will "map" onto the same location in map space in a many-to-one correspondence. At the same time, the power and effectiveness of maps are based upon this simplification. It is through portrayal of some elements and not others that the complexities of the real world are filtered and controlled, allowing the viewer to focus attention on particular aspects. Thus, as with all images, maps are necessarily simplifications, or generalizations, of reality. There is an often-cited scene in Lewis Carroll's *Sylvie and Bruno Concluded* about the "perfect" map. In order to show all the detail of the real world, it was made at one-to-one scale. Because this map could never be unfolded, for it would block the sun from the farmers' fields, the characters used the real world as its own map, and this served *almost* as well. This statement alludes to the inescapable fact that any map is an abstraction of reality. Paradoxically, the more detail provided in the map, and the more complete the information represented, the more redundant and confusing the map becomes in its representation of perceived reality. It is through simplification of the complexity of the real world that maps derive their expressive power.

Maps are not just a means of storing and communicating geographic information. As images, they allow new knowledge to be uncovered. Thus, the map *becomes* the territory, with the opportunity for new discoveries even in territories already well-known. Not all maps do this, however. This is the difference between what Bertin (1981) has called a "reading" map and a "seeing" map. The former is a compilation of facts. The latter allows pattern and general relationships to be seen. This relates back to the various levels of information Bertin described in relation to images in general.

"Reading" maps allow information to be retrieved at the elementary level. At this level, the value of a map element can be read to answer questions such as "What is the land use at 234 Broad Street?" or "Is Baltimore north of Washington, DC?" This is learning via indirect experience: to know the information by reading the map. At the overall information level, we can answer questions such as "What is the overall pattern of commercial land use in the city? Is it linear or clustered?" A "seeing" map allows such patterns to be easily determined. This is also indirect experience, but it involves imagination. To gain this type of knowledge through direct perception of the real world would require many individual experiences to derive the required generalization.

In stressing the scientific approach in cartography, the focus has been on the accurate reproduction of information—portrayal of what is already known. An unfortunate consequence of this approach is that the use of the imagination in finding *new* information (i.e., new patterns and relationships) has been ignored, if not actually repressed (Peuquet and Kraak 2002). With the adoption and extension of Bertin's semiology over the past 30 years, systems of symbology have been codified, so that overall patterns can be most effectively portrayed by the cartographer in getting an intended message across to the map user. The use of the imagination in cartographic portrayal has in large part been left to the artist. The map-as-metaphor is one such "artistic" form that is also a very powerful communication form, stretching the ability to go even further beyond the individual elements portrayed—beyond the power of the map or of metaphor alone.

There are many examples in which the map is used as a visual metaphor in making various political statements. These maps visually superimpose other figures that are not geographic features. One of the best examples is Leo Belgicus (the Belgic Lion), shown in Figure 7.3. This metaphorical map was first conceived by Michael von Aitzing around 1579, when the Lowland provinces (now The Netherlands and Belgium) were fighting for independence from Spain. Seeking to create a symbol of national pride and identity, von Aitzing superimposed a lion onto a map of the Lowland provinces. The lion appeared on the coat of arms of almost every province, so that in addition to implying lion-like strength and dignity, it also served to imply unity. The artistic blending was not forced, in that the geographic accuracy and detail of the map were not compromised in order to superimpose the lion, making the intended correspondence very clear. This map appeared in various editions and forms for over two centuries. Figure 7.4 represents a modern example. A dart board cut into the outline shape of the United States was used by the Memphis Area Chamber of Commerce

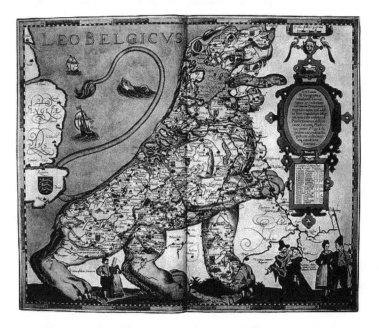

FIGURE 7.3. Leo Belgicus from Germania Inferior by Petrus Montanus, engraved and published in Amsterdam, 1617.

as an advertisement—Memphis as "America's distribution center" (Holmes 1992).

Regardless of whether the map is viewed from a scientific or an artistic perspective, the modern computer has fundamentally changed the predominant way maps are produced and used. Instead of the traditional paradigm of professional cartographer designing a map product for the map user, map-making has become "democratized." Widely available, automated mapping tools and GIS, combined with the advent of readily accessible and shared geographic databases, have given map users the ability to make their own maps. Moreover, the map itself has often become an ephemeral object, seen only on a computer screen perhaps for less than a minute, until another map or table is generated. Instead of being a data storage medium and communication tools, maps are being used in a fundamentally different way as exploratory tools for gaining new information and insight from large, digital databases.

The combination of modern computing technology and exhaustive digital coverage of the earth at multiple scales is a particular form of cyberspace, blurring the distinction between reality and a representation of it. Of course,

FIGURE 7.4. Memphis Area Chamber of Commerce advertisement (Holmes 1992). Reproduced by permission of Memphis Area Chamber of Commerce.

small-scale maps, as indirect experience, have always been how we construct our perception of geographic space on a scale beyond what we can experience directly through our senses, such as our view on a global scale. Although at one time, a main purpose of world-scale maps was to record and present known knowledge of the larger world, they were also of necessity abstractions and simplifications. Both types of maps are now being replaced with what may be described as imitation of experience through exhaustive coverages in cyberspace, complete with realistic images and imagery.

Unlike traditional forms of indirect experience of geographic space through textual descriptions and graphics, the geographic experience through cyberspace is quickly taking on the character of the holodeck on the spaceship *Enterprise*. The boundaries between direct and indirect experience of the environment are becoming so blurred that it is difficult in a cyber world to distinguish between the real and the created—past, present, or future. But will this ever replace textual and graphic representations of geographic space? The answer, clearly, is no, but maps need to change, as mapping has already changed. We will always need simplified representations of reality in order to comprehend its complexities. What we need, then, is research into how maps can be best used as visualization tools for exploring digital geographic databases and as interactive aids in experiencing the world, deriving decisions, and solving spatial problems.

Computing thus presents both a need and an opportunity. There is a need for development of strategies to ensure that cartography and related displays are appropriately employed by the user. There is also a major opportunity that the very different medium of the computer provides for cartography in developing new kinds of displays for interactively exploring virtual worlds.

Computers have capabilities that traditional cartographic media do not have. No special skill is needed. Both the map design decisions relating to default projection, symbology, and other elements, and the actual rendering are handled by the computing software and hardware. Besides being able to draw a new map literally in the blink of an eye, the computer is capable of displaying time dynamically. Whereas there are a number of strategies for displaying change through time on traditional maps, these are inherently limited within a static medium (Peuquet 1994). Just as we can establish a "mapping" of locations from perceived geographic reality to any two-dimensional surface, we can map time onto a dynamic computer display. Certainly, in today's rapidly changing world, a natural and intuitive representation of space–time dynamics has become more important.

These capabilities have already allowed the easy generation of "map movies," and cartography in general has become an interactive tool. So far, however, computer cartography only represents an extension of traditional cartographic techniques. If the computer is to be used to its full potential as a cartographic medium for data exploration and problem solving, cartographic techniques need to be developed that "spark the imagination," rather than using the traditional paradigm of storing known information and getting an intended message across. In data exploration, no one knows what the "message" is when the map is drawn.

As discussed in Chapter 6, researchers in cognitive psychology have already suggested some preinventive properties of images that encourage imaginative thought, specifically, novelty, abstraction, incongruity and ambiguity. In interactively manipulating the map, the user of cartographic software is (unknowingly) directly manipulating these properties. For example, zooming in and out (i.e., changing the scale) varies the level of abstraction. The level of incongruity is increased by overlaying themes (i.e., combining variables) on the map that would normally not be associated with each other.

MacEachren (1995) has noted that as the symbology used in a map becomes less mimetic (i.e., imitative of the thing symbolized) in form and more arbitrary, the likelihood of multiple interpretations escalates. In other words, the map itself becomes more ambiguous. Although there is pragmatic necessity in making map symbology less mimetic and more general as the scale decreases, and the map as a whole becomes more generalized (cf. Figure 7.5), this can also serve to spark the imagination to see unanticipated interpreta-

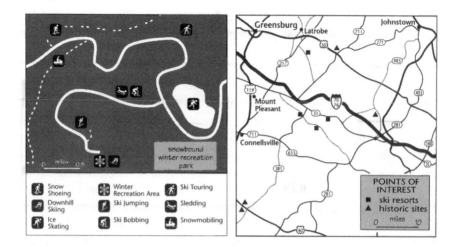

FIGURE 7.5. The tendency toward more general symbols as map scale increases is evident when we compare a winter recreation park map and a regional map used to locate such parks and other points of interest. From MacEachren (1995, p. 322). Copyright 1995 by The Guilford Press. Reprinted by permission.

tions by deliberately utilizing arbitrary symbology. Thus, ambiguity can be introduced in a controlled way in any map, through the deliberate relaxation or violation of semiotic rules.

Other properties that are more critical to how we construct maps than imagery should therefore also be considered in the cartographic context as preinventive properties. Framing is certainly one of these properties. What portion of geographic space we choose to delineate as being within the map instead of outside is a quite different form of element selection than abstraction or thematic portrayal. It is very much like what we see out a window and is similar to the town portraits of former centuries. Given that the presence or absence of kinds of features can vary considerably over space and through time, the framing we choose has epistemological implications. Framing selects one interval of space–time, while ignoring the rest. A unity and totality of space are defined within the map frame (Cosgrove 1999). If contained within the frame, perhaps excluding confounding or distracting features, patterns can become visible, but they can just as easily become invisible if not contained, or only partly contained, within the frame of the map.

We are already manipulating some of these properties instinctively in using maps and other forms of graphics to explore geographic databases interactively and solve problems. We pan the map to change the framing, zoom in and out to change the level of abstraction, change colors to make certain features more visible, or choose a monochromatic scheme to look for overall pattern, all of which involve "playing" with the map to let latent relationships emerge. There are other methods of manipulating maps for this purpose that we would not usually use, given how we normally think of maps—turning the map upside down or sideways, for example. Other methods of aiding the mapmaker/user to see emergent features and to allow computer mapping to become a truly exploratory tool need to be investigated.

Insight can be gained by investigating historical maps that "spark the imagination." A classic example is Charles Minnard's map showing Napoleon's 1812–1813 Russian campaign. Minnard himself did not call it a map, but rather a *carte figurative*, acknowledging that he took significant liberties with cartographic convention. But in so doing, he was able to display many variables and in a way that allows the reader to see immediately and clearly several interrelated patterns. The map shows the overall spatial pattern of the route, both to Moscow and back, the number of troops, the temperature on the way back—all within a temporal sequence. Minnard also shows essential details for all of these variables—all in a manner that is very accessible and elegant. Certainly, this simplification was by design. The actual troop movements of Napoleon's army were much more complex. The reason that Minnard was able to achieve this visual *tour de force* in this particular instance is that the overall spatial movement coincided with a temporal progression. So spatial movement over time is collapsed into a single east–west, space–time dimension.

What else can we see by "playing" with Minnard's map? Plenty (Peuquet and Kraak 2002). If we "pick apart" space and time in his combined east–west axis, other properties are seen in the data. In the example in Figure 7.6a, where the horizontal axis is a complete time line, we can see that Napoleon's army was not constantly on the move, as the original graphic leads us to believe. The army stopped at several locations along the way, sometimes for significant periods of time. If we allow the spatial dimension to dominate within a consistent x–y grid, as in Figure 7.6b, we see more clearly that Napoleon took the same route back part of the way. As in Minnard's original graphic, the timing relative to the group that separates and takes a circular route on the left side of the graphic is unclear. In the time line of Figure 7.6a, however, it is quite clear that after branching off from the main force, the group arrived in Polock in mid-August and remained there until October. They later joined the main retreat in early December (Roth, Chuah, Kerpedjier, Kolojejchick, and Lucas 1997). Of course, the pages of this book have the same limitations as the paper map. On a computer, where the graphics in Figure 7.6 were de-

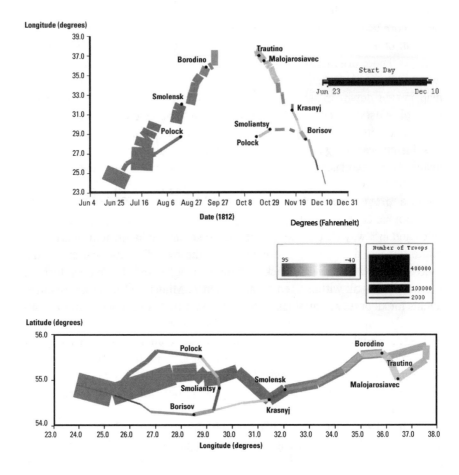

FIGURE 7.6. Alternative visual representations of Minnard's map. From Roth, Chuah, Kerpedjiev, Kolojejchick, and Lucas (1997). Reprinted by permission.

signed to be displayed, we could examine various components of the image for further information or zoom in for a closer look. The horizontal grab bar labeled Start Day could also be "grabbed" and used to steer the dynamics of the display, playing them forward and back, slow or fast, in the temporal dimension.

This is only a brief example. Future research should reveal some new forms of cartographic representation that are very different from the maps that we currently use, although computer displays have some constraints given the current technology. Projected holographic images are not yet available, so computer graphics are limited to flat screens, with limited choice in size. The dynamic and interactive capabilities of computer representation of graphic images greatly enhance the potential of the map as a tool for creative thinking in ways that before were only possible via mental imagery: They instantaneously create, modify, and manipulate images. Images can be set in motion and speeded up or slowed down. Indeed, whole worlds can be created that would be physically impossible in the real world—Einstein riding a beam of light . . . with a grab bar!

The primary challenge in understanding maps as representations of geographic space lies in the inherent melding of data, knowledge, and personal and cultural perspectives. This melding is also intertwined with the dual image/structure nature of maps and is certainly more difficult to untangle than textual descriptions of geographic features or phenomena. On the one hand, personal impressions and interpretations are usually expressed in more generalized terms than observational facts. Observational facts, on the other hand, are frequently expressed in a more detailed and concrete manner, which reflects how people cognitively store knowledge. The dual nature of maps, and the complex web of symbolic devices from which they are constituted, offers the capacity to intertwine objective data and subjective, cognitive impressions in ways that at times may frustrate the mapmaker's attempts to control interactions between such elements.

THE POWER OF LANGUAGE

In addition to using real-world experience and graphic images, people acquire information about spatial environments through language by reading text, or by hearing spoken descriptions. Language can be happy or sad, beautiful and eloquent or simply informative, and evoke images and memories of places and events. The combination of these capabilities has given rise to cultural mythologies throughout history. When people were unable to explain phenomena, they invented stories in terms that they could understand. Thus, earthquakes and erupting volcanoes were explained in terms of angry gods; if these gods were appeased appropriately, tranquility and prosperity would return (Bang 1993).

Language is used in an accumulation of knowledge as a collective project within a culture. This is perhaps even more important in the case of language than in cartography, because language is used to represent all knowledge domains, not just the geographic domains or ideas that lend themselves to expression in spatial form. As was the case for maps historically, the knowledge represented in geographic as well as textual form concerning other domains has been built cumulatively, through successive human iterations to others, both through verbal and written means, in addition to personal experience. The use of references in academic books and journals is an institutionalized acknowledgment of this cumulative process.

Before writing was developed in the third millennium B.C., as well as in more contemporary cultures that lacked writing, knowledge was passed down to younger generations verbally. Certainly, practical skills were passed along by both demonstration and verbal instruction. Throughout history, telling stories has been a way to preserve and perpetuate the morals and values of a culture, as well as to relay accounts of past events. The accuracy of the story was dependent on the memory of the individual telling the story, and certain individuals became storytellers. Presumably, these people had not only a talent for dramatizing stories to make them entertaining but they also had very good memories. When storytellers died, their memories, and their versions of stories, died with them. With the development of writing in the third millennium B.C., a second means (in addition to drawing) became available to represent information externally. The use of text as an external knowledge representation meant that the storage of the information contained in language became independent of any individual. The fundamental idea behind libraries is to collect existing written knowledge, including that from preceding generations, and to make it more available as external, collective memory. All one needs is to know how to read the text in order to retrieve the information. Thus, as with maps, language available in written form serves as a source for both current and previous knowledge. The evolution of knowledge can be tracked retrospectively by studying the textual archives, thereby providing additional insight.

It has been demonstrated that language alone not only conveys geographic information but it also allows people to construct integrated mental representations of places and environments that they have never visited, without the aid of graphics or any direct experience of a given place. This is why Taylor and Tversky have claimed that "language is a surrogate for experience" (1992a, p. 495). In one of two studies, subjects were asked to study maps and subsequently convey that information back in the form of either a drawing or a textual description (Taylor and Tversky 1992a). In the other, subjects read descriptions of environments and then drew maps of them (Taylor and Tversky 1992b). The authors found that regardless of the form of the external representation presented in the experiment, the information subjects reported back was quite similar. Experimental subjects have also demonstrated the ability to infer object relationships not explicitly described within textual descriptions (Bryant, Tversky, and Franklin 1992; Ehrlich and Johnson-Laird 1982; Franklin 1992). The structure of how the information is represented in the text (e.g., ordering and relative emphasis) also affects the construction of mental representations (Denis and Cocude 1992; Denis and Denhiere 1990; Ferguson and Hegarty 1994; Franklin and Tversky 1992; Oakhill and Johnson-Laird 1985), as do graphical images.

All three types of representation—graphic, linguistic, and cognitive—are therefore interlinked. Because of this, the use of these forms of external knowledge representation for experimental research on understanding cognitive knowledge representation presents some particular difficulties. Such experiments must be carefully designed, so that the results can be correctly interpreted as to whether the observed effects result from the structure of the graphics or language used in the experiments, or are truly representative of cognitive structure. The literature that reports on experiments using only language to understand cognitive structure is large, although much of it fails to maintain an adequate distinction between linguistic and cognitive structure. Certainly, some researchers, as I describe in Chapter 8, espouse the theory that people store cognitive knowledge in linguistic and/or imagistic form. But this presumption in experiments using either language or images alone to provide insight on cognitive structure is circular reasoning. Because of this, I deliberately avoid much of this literature in the current discussion. I maintain a metalevel and cross-linguistic approach in looking at language as a form of knowledge representation, and subsequently, at the implications that can be drawn about the form of cognitive knowledge representation from the commonalities derived from the study of various languages.

LANGUAGE AS A SYMBOLIC SYSTEM

Natural language—the kind of language people use every day—consists of sequential arrangements of symbols. These symbols always occur in spoken form, and most also have a written form. The individual symbols and their sequences have meaning, as well as a complex, hierarchical structure. A collection of elemental sounds (phonemes) characterize each individual symbol that makes up words. There are various types of words with different symbolic functions: Nouns symbolize things; adjectives symbolize attributes of things; verbs symbolize actions or events; adverbs signify attributes or modify verbs. Words are grouped into sentences; sentences are grouped into paragraphs. Noting this structure, Saussure (1986), one of the founders of what has since become the field of semiology, viewed language as the model for all other sign systems.[1]

There is a lot of variability in how the meaning of words is interpreted within any given language. Perhaps one of the better studied examples is the use of spatial prepositions. The following are all legitimate uses of the English word "on," but each is used in a different way:

The cup on the tray
The picture on the wall
The town on the border
The house on the lake
The road on the left

English speakers, if asked without first being given such a list of examples, would tend to say that the meaning of the word "on" has something to do with support, as in the first two examples. A further analysis shows that this is usually true, but that the most essential element central to the meaning of "on" is contiguity (see Herskovits 1986 for a more extensive discussion). The variability of symbols denoting different shades of meaning for the same object, event, action, or attribute, as well as the variability within the rules of proper ordering (i.e., syntax), provides a significant amount of flexibility in natural language for expressing and communicating a wide variety of ideas.

The rich system available in language for combining words in a variety of ways also provides the capability for shades of meaning beyond what the individual words allow (Landau and Jackendoff 1993). As is the case with graphic images, the power of natural language is also greatly extended through the use of imagery and metaphor. But as previously stated, imagery and metaphor are not simply linguistic (or graphical) devices. They are fundamental mechanisms of thought. From this perspective, it can be said that the use of imagery and metaphor in natural language is unavoidable. These mechanisms extend the literal meaning of language by invoking the imagination, thereby extending the use of cognitive knowledge by associating differing knowledge domains and conjuring up mental scenes in the "mind's eye."

From the objectivist view, linguistic models are considered vague, inexact, and not internally consistent. This is the purpose of mathematical/algebraic models: a type of language specifically designed to be literal and not open to variations of meaning. In this sense, creative thought is avoided. Relationships among mathematical model components are explicitly set according to mathematical rules (Willmott and Gaile 1992). As a result, they more easily convey information unambiguously and permit verification. As such, mathematical/algebraic models have become the generally favored representation within science. This has been particularly true within the physical sciences, since the work of Newton. Accepted conventions (i.e., rules), structures, and control mechanisms (e.g., peer review) for academic discussion and publication are also specifically intended, if not to minimize, at least to control ambiguity in the use of natural language. On a more general level, the genre of academic discourse combining text, mathematics, and graphics, has continued to evolve ever since the founding of Plato's Academy. Academic discourse, as a whole, utilizes not only the imaginative power of linguistic models but also the literal clarity of mathematical/algebraic models whenever exactness and nonambiguity are desired in the ideas being represented.

HOW LANGUAGE CONVEYS SPATIAL KNOWLEDGE

But why, then, if there is so much inexactness in natural, everyday language, can an intended message be consistently and effectively conveyed, as demonstrated in the Taylor and Tversky studies? One factor is that we all interact with the world using the same basic physical systems for both perceiving and communicating. A significant amount of evidence shows consistency at a basic level in denoting the same general idea with the same type of sound. In particular, evidence was found early on that people associate meanings to sounds, independent of language. When subjects were asked to evaluate various vowels on the basis of relative size, they suggested that *i* and *e* denote small sizes, while *o* and *u* denote large sizes (Newman 1933). It was suggested that this has a direct physical connection and relates to the relative openness of the mouth when forming these vowels—closed for *i* and *e*, and open for *o* and *u*. It has also been demonstrated that there tends to be a shift from one sound group to another in the characterization of spatial position (nearness and farness) relative to the speaker. The dull-sounding vowels (*o*, *u*) are generally used to designate that the place of the person addressed—*you, vouz, tu, du, dort,* and so on, is away from the speaker. The sharper vowels (*i, e*), in contrast, are used to denote proximity to the speaker—*I, me, je, ich, here, ici,* and so on (Ivic 1965).

Our brains are hardwired to learn language, and we start this learning process at a very early age. Infants begin by babbling—mimicking individual sounds they hear others make. As is the case for knowledge in general, learning elements of language begins by directly relating them to the self in an egocentric world view: The child learns that making sounds—beginning with crying—is an effective means of gaining the mother's attention to attend to its personal needs and desires. By age 11 to 12 months, the small child is usually speaking single words and a few short phrases. Positive reinforcement in the form of attention from smiling parents and getting sounds and words uttered back to them provides children with initial encouragement. Later, the discovery that words have meaning—that they symbolize things—provides more encouragement. The child discovers language as a much more effective means of communicating specific wants and needs than crying and screaming. Until about age 15 to 18 months, a child learns about one word every 3 days. Then, this suddenly accelerates to about 10 words a day. Gleitman and her colleagues (Gleitman and Gillette 1999; Lederer, Gleitman, and Gleitman 1995) have found that the sudden acceleration at this age is the result of gaining some understanding of the overall structure of language and being aware of some rules of grammar. The ability to interpret a new word within a structural context helps significantly in discovering its meaning. This process of deriving meaning through the pattern of linear sequencing and context actually begins much earlier (Saffran, Newport, and Aslin 1996). Gleitman claims that this wholesale learning of words continues until about the age of 30, when the average American has a vocabulary of between 80,000 and 100,000 English words.

The overall process by which children learn language is in accordance with two Piagetian principles regarding early development. First, knowledge within the first 2 years of life is centered around personal interaction with the environment. This forms the basis for understanding language, as well as the construction of higher level knowledge structures. Second, development of higher level knowledge involves the revision and the conceptual reorganization of existing knowledge. The essence of learning language is a gradual discovery of structure, and an increasing understanding of the generalized rules and patterns and ability to think in abstract terms (Clark 1973; Sinclair-deZwart 1973). For language, this means gaining an understanding of the relationships between the symbols of language, and between the symbols and the world as perceived.

Learning to attach meanings to words by the young child involves a progressive refinement of the meaning attached to specific words based upon sensory perception. The child initially seems to utilize very simple criteria for attributing names to perceived objects. For example, he or she may use the word "dog" (or "doggie") for any animal with four legs (e.g., cat, cow, horse, etc.). This reflects an *overextension* of the conceptual categories corresponding to these words (Clark 1973). Gradually, the child learns to discriminate more finely these perceived objects, utilizing additional sensory criteria (sound, color, shape, etc.), until eventually the child's categorical definition of these basic-level objects coincides with that of an adult.

The next step in the acquisition of language is the ability to put words together in phrases and, later, into whole sentences. Eventually, in developing conceptual hierarchies, the child can understand and compose original metaphors, as the ability to think in abstract terms is manifested in language. Although acquisition of language and development of conceptual structure seem to be intertwined, there can be no equivalence implied in comparing the timing of the acquisition of language compared to learning and development in general.

In the development of language from a cultural perspective, early studies of some primitive languages, as cited early on by Cassirer (1964), also show use of the human body as a metaphor for expressing spatial relationships. In Mandingan languages, "in front of" is expressed by a word literally meaning "eye," "in," by a word that translates as "belly," and so on. This is a clear reflection of the acquisition of learning through interaction with the environment. More specifically, it corresponds to Cassirer's *symbolic* stage of learning, when specific sense data begin to become cues for something else within an entire process of transferring and integrating experience into symbolic forms within a process of cultural evolution (Cassirer 1973). It also corresponds to an earlier stage of how learning progresses, espoused by Cassirer, Piaget, and others, in which notions of space–time relationships begin as the child directly relates them to the self in an egocentric world view. Thus, there is evidence that early development of language, like learning, is not only based upon common, biological attributes, but it also develops within a shared world view.

Language encompasses an agreed-upon system of symbols that is also set within a specific cultural context. Whorf (1956), one of the earlier researchers on the relationship between language and culture, stated this relationship succinctly:

> We cut nature up, organize it into concepts, and ascribe significances as we do, largely because we are parties to an agreement to organize it in this way—an agreement that holds throughout our speech community and is codified in the patterns of our language. (1956, p. 97)

Thus, language within different cultural contexts partitions the many possible, observable, space–time relationships in the environment in various ways. Expressing geographic direction in certain island cultures as "mountainward" and "seaward," instead of the Western tradition of east, west, north, and south, is one example. Some elements of a relationship are also variably included or excluded within specific languages, with some elements seemingly ignored altogether. Bowerman (1989; Choi and Bowerman 1991) contrasts English with German and Korean in terms of how some specific spatial relationships are expressed. Thus, in the English expressions "The book is *on* the table," "The picture is *on* the wall," and "The shoes are *on* his feet," in which the relation is horizontal, vertical, and surrounding, respectively, there is a separate word in German. Choi and Bowerman (1991) cite another example in which there is one word in Korean for placing one object in another when there is a loose fit ("The apples were put into the bowl") and a different word for when there is a tight fit ("The pencil was placed into the pencil sharpener").

Cultural and environmental contexts, and experiences within those contexts, influence how we group all concepts, not just spatial relationships. Levy (1973) reported that Tahitians have no word for sadness. They also seem to have no concept of it, and no culturally dictated ways for dealing with various feelings of sadness, such as bereavement. Lakoff (1987) provides an extensive discussion of cultural variations in categorical designations and groupings. For a more extensive discussion of the geometry implied in specific locative terms, see Herskovits (1985) and Talmy (1983, 1987).

Regardless of cultural variation, it has been shown that there is still a structural commonality in spatial expressions among languages. Among more recent studies, Munnich, Landau, and Dosher (2001) compare naming of spatial relationships in English, Japanese, and Korean. They found similarities in expressions relating to axial structure (e.g., "above," "below," "right," "left") as well as differences in expressions relating to contact/support, as did Bowerman (1989), and Choi and Bowerman (1991). There is evidence that the lack of perfect correspondence is due to differences in the way the two systems develop and are used (see Crawford, Regier, and Huttenlocher 2000). The commonalities, however, imply that perception and linguistic representations must both draw upon a common cognitive structure of spatial knowledge at some deep and fundamental level.

Talmy has perhaps done the most work on comparing spatial expressions in languages in differing cultural contexts. He has pointed out a number of universal structural mechanisms within language for representing, and thereby conveying, spatial and temporal information. These mechanisms are both similar and dissimilar to how images convey information, which I discuss later in this chapter. Besides the organizational structure of language that is readily apparent on a surficial level (hierarchical arrangement of words, sentences, paragraphs, etc.), two fundamental subsystems present in all languages are used to convey meaning: the grammatical and the lexical. These two subsystems have complementary functions in any given sentence, with the grammatical elements providing the conceptual structuring system or framework, and the lexical elements, the majority of its content (Talmy 1987).

The lexical subsystem, for the most part, provides the vocabulary of a language and includes nouns, verbs, and adjectives—what in linguistics are known as *open-class* elements. A word-class in a language is called open-class if it can be easily added to, and because of this, such words within any given language tend to be numerous. Open-class elements, particularly nouns, are easily extended within any language. New nouns are generated continually, as needed, to provide terms for new things and contexts as they are encountered. Certainly, the proliferation of new words in English, such as *Internet*, *petabyte*, and *telecommuting*, with the advent of the information revolution over the past 15 to 20 years, is a very noticeable example of this.

In a general sense, the grammatical subsystem determines the conceptual structure that, in a spatial context, provides the main setting for the scene, situation, event, or idea being communicated. The grammatical elements of a given language include all the other word classes, including prepositions, conjunctions, articles, and demonstratives (e.g., "this" and "that"), as well as prefixes, suffixes, and intonation. These elements, taken as a whole, also specify a central set of concepts contained within a given language that are hard to add to and limited in number. As such, grammatical elements of a language are called *closed-class*.

As Jackendoff and Landau have pointed out, there are few words to express spatial relationships relative to the number of words available for things. For example, an average adult English speaker knows on the order of 15,000 names for things—count nouns—that label different kinds of objects, but there are only 80–100 prepositions in the English language for expressing all possible spatial relationships (Jackendoff 1992; Jackendoff and Landau 1991). There are English words to name certain relationships, such as "above" and "centered"; there is no word "cenabove" to express the combination (for their complete description of this example, see Jackendoff and Landau 1991). As discussed in Chapters 4 and 5, familiarity and depth of knowledge also bring with them expanded vocabulary for expressing finer levels of distinction or precision.

Through examination of various languages across different cultures and areas of the world, Talmy and others have shown that the concepts that are part of the grammar can vary in importance and prominence from language to language (Talmy 1987). Some core concepts seem always to occur within the grammar of language, and others possibly never do. Any concept that is excluded as part of the grammatical structure, however, can be expressed within the lexicon. As Talmy (1987) has described, notions of spatial and temporal relationships seem always to be included within the grammatical structure of a given language, but these tend to be relativistic, not absolute, in nature. Thus, the verb suffix -ed in English indicates past tense, but no notion of exactly how far in the past is included within the meaning. English has spatial prepositions, such as "above" and "below," "near" and "far," that are topological, but metric relationships (e.g., 50 meters below the bridge) must be expressed lexically. Besides distance, magnitude in general—including size—seems limited to the topological within the grammatical subsystem.

Other spatial relationships that seem limited to relative, topological notions are direction and shape. The same is true for temporal relationships. In English, we can speak of events occurring *before, after* or *during* another event or state of affairs, but exact temporal distance or duration can only be expressed by phrases such as "He resigned from the military 4 years after enlistment," or "The storm lasted for 3 days." Thus, lexical elements are used to provide specificity as well as nuance of meaning. As can be seen from these examples, despite the number of words included within the closed-class elements for expressing such relationships, it is the much more numerous and varied words included as lexical elements that provide the ability for a very rich range of variation in expressing ideas.

The dual grammar/lexicon systems of language are also subsumed within a group of much less surficial mechanisms for conceptually structuring space and time, including: schematization, frame of reference, and focus of attention. Schematization is the mechanism used for defining the basic spatial and temporal relationships grammatically within any given language, thereby also judging their appropriateness within specific situations. Frame of reference defines the spatiotemporal referencing system within which the various elements of a scene are set. Focus of attention involves directing attention to specific elements within a scene and shifting attention from one element, or group of elements, to another. While the fundamental meaning of relationships is defined through schematization, frame of reference and focus of attention provide the means to adjust or modify a representation to suit a particular situation.

Schematization involves the systematic abstraction, selection, and idealization of spatial objects in the use of spatial prepositions (Herskovits 1998). Abstraction is important in all linguistic meaning in spatial and temporal expressions, ignoring irrelevant attributes. For example, in "The bicycle is next to the church," color is completely ignored. Selection goes beyond abstraction in using only one aspect of an object to represent the whole. Thus, in the preceding expression, only one side of the church is used to represent the entire church. Idealization is the imposition of a standardized geometric structure on the particular objects and/or scene that may or may not reflect the geometric structure of the actual objects. In general terms, then, schematization defines the patterns through which interrelationships are expressed.

One element of schematization at work in expressing all spatial relationships involves a distinction between a *figure* object and a *reference* object (Talmy 1978). In everyday speech, expressions of spatial location for the object of interest (the figure object) generally are relative to some other object (the reference object). Thus, the bicycle is *beside* the house, *in* the garage, or *under* the tree. All spatial prepositions seem to rely on this figure–reference structure.

In his analysis of the structure of spatial schematization in language, Talmy posited that the figure object can generally be considered a dimensionless point, and that size and shape attributes are only considered for the reference object. Herskovits has subsequently claimed that this is not necessarily the case, that there are instances in which actual, but still generalized, shape of both objects and other details about relative placement comes into play in determining the applicability of a preposition (Herskovits 1998).

Nevertheless, there is still an asymmetry involved in most linguistic descriptions of space. In the expression "The bicycle is beside the house," the constraint in the relationship geometry is that the figure object must be located along a dimension of a reference object that has some length. The reference object is also usually at least as large as the figure object, and relatively immovable, or at least cognitively preexisting. Thus, reversing the expression to say "The house is beside the bicycle," although the relationship itself is still true, is not the way the relationship is expressed. This also holds for one's self and other people as the figure or reference object. Thus, George is standing *beside* the car, I am standing *in front of* the church. Reversing the latter phrase to say the church is *in front of* me is syntactically correct but has a different implication of motion along a path. In this case, the ground in the relational expression is the moving object, serving as the reference for things encountered by that object along the path in space–time (Talmy 1983).

While most spatial prepositions express a one-to-one relationship (figure vs. a single ground element), such as "on," "in," or "next to," a few also express a one-to-two (e.g., "between") or a one-to-many (e.g., "among") relationship. All of these are topological relationships expressing a relative position between objects in space. There are also a few prepositions that express a space–time relationship as motion. Examples in English include "across," "through," and "into." These also imply a more complex geometry for the reference object. "Into," for example, utilizes a volume of space as the schematization for the ground object. For expressing a strictly temporal relationship, spatial prepositions are often employed with the temporal context revealed by the nouns used, such as *in* the month of January, or *before* evening. More complex relationships can be expressed by combining expressions. Some languages also denote additional geometric characteristics beyond those implied in propositions. But regardless of the order of the relation, one-to-one or one-to-many, or the dimensionality, no spatiotemporal preposition is metric or based upon an absolute coordinate system that uses space, rather than objects, as the frame of reference.

Furthermore, no exact geometric limitations qualify the relative positions of objects as within a particular relationship or not. For example, "The bicycle is *near* the garage" implies proximity, but there is no exact distance boundary. If the garage were a block away from the bicycle, this relationship would no longer make sense. But it does make sense to say that "Delft is *near* Amsterdam," if I am in New York, even though Delft is many kilometers from Amsterdam. If I am in Delft, however, the hour-long train ride necessary to get there may make Amsterdam seem not very near. So the scale of reference (among other factors) greatly impacts the relation.

Frame of reference provides the coordinate system and overall referencing structure as the context for linguistic description of spatial scenes, and judgment of relationships within those scenes. Conceptually, frame of reference defines the specific, abstract spatial schema to be used. Thus, the linguistic frame of reference employed in any statement drives the mapping onto conceptual space (Fauconnier 1994). The various frames of reference used in language have been defined with varying levels of distinction (Jackendoff, 1996, has asserted that there are at least eight). On a general level, three fundamentally distinct frames of reference are used in language and described by a number of researchers: relative, intrinsic, and absolute. These have also been called ego-centered, object-centered, and environment-centered, respectively (Landau and Munnich 1998; Levinson 1996), and are shown schematically in Figure 7.7.

With the relative frame of reference, the coordinate referencing system is centered on the observer of the scene, as shown in Figure 7.7a, and defined on the basis of half-planes through the body of the observer (up–down, left–right, front–back), forming a polar coordinate system. Employing a relative frame of reference in the example expression "The bike is in front of the house" means that the bike is between the viewer and the house. Thus, relationships such as "left of" can change with a change in the position of the observer. Directional relationships also change, depending on whether the viewer is facing the objects referred to in the relation. Thus, a guide on a tour

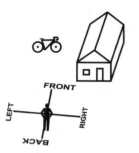

"The bicycle is to the left of the house"

a.

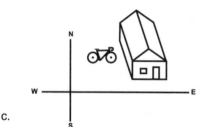

"The bicycle is on the right side of the house"

b.

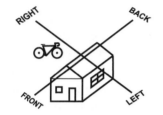

"The bicycle is west of the house"

c.

FIGURE 7.7. Frames of reference in language. Adapted from Levinson (1996). Copyright 1996 by MIT Press. Adapted by permission.

bus would say in English, "Look at the building on my right—your left." The relative coordinate systems used by the guide and the tourists vary by a 180-degree rotation, because they themselves are facing in opposite directions. In some languages, the coordinate system may be translated rather than rotated to express this situation (see Hill 1982).

The intrinsic frame of reference utilizes a coordinate system anchored on the ground object in the scene. The relationships are defined via its various "facets." In English, we would talk of the house in Figure 7.7b having front, back, left, and right sides (Levinson 1996). Thus, to say "The bike is on the right side of the house" using this reference frame means that the bike is on that side because of what we define as the front (usually the side with the main entrance in the case of a building). Coordinates within this frame of reference tend to be polar—radiating from the reference object at the center.

In the absolute frame of reference (cf. Figure 7.7c), the coordinate system derives from the environment in which the figure and ground are situated. These are fixed bearings usually based upon planetary orientation in defining east, west, north, and south in the case of English, for example. However, the coordinate system is not always Cartesian. The native Polynesian use of what translates to mountainward and seaward is one example of this. The absolute frame of reference seems, in a general sense, to rely on some environmental gradient (planetary orientation, elevation, river drainage, etc.) (Levinson 1996). The use of an absolute frame of reference does not vary, regardless of the position of the point of view or relative location of objects. This is the coordinate framework used for "dead reckoning" in moving through the environment and locating one's self (and other objects) within the entire space in terms of direction, distance, and orientation.

While these seem to be the only three distinctive types of reference frames used in language, not all languages use all of them. Some languages may predominantly use either absolute or intrinsic frames of reference, and some may extensively use two of them, which can be intrinsic and relative, or intrinsic and absolute (relative seems to require intrinsic). Levinson (1996) also showed that the predominance in the use of a particular frame of reference in a particular language co-occurs with the preference for these same frames of reference in nonverbal tasks.

Speakers frequently switch from one framework to another, and listeners are able to follow (Schober 1998). Guidance is often provided, but not always by linguistic cues. The degree to which such cues are employed also varies from language to language (Herskovits 1998). Experimental evidence indicates that frequent switching of frame of reference is a basic characteristic of the human attention system (Logan 1995).

As can be seen from the previous discussion, there seems to be a number of universals in language that, as such, provide insight into how we cognitively organize spatial knowledge, summarized as follows:

1. Through schematization, spatial language disregards many geometric details. This parallels cognition in generalizing knowledge, disregarding unneeded information. It also appears that the classificatory system in language parallels that used cognitively as a means of both organizing and minimizing information. Talmy (1983) suggests that such schematization in natural lan-

guage is evidence that linguistic relations are *representative* of cognitive relations between objects.

2. The separation of relative and absolute–metric within the distinction of grammatical (closed-class) and lexical elements of language, respectively, parallels the relative and absolute views of space and time seen in cognition and, historically, in various cultures. These are also reflected in the frames of reference used. Spatial relationships expressed within the most basic elements in language, the closed-class elements, are the relative relationships associated with early learning and direct interaction with the world. The use of a figure–ground framework in linguistic expressions of relative spatial relationships also parallels human conceptualization as relationships between objects; relative relationships are viewed as a property of objects that relates them to other objects.

3. The mechanisms of focus of attention and shifting of attention in language parallels the cognitive process of an observer experiencing a visual scene. This cognitive process involves both a sequential process of examination/exploration of elements in the scene, as described by Neisser (1967, 1976), and a hierarchical strategy, as described by Marr (1982).

A number of researchers have argued that the basis for these language universals is the commonality of our basic physical system and of seeing, touching, and generally interacting with the physical world, which is also the foundation for all learning and understanding.

THE RELATIONSHIP BETWEEN LANGUAGE AND COGNITIVE STRUCTURE

Because natural language and cognitive structure are interconnected, the study of language has become a means to understand how we store and use knowledge. The study of the structure of linguistic categories, in particular, has shed light on cognitive structure. The study of how categories are applied in attributing meaning to words provides insight into certain aspects of how we think.

As Hayward and Tarr (1995) point out, this is similar to the prototype theory of categorization (Rosch 1978), or to what Lakoff (1987) refers to as radial categories, in which the new sensory stimulus is compared to a prototypical example. With this view, there are no clear boundaries between what is included within a particular category and what is not. They are also not mutually exclusive. Rather, there is a set of "goodness-of-fit" criteria for each classification. The stimulus is compared with the prototype according to those criteria along a graded continuum. Thus, the new stimulus can be judged as anything from a good example, closely resembling the prototype, to marginally acceptable but still within the relation, to marginally outside of the relation, to clearly not within the given relation. There is experimental evidence that the prototypical relation is something akin to a template (Logan and Sadler 1996), which in cognitive terms would correspond to an idealized pattern or schema. Such idealized patterns or schemata would also not be tied to any particular scale.

Words for entities, properties, and relations are symbolic tokens for the things themselves; therefore, the process of attaching meanings to words is essentially the same as attaching meaning to objects. According to Jackendoff (1983), at least three sorts of conditions are needed to describe word meanings adequately:

1. *Necessary conditions*—For example, "lake" must contain the necessary condition of also being a hydrological feature.
2. *Centrality conditions*—These graded conditions specify a central or threshold value for a continuously variable property, such as elevation for "mountain."
3. *Typicality conditions*—These are conditions we would normally associate, such as fresh water with "lake" (although some contain saltwater).

The following *functional* characteristics must apply to how the representation behaves when known, necessary conditions alone are often not sufficient to classify objects and spatial relations between them (McClelland et al. 1986):

1. *Graceful degradation*—This is, effectively, "benefit of the doubt." Given that there is often no entity that satisfies the relational qualifier in the strictest sense, all entities that marginally qualify are accepted. Thus, in the question regarding a spatial relationship "Is there a city in central Pennsylvania?" we may answer in the affirmative, saying that State College (definitely in the center of the state) qualifies. Someone living in Philadelphia or Chicago, particularly outside the context of this specific question, may not agree that State College is a city. Similarly, we may say that the presence of Altoona, more or less in the center of the state, also would mean the question could be answered affirmatively. Graceful degradation becomes particularly important when there is a combination of several qualifiers that must be simultaneously satisfied and must rely on a combination of centrality and typicality conditions.

2. *Spontaneous generalization*—Objects or object classes tend to "inherit" properties of a higher level object in the hierarchy. For example, lawns, trees, and shrubs all tend to be green. Thus, in filling in unknown properties concerning a partially known object or object class, certain properties and property values pertaining to its parent class, or other class at a higher level to it, are used.

3. *Default assignment*—This is the assertion that we can also assume some properties and values of properties for partially known objects on the basis of similar objects, or work on the basis of typicality conditions. In this context, properties of an unknown location or group of contiguous locations (i.e., an area) can be inferred or interpolated from the value of that property at adjacent locations.

In this general mechanism, we see that the necessary, centrality, and typicality conditions function throughout as how judgments can be made for the definition and classification of objects and locations, and for the relations between objects and between locations (left of, near, etc.). These follow Lakoff's (1997) notion of radial category structure and rely on mechanisms for cognitively navigating category structures. Most of these conditions and characteristics cannot be applied if the category structure within a given domain is not sufficiently developed. And most also are often not used when metaphor and other tropes come into play. The mechanism involved in tropes is primarily the extension of *spontaneous generalization* into category structures of very different knowledge domains, as well as the use of imagery schemata. The mechanisms of thought involved in attributing meaning to metaphor were discussed in the previous chapter.

Given work in cognitive linguistics by Lakoff, Talmy, and others (Hayward and Tarr 1995; Herskovits 1986; Lakoff 1987; Logan and Sadler 1996; Talmy 1983), it does seem to be the case that spatial language encodes the world utilizing a prototype categorization scheme based on spatial schemata that reflect cognition. There is also evidence of a correspondence in the cognitive separation between relative and metric, or absolute, as well as "what" versus "where" information (Kosslyn et al. 1989; Landau and Jackendoff 1993). The conclusion to be drawn from this is that there is a connection between how language in general encodes and structures spatial knowledge, and how that knowledge is cognitively encoded and structured. Because language—as do maps and images—provides a means of indirect experience, it can be viewed as circular reasoning to attempt to claim a causal relationship and say that linguistic structure is the result of cognitive structure or vice versa. Still, assuming that there is a correspondence, much insight can be gained into the cognitive structure of spatial knowledge by studying linguistic structure. Such insights are discussed in more detail in Chapter 8.

There is now general agreement among philosophers, cognitive scientists, and linguists that, at a fundamental level, the main reason language is an effective mechanism for representing and communicating knowledge is that the users of any particular language also share a common world view (Glenberg and McDaniel 1992; Jackendoff 1983; Lakoff 1987). In other words, language is *dependent* on reference to a world view. The basic structure of language is a reflection of cognitive knowledge structure. Moreover, shades of meaning are determined through reference to this structure. Language does not stand alone in conveying geographic information. Rather, like all other forms of experience, both direct and indirect, it relies on cultural context and the sum of previous experience and knowledge base for interpretation.

This, however, is a relatively recent perspective, beginning particularly with Wittgenstein's work on the correspondence between language and reality (1953). It was also the starting assumption in Saussure's earlier work in the development of what has become known as semiology. In Chomsky's work, starting in the 1950s, it led the development of what is known as generative linguistics. Generative linguistics is based on the notion that language is a separate cognitive system, independent of other knowledge. Chomsky's theory of language was influenced by the philosopher Gottlob Frege, who, in effect, developed the first predicate calculus. In the pursuit of his long-term (but unsuccessful) project to demonstrate that all mathematics is reducible to logic, Frege did not limit his examples to mathematical statements but also included linguistic inference. He based his theory regarding language on a strict distinction between a symbol, the meaning it conveys, and the idea that it represents for the speaker/writer. The exact idea represented can only be known by the speaker or writer, because all ideas are based on a unique set of experiences. However, a word, as a symbol, refers to something in the real world and, as such, it has an intrinsic meaning. Thus, he reasoned that correctly interpreting the meaning of symbols in a given language only requires being familiar with that language (Frege 1980).

This was in line with classical category theory, namely, that categories share common properties that are inherent to all objects or things within the same category. Thus, according to classical category theory, how things are grouped into categories—and thereby identified by name within a language— is determined by those inherent properties. Through this line of reasoning, categories are equivalent to sets, and sets of sets. This is extended further in generative linguistics. If words have meaning that is inherent in the things in the world to which they refer, then the vocabulary of any language is objectively and externally defined. As such, a grammar can be defined as a set of combinatorial rules that manipulate words as symbols, without regard to their meaning. Without a cognitive context, and independent of semantics, a system of grammar for any language is also internally consistent and complete (Chomsky 1957, 1965).

Generative linguistics, developed before Rosch's work on the theory of prototypes and basic-level categories, was certainly appealing as an objective, formal, closed system. Chomsky's work was widely used in the earlier days of artificial intelligence, with the goal of developing algorithmic definitions of language for use on computers. Generating and understanding natural language on computers, however, remains one of the "grand challenges" of computer science.

We now know that any natural language is structured in terms of a complex system of categories of many kinds, including situational and metaphorical, as well as morphological (Lakoff 1987). Individual words for physical objects may be fairly well, although not perfectly, defined in objective terms. Even within the same culture, what is a "lake" to one person may be a "pond" and not a "lake" to another. The tomato is a vegetable in everyday terms, but biologically, the tomato is categorized as a fruit. Entities in nonphysical domains and higher level abstractions have even more contextual interpretations. Interpretations also often depend on whole phrases, as well as general context. Perhaps even more problematic with respect to this approach is that language often conveys far more meaning than the mere sum of the words—in evoking mood, emotion, and imagery, as well as other meaning that may not be represented in any literal sense: thus, the expression "Reading between the lines."

It is these gestalt properties of natural language that make the treatment of natural language as essentially a form of mathematical/algebraic model too limiting, which in turn has focused interest in what has become known as cognitive linguistics. But perhaps from the perspective of understanding natural language on computers as a learning process, taking the absolute (i.e., literal) view made sense in retrospect as a place to start.

What the preceding discussion indicates is that language does not directly symbolize the external world. Rather, language is a *representation* of our shared mental model of the world, as Talmy (1987) asserted. Perhaps the lateness of this realization relative to the same, long-held realization with regard to maps is that gaps in geographic knowledge, and the changing form of that knowledge (in a spatially literal sense), was much more obvious in graphical form. The correspondence between the changing form of geographic knowledge and maps was, for example, undeniable during the Renaissance and the Age of Exploration.

If any specific language reflects a specific world view, the question arises as to how it is then possible for anyone to learn a different language, particularly one from a very different culture than their own. Lakoff (1987) has explained that part of learning a foreign language includes learning the cultural context (i.e., the world view) of that language. Indeed, it is part of any foreign language course to learn about the customs, history, and literature of that people.

As with the perceptual cycle portrayed in Figure 4.3, our schemata, cognitive category structures, and the contents of those structures (derived from the sum of our experiences within a cultural context), are used as a reference to interpret language. Reading text and hearing speech in turn modify our schemata and category structures as indirect experience. We also use our knowledge to formulate language as a means of expression and communication. This means that language (and graphical images) as a communication device also can mediate between differing individual world views. The basis for this lies in the commonality of our physical perceptual system and the learning process. This in turn has resulted in a number of universals in the structure and function of language that transcend cultural context.

GRAPHICS AND LANGUAGE INTERTWINED

Although graphics and language can each alone convey information, these forms of external representation are often combined. Atlases contain descriptions that are intended to help interpret and elaborate on the maps they contain. Geographic texts and tourist guides usually contain maps to supplement the text. Video documentaries combine pictures and narration. As mentioned in Chapter 6, a significant body of evidence indicates that constructing a mental model is facilitated when language and graphics are combined (Glenberg and McDaniel 1992; Glenberg and Grimes 1995; Hegarty, Carpenter, and Just 1990; Mandl and Levin 1989; Mayer and Gallini 1991), and specifically when text is combined with maps (Ferguson and Hegarty 1994).

According to Glenberg and McDaniel (1992), the use of graphics with text helps to overcome two inherent shortcomings of language when conveying spatial information. First, the linear (i.e., one-dimensional) nature of language is ill-suited to represent the higher dimensionality of spatial information. As mentioned in the earlier discussion on graphics and imagery, there is a direct mapping of elements of the perceived world in the graphic representation. The human visual system is also particularly well suited for detecting form and pattern when displayed spatially. Second, the number of terms in natural language for expressing topological spatial and temporal relationships is hard to add to and very limited. Nevertheless, English-speaking adults are certainly aware of a finer level of distinction in spatial relationships than that represented in English grammar. Talmy (1983) and Glenberg and McDaniel (1992) have pointed out a pragmatic reason for this: It would simply require far more terms than humans are capable of remembering to give distinct and discrete names to all possible topological spatial relationships. For instances where precision is required in expressing a particular distance or directional relationship, we rely on some culturally accepted metric (e.g., the train station is *900 meters* from his house). Nevertheless, many spatial descriptions cannot be adequately expressed. For example, try verbally describing the shape of Canada or the United States.

Given the similarities of language and graphics, it is indeed the case that one can substitute for the other (although there are trade-offs, as I will discuss in more detail in Chapter 8). As discussed in the section, "The Legacy of Maps and Mapping," the history of maps and spatial information represented in the form of text is intertwined; many geographic texts through history have included maps, and maps have been accompanied by textual descriptions. Maps have also in large part been compiled from texts (in both linguistic and algebraic form). The *Description of the Earth* by Dionysius in the second century A.D. was intended as a guide to the geography incorporated within the stories of the Greek poets, such as Homer. The Latin translation of his geographic discourse became popular during the Renaissance and into the 18th century as part of the general interest in rediscovered ancient texts and the discovery of new lands (Jacob 1999). Through the use of metaphors and imagery, Dionysius quite successfully conveyed a view of the entire known world. The text was divided into regions, and the description for each region—as the author informed the reader—was described as a verbal map in a systematic and organized fashion, building up successive types of geographic features. This allowed the reader's mental representation of the geographic space portrayed also to be built in a systematic fashion, including a wealth of detail about geographic features, and peoples and their interrelationships, but without the implied exactness of lines and points on a map that may not always have been appropriate.

Although graphics and language seem to be very different forms of external knowledge representation, as the discussion in this chapter has shown, they also have a number of fundamental similarities (Andrews 1990; Robinson and Petchenik 1976):

1. Both are symbolic systems with a vocabulary and a grammar.[2] Nevertheless, there is no necessary resemblance between the symbol and its referent, nor is there a necessary correspondence between the distinguishable parts of the symbol system and the separate parts of the elements in the real world.
2. Both have a hierarchical structure, with groupings and linkages among elements.
3. Both convey extrinsic as well as intrinsic meaning. Meaning is conveyed intrinsically on the basis of the combined definitions of the lexical and grammatical elements employed. Extrinsic meaning goes beyond the individual elements portrayed and must be interpreted as a whole. For both graphics and language, extrinsic meaning relies to a very great degree on imagery, metaphor, and cultural context.
4. They do not represent reality directly, but rather are representations of our cognitive knowledge of the world.
5. Both are selective abstractions of cognitive knowledge.
6. Both are *reflections* of cognitive structure.

Taylor and Tversky (1992a) assert that the commonality of mechanisms involved and the ability to communicate an intended message are due not to one but two factors: commonality in how our world knowledge is organized, and a common underlying goal of communication. I have already discussed evidence of how graphics and language each succeed as means of communication and external representation of knowledge through reference to a shared cognitive representation. As to the latter, Taylor and Tversky argue that some common properties inherent in graphics and language, specifically, hierarchical organization, grouping, and linkages among elements, contribute to ease of comprehension and memorability. It is only logical that a structural correspondence between external and cognitive knowledge representations would facilitate these functions.

THE ROLE OF INDIRECT EXPERIENCE

In this chapter, I have examined how we acquire knowledge through graphics and language, the two primary means of indirectly experiencing the world around us. Although the knowledge thus derived has been called secondary knowledge, as opposed to primary knowledge, which is gathered through direct sensory experience, both are essential to learning. Knowledge, regardless of how it is gained, is stored cognitively as an integrated whole.

This integrated whole includes both internal knowledge stored within the mind and external knowledge. In previous chapters, I have discussed how the environment itself provides cues, triggering associations and memories in a process of cooperative cognition. This also holds true for indirect sources. Besides helping individuals to learn new information, maps and written language serve as memory aids. Most of us write down shopping lists, lists of things to do, and appointments as reminders. Images and language are also important tools for thinking. We may often take along the sketch map to a friend's party, not because we are unfamiliar with the area, but because the path may be complex.

Maps and written language are important tools for thinking as well as remembering. In multiplying two multiple-digit numbers, for example, it is much easier to perform the calculation with the aid of pencil and paper. In a simple experiment, the aid of paper and pencil was found to reduce the time it takes to perform this task by a factor of five (Card, Mackinlay, and Shneiderman 1999). Writing the numbers down as the calculation progresses serves to hold the intermediate results externally, until they are needed at a subsequent stage.

Both maps and written language provide an invaluable record of man's evolving and varying world views through the centuries and across different cultures. Maps, in particular, provide a very traceable evolution; mapmakers themselves rely on previous knowledge of others in what is often a long chain of preceding maps and textual compilations that are literally visible in maps. Writings on how to make maps that have survived from successive time periods provide added insight.

Not everyone within a culture has the same world view, and views change within the context of the task at hand. In addition, because both maps and language are fundamental tools for conveying messages (as well as information), the view of reality portrayed may be deliberately modified to fit the message. A number of authors have recently provided extensive discussion on "how to lie with maps" (Monmonier and de Blij 1991; Wood 1992).

The use of images, maps, and language as means of storing and communicating information is essential because of the scale and complexity of the real world. Internal and external representations are inextricably interwoven in thought and in the general way we store and use environmental knowledge. Because of this, both maps and language provide significant insights into cognitive representations of our world.

In the preceding chapters, I have examined the various means and the overall process by which we acquire knowledge about the world and build an integrated world view. In Chapter 8, I examine how spatial knowledge is cognitively encoded.

NOTES

1. As previously noted in Chapter 2, semiotics was developed independently by C. S. Peirce, from the perspective of logic, and by Saussure, from a linguistic perspective.

2. Robinson and Petchenick (1976) asserted that maps have no syntax. They based their view on the observation that, unlike text, maps need not be read in any particular order. As both Andrews (1990) and MacEachren (1995) have subsequently pointed out, however, the well-designed map can lead the map reader through an intended sequence, from overall features into specific details. So although maps are not read in a linear fashion, there is a distinct ordering, or syntax.

CHAPTER 8

HOW SPATIAL KNOWLEDGE
IS ENCODED

MEANING AND UNDERSTANDING

As with any aspect of knowledge, meaning is totally individualistic and relative. It involves what is meaningful to us. Nothing is meaningful in itself. Meaningfulness is relative to our own needs and experiences within an environmental and cultural context. Basic-level concepts are meaningful to us because they derive directly from the physical characteristics of the organism and the way it interacts with its environment. Recalled experiences and known procedures can thus be meaningful only because they are grounded in basic-level concepts and extended on the basis of such things as biological function and purpose.

What, then, is the difference between meaning and understanding? Based on what I have discussed so far, understanding can be described as relating meaning to an integrative conceptual structure. From an experiential perspective, we may attribute meaning to a particular sight or sound, but we must have an integrative conceptual structure in order to have a broader understanding of the *situation*. For example, we may recognize a new building as a supermarket simply by its visual characteristics—its large size, food advertisements in the front windows, and shopping carts in front of the building. We are able to attribute *meaning* to the visual experience because knowing about supermarkets as a type of object in our store of declarative knowledge, we can not only identify it on the basis of visual cues, but we also associate with it a particular function that has significance in our everyday lives. We may also *understand* that this particular supermarket may cause the mom-and-pop grocery store 3 blocks away to go out of business. This is meaning within the broader, overall context of how elements in the environment are interrelated. It first requires awareness of where this particular object (the supermarket) is located in relation to other objects within the environment. Second, it requires knowledge of how supermarkets, as a type of element that can occur in the environment, interact functionally with other elements.

Similarly, if I am planning a trip to New York City, on the basis of my procedural or route knowledge relating to the area, I may know that I can drive to New York via I-80. However, I decide to take the train, instead, because snow is predicted. I make this choice on the basis of integrative (configurational) knowledge that provides a broader *understanding* of the situation: I know that snow causes roads to become slippery, which in turn would make such a drive more hazardous and probably take longer.

When reading a book on physics, the reader may know the meaning of many mathematical symbols, many individual words on the page, and, indeed, some whole sentences. Nevertheless, the reader cannot understand what he or she is reading without a sufficient background in physics (i.e., sufficient integrative knowledge relating to the domain of physics). These concepts are initially built on percepts and subsequently grouped together with a hierarchical structure, and integrated with internal links as increasing knowledge on the subject is acquired.

But how is this knowledge cognitively encoded? How are the various elements represented and related to each other? Unfortunately, we cannot directly observe how knowledge is encoded by simply probing or dissecting the brain but must rely on behavioral evidence. Based on the discussion in previous chapters, it is apparent that the structure of knowledge is multifaceted, highly interrelated, dynamic, and intertwined with an environmental context in a number of ways.

In the remainder of this chapter, I draw on the discussion so far and address how human spatial knowledge is cognitively encoded. As we have already seen, because vision and other forms of sensory perception play a critical role in interacting with and learning about our environment, I begin by discussing imagery and language as mental constructs.

THE FORM OF IMAGERY AS A MENTAL CONSTRUCT

As mentioned in Chapter 6, imagery is sometimes described as pictures "in the mind's eye." The very idea of the mind's eye gives an indication of how closely connected imagery is with visual perception. Whether imagery is actually stored in pictorial form remains a topic of debate (Thomas 1999).

Classical and medieval thinkers, including Aristotle and Aquinas, gave imagery a central position in describing the process of human thought. In the 4th century B.C., Aristotle asserted that there is no thought without an image. In the 17th century, Descartes argued that not all thought is imagistic, by differentiating between imagery and concepts. These thinkers were followed by the British empiricists in the 19th century (Berkeley, Hobbes, Hume), who explored the connection between sensory experience and the acquisition of knowledge. A consensus developed from these explorations that if knowledge is based upon experience, then how we mentally represent that knowledge must reflect the form of those original perceptions (Cornoldi and Logie 1996). The relationships of mental imagery to perception and the representation of knowledge became one of the most hotly debated areas in the field of psychology. There is now considerable evidence, both experimental and anecdotal, to support the view that much of our thinking is indeed based on the formation and manipulation of mental images (Cooper 1990; Denis 1990; Finke 1989; Finke and Shepard 1986).

Most people would not question that mental images exist or that we use the same thinking strategies on them as we do on perceived images. The major imagery debate within cognitive psychology continues to focus on the form of these mental constructs: To what degree is there an isomorphism between visual imagery and sensory perception of visual scenes? We often describe our mental images as pictures in the mind's eye. However, many have questioned whether mental images actually take the form of pictures in the brain, and claim that their form is more like verbal descriptions. A variety of empirical findings support the idea that there is indeed an isomorphism between seeing and mental imagery, at least at some level. Applying some of the same research techniques developed for sensory perception to imagery, these experiments tended to employ visual problems that depend on size or shape comparison, or some sort of mental image manipulation (e.g., rotation, folding, assembly or disassembly) for solving the problem.

The first objective, quantifiable experiment supporting the pictorial representation view involved mental rotation. Shepard and Metzler (1971) asked individuals shown line-drawn figures differing in their spatial orientation whether pairs of figures were identical. They found that reaction times had a direct, linear relationship to the amount of rotation needed to "line up" the figures. Subsequent mental rotation experiments using other stimuli, such as letters of the alphabet, have had similar results (Cooper and Shepard 1973; Jordan and Huntsman 1990; van Selst and Jolicoeur 1994). These findings suggest that people generate pictorial images of the objects, which they then mentally rotate in a fashion analogous to having the figures in their hands and manipulating them physically.

Kosslyn (1975) investigated the size of elements within images, as well as other functional characteristics of imagery. If mental images are pictorial, then people should take longer to find details on a relatively small image, as opposed to a large one, when asked questions about them. He forced the idea of relative size by dealing with imagery of two things of very different size, for example, a rabbit and an elephant, for a small rabbit image, or a rabbit and a fly, for a large rabbit image. Indeed, people's judgments regarding image details were consistently faster with large rather than small mental images.

In a similar vein, Moyer (1973) used pairs of names for animals and asked which animal was larger. As expected, it took longer for people to respond when asked to distinguish between pairs of animals close in size (e.g., an ant or a bee) than for pairs very different in size (e.g., an ant or a elk). Both of these investigations are examples of what is called the *symbolic distance effect*. Interesting results concerning auditory distance by Intons-Peterson, Russell, and Dressel (1992) showed that it took longer for participants to mentally adjust a pitch to a desired pitch when the two pitches were close together than when they were further apart.

Experiments involving judgments about shape differences have shown a relationship similar to those regarding size differences, with people taking consistently longer to determine differences when the shapes were similar than when they were very different. This held for judging very basic and controlled shapes, such as the angle formed by the hands of a clock (Paivio 1978), and for judging differences in irregular and uncontrolled shapes, such as the shapes of various states in the United States (Shepard and Chipman 1970). There is also evidence that people rotate familiar or sharply defined figures more quickly than unfamiliar or fuzzy ones (Duncan and Bourg 1983; Jolicoeur, Snow, and Murray 1987), again pointing toward a pictorial representation.

Other evidence involves mental scanning experiments. The presence of a mental image in pictorial or similar imagistic form suggests that the time it takes to scan between two objects in an image would be a function of their distance from each other. To test this hypothesis, Kosslyn, Ball, and Reiser (1978) asked students to learn the exact locations of the objects on a fictional map. The map was then removed, and the students were asked to focus on a particular named object. They were then given the name of a second object and asked to mentally scan the map (in the manner instructed), until they reached the second object. As predicted, there was a close, positive relationship between distance and the amount of time it took to mentally scan from the first location to the second. This relationship was also shown to hold when participants were given verbal descriptions only (Denis and Cocude 1992). Farah and others have also presented evidence that visual imagery tasks involve the same portions of the brain as visual perception (Farah 1988; Farah, Peronnet, Gono, and Giard 1988; Kosslyn and Koenig 1992).

Despite the significant weight of experimental evidence that visual imagery is mentally represented pictorially, a number of researchers have argued that we represent imagery internally in terms of a propositional code (i.e., in a language-like form). This position had been favored by those concerned with computational simulation of intelligence, most notably Pylyshyn (1978, 1984, 1989). This view holds that our representation of imagery is more analogous to language or verbal descriptions than to perception. The arguments supporting this view focus on theoretical rather than experimental issues, or on a fundamental methodological argument against all imagery experiments.

Before continuing this discussion, I need to address the terminology used in the imagery debate. The first authors to discuss each of these representational forms for imagery referred to analog and propositional forms, respectively (Pylyshyn 1978; Sloman 1971). Kosslyn refers to this opposition as "depictive versus descriptive," and Jackendoff refers to it as "geometric versus algebraic" (Kosslyn and Koenig 1992, Jackendoff 1987). I agree with Jackendoff's objection to the original terms, that "analog" connotes noncomputational structures, and "propositional" suggests a truth value. (Another problem with this term will become evident later in this chapter.) Jackendoff uses the terms "geometric versus algebraic." I feel that "geometric" itself connotes a level of formalization and discreteness that is also not really appropriate for encompassing all imagistic forms, including naturalistic visual scenes. "Algebraic" does convey the notion that objects are represented through the use of tokens that may have no physical resemblance to the objects represented, and that there are formal relationships among symbols. In the continuing search for neutral terms, I use the following terms: "pictorial," to denote any imagistic form, and "linguistic," to denote any form using sequences of symbols, whether formal (e.g., algebraic) or natural language.

Some of the main arguments against support of the pictorial form of representation are as follows: From a methodological perspective, it has been claimed that all tests of visual imagery are distorted because of the subjects' tacit knowledge and expectations. Pylyshyn argues, for example, that mental scanning experiments are subject to the participants' utilizing knowledge-driven cues in a way that can be interpreted externally according to the expectations of the experimenters (Pylyshyn 1981). People usually recall a visual scene not as an evenly distributed surface but hierarchically. They recall detail by using the whole scene as a cue (Brann 1991). Reaction times of the knowledge-driven process are similar to those that would be generated if the task were performed in the manner of an actual scanning process. Most of these criticisms against the pictorial view have been answered experimentally (for a discussion, see Brann 1991; de Vega and Marschark 1996).

From a theoretical perspective, proponents of the linguistic approach claim that everything that can be represented pictorially can also be represented in a linguistic sequence. There are also completely abstract concepts that can be represented linguistically but not pictorially. Part of this side of the debate includes the argument that pictorial representations have limitations, in that they can be fuzzy, incomplete, or ambiguous (Dennett 1969). But imagery is well known to have these properties.

As would be expected, there are others who do not go to either extreme: pictorial versus linguistic should not be thought of as an either–or choice. Some researchers believe that we cognitively use both linguistic encodings and pictorial imagery. Paivio (1971, 1986), the primary architect of this "dual encoding" hypothesis, explored the relationships between pictorial and linguistic representation of visuospatial information empirically, focusing on the role of mental images in memory enhancement for specific learning tasks. Two of his primary findings are consistent with the dual-coding model: the *concreteness effect* and the *imagery effect* (Paivio 1971, 1983, 1986). He found that concrete words (words that refer to an object, material, or person and can be experienced by the senses) are generally easier to remember than abstract words: the concreteness effect. He also found that the use of imagery improves memory recall of verbal materials: the imagery effect.

Johnson-Laird (1983) proposed yet another hypothesis, describing the relation between sensory perception and mental representation as a "weak isomorphism," or "homomorphism." This is a more relaxed claim of isomorphism, a model-oriented view in which some intermediary model preserves entities and relationships but does not preserve the form of sensory sensations. This maintains an operational correspondence, but not a structural correspondence between forms. For example, pictorial information may well be projected into a something like Marr's 2½-D sketch (Johnson-Laird 1996). It is this form that can be rotated into various points of view and manipulated to construct alternative possibilities, but it is not necessarily an encoding analogous to a photograph or 3-D display. Johnson-Laird's view is thus a triple encoding (picture-like, language-like, and an intermediate form between the two), in which the details of the encodings themselves are unknown. Various experiments have supported such a model for visual information (Finke and Slayton 1988; Metzler and Shepard 1982).

An alternative to this view is Kosslyn's computational imagery model, which includes a "visual buffer" in working memory that is generated and refreshed from long-term memory (Kosslyn 1980). More recently, Thomas

(1999) has espoused a perceptual activity approach to imagery, which was first proposed in the late 1960s by Hebb (1968) and Hochberg (1968), and is now receiving some renewed attention. In this approach, no permanent cognitive image is ever created. Rather, imagery results from updating the schemata through active and purposeful perception of, and interaction with, the environment. This very Gibsonian approach relies on active perception and neglects imagery's role in pure reflection.

Anderson has pointed out that part of what fueled the pictorial versus linguistic debate in the first place is that the very term "mental imagery" is ambiguous. No one really said precisely what an *image* is (Anderson 1978). Anderson demonstrated that wide classes of different representations, and particularly linguistic and pictorial representations, can be made to yield identical behavioral predictions. He thereby concluded that it would be impossible to establish which representation is correct. Johnson-Laird (1996) later explored this question, citing the following quotation from Fodor (1975):

> There is an indefinite range of cases in between photographs and paragraphs. These intermediate cases are, in effect, images under descriptions; they convey some information discursively and some information pictorially, and they resemble their subjects only in respect of those properties that happen to be pictorial. (p. 190)

Johnson-Laird noted that Fodor was not really arguing for the existence of an infinite variety of representations, merely that there are mixtures of pictorial and linguistic representations in infinitely varying proportions. The implication here is that a pictorial type of representation is better for some things, whereas linguistic representation is better for others.

As already mentioned, there is no way to prove observationally how people mentally represent what they sense and know. All evidence is indirect, through observed behaviors and introspective reports. It seems that in relying on the perceptual side of the process for experimental evidence, many researchers have also remained focused on the perceptual side in contextualizing their explanations. In doing so, they have lost sight of the bigger picture: the essential difference between percepts and concepts, and the entire, interrelated process of interacting with the environment and learning (cf. Figure 4.3). I summarized earlier the role that visual perception plays in developing knowledge, with the visual perception portion of the brain "hardwired" to process shape (form), spatial relationships, and overall pattern.

Within the broader context of the imagistic versus propositional debate, the best working assumption is that we indeed use both forms of representation, although their exact form or forms is most likely impossible to determine. In other words, we cannot determine if there is indeed a structural correspondence between concepts and percepts (i.e., a first-order isomorphism). We should be able to say, however, that there is at least a functional correspondence, or a second-order isomorphism. If we accept this assumption, then two questions immediately follow: First, what are the functional characteristics of each type of representation, and what are the expected advantages and disadvantages of each? Second, how do these representational forms functionally interrelate?

PROPERTIES OF PICTORIAL
VERSUS LINGUISTIC REPRESENTATIONS

Sloman (1975, 1971) was first in the modern literature to compare the properties of pictorial and linguistic representations within an imagery context. He described pictures, maps, and scale models as all being examples of what he called anological (i.e., pictorial) representations, while the predicate calculus, programming languages, and natural languages were examples of linguistic representations. The key difference, he claimed, concerns the manner in which the elements and their interrelationships and properties are expressed. In a pictorial representation, the structure of the representation itself conveys information about the structure of the thing represented. For example, relative distances and positions between elements are inherent in the scene depicted in Figure 8.1, such as whether a steeple is "on top of" the church, or "above" the church, or whether the house is "to the left of" or "to the right of" the church. These correspond to the relationships of their real-world referents. Such inherent interrelationships can be very complex.

In a linguistic representation, on the other hand, the structure of the representation (e.g., in the expression "the door is next to the window") is independent of the structure of the thing or situation it represents. The real power of linguistic representations derives from this property—in that the same rules of syntax, semiotics, and inference can be applied to very different subject domains, and indeed, associations and inferences made across differing contexts. Entities, relationships, and properties can be denoted unambiguously by any symbolic form, as long as the rules of syntax for combining and manipulating them are complete and consistent. It is this generality of algebraic representation that Sloman (1971) claimed may account for the extraor-

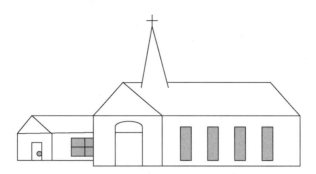

FIGURE 8.1. A simple pictorial representation.

dinary richness of human thought. Linguistic representations also allow the storage of quantitative information. In contrast, the spatial relationships portrayed in pictorial representations are relative in nature. Quantitative measurements, such as the distance between the door and the window in Figure 8.1, would take conscious effort in an attempt to derive them, if they could be derived at all, with any degree of exactness.

The power of a pictorial representation derives from both the arrangement of elements and the implicitness of the interrelationships inherent in the representation. The nature of a pictorial representation allows a level of parallelism, and therefore, of processing efficiency that cannot be attained in a linguistic representation, where elements must be processed in a much more sequential manner. Using a map, for example, interrelationships spatially adjacent to an element of interest can be accessed by locating that element. In problem solving, information for a single step in the process is often found at a single location or contiguous area, and information for the next step is found at an adjacent location. This provides an ease and economy of effort for information retrieval. The implicitness of the information stored also provides significant economy in manipulating the stored information: Moving a symbol, or warping an entire scene in some way, would be much simpler in the pictorial than in the linguistic form of representation.

The associative property of pictorial representations just described also allows new pieces of information to be discovered and inferences to be made at a later time through subsequent inspection and/or manipulations (rotation, deformation). All entities, properties, and relationships must be explicitly stated in a linguistic representation. Thus, given only the expression "The door is next to the window," there is no way of knowing which side the window is on in relation to the door, how big the door is, and so on. Nevertheless, it must be noted that pictorial representations are still typically nonexhaustive. With linguistic representations, we know that only the relevant elements for a given context are stated, but for pictorial representations, elements not included are treated as if they simply do not exist within the depicted area.

The distinction between pictorial and linguistic representations is not completely clear-cut. Linguistic representations, such as natural languages and programming languages, can be least partially pictorial, in that the ordering of elements in a linguistic expression can be a deliberate reflection of the spatial or temporal ordering of their referents, for example, "She put the car in gear, and drove away" or "First the highway was built, then a significant amount of commercial development followed around the interchange." An association must also be made at some point between the linguistic and pictorial representations, in order for meaning to be attached to those symbols. Of course, concepts can be learned entirely without reference to pictorial representation, but this is certainly not as efficient—giving rise to the famous saying "A picture is worth a thousand words."

Given these properties, the generality and more context-independent nature of linguistic representations allows for complex thought and inferences and associations to be made. As Sloman (1971) points out, this comes at the price of efficiency within any specific domain where concrete and specific solutions are needed. An example of this is real-time route selection as one navigates around traffic jams on the way to a downtown appointment. For specific problem contexts, such as this example, it might be more effective to utilize a pictorial form of representation. With a pictorial representation, implicit interrelationships and the property of "discovery" of information, as well as the more cognitively primitive manipulation procedures available for imagery and other forms of pictorial representations, can be brought to bear for quick problem solving in ad hoc situations. Because of these characteristics, this form of representation would also be advantageous for storing specific information that relates to frequently encountered situations.

It follows from this interdependency of representational types that combining both in an overall knowledge structure would be advantageous. Embedding pictorial and linguistic representations within an overall symbolic system makes it easy to switch flexibly between different representations, according to the needs of the problem. Following this assertion, information stored in pictorial form within long-term memory would tend to be memories combined with constructions of specific, domain-dependent information, whereas information stored in linguistic form would tend to be derived, abstract knowledge relating to generalized rules and relationships.

FROM OUTER TO INNER SPACE:
BASIC COGNITIVE MODELS

So, what can be said about the form of spatial knowledge? We must first recognize that we are dealing with a complex and highly interrelated *process* of visual perception, visuospatial cognition, memory, reasoning, learning, and progressive abstraction. Moreover, knowledge in general is an open system, with both direct (sensory perception of environmental scenes) and indirect experience (photographic images, drawings, text, etc.) providing a wide variety of forms.

Without the ability to examine cognitive spatial–temporal representations physically, we can still gain much insight about the form of representations and how they function. While there may be some comfort in the notion that our internal, cognitive representations are the same as the external representations that are, in a sense, tangible, and certainly familiar, we must take care to keep in mind that this view of cognitive representation is a simplification for the convenience of description. In more precise terms, we should think of the form of spatial knowledge as having at least a general homomorphic relation to pictorial and linguistic forms—having some of the same properties, but not being pictures or strings of symbols in the head, as such. It is, in any case, more informative to view cognitive knowledge representation from a more fundamental level. From the perspective of basic types of models, such models should be defined on the basis of what informational elements are being represented, and how the interrelationships among those elements are derived and structured.

Both imagistic and linguistic perspectives on cognitive representation are grouped together as symbolic types of representation. A long-standing debate between proponents of symbolic types of representation in general and proponents of connectionist models has gained a significant amount of renewed attention recently (Bechtel and Abrahamsen 2001; Frasconi and Sperduti 2001; Goonatilake and Khebbal 1995; Medsker 1995; Wermter and Sun 2000).

As described in Chapter 4, connectionist (parallel distributive process, or PDP) models from a functional perspective emphasize the evolution of connections within a distributed network. The basic idea is that structuring occurs among neuron-like elements on the basis of weighted connections. Complex associations can be built up, out of multiple connections across elements. Connectionist models exhibit characteristics that are both consistent with how the brain works and needed for solving complex pattern recognition problems; they are highly parallel, and connections are highly adaptable. These last two characteristics show advantages for robust fault-tolerant processing and gradual plausibility learning.

Symbolic types of representation can be generally characterized as consisting of discrete symbols that are grouped and interrelated in specific ways. All groupings and interrelationships are governed by rules of combinatorial syntax and semantics. Symbolic models are therefore much more adept than connectionist models at structured reasoning tasks and structure-sensitive problems (e.g., routing problems).

Computer implementations of symbolic representations have been more successful in simulating high-level cognitive tasks, as demonstrated in expert systems, whereas computer implementations of connectionist representations have been more successful at low-level cognitive tasks, such as computer vision. Although earlier criticisms focused on what each of these two representational types could not do well (Churchland 1992; Fodor and McLaughlin 1990; Fodor and Pylyshyn 1988), recent attention has focused on developing hybrid approaches, the result of a recognition through experience that differing representational approaches each have their own inherent advantages and disadvantages. Recent advances in the area of formalized knowledge models are discussed in Chapter 14. It is more relevant to our current discussion to focus on the overall organization of space–time knowledge within the mind.

DIFFERING VIEWS WITHIN THE MIND

The human organization of knowledge utilizing only four structuring principles serves to provide a level of unity in what is otherwise a very complex structure. There are a number of different, yet interdependent, views of our environment within the mind. As previously described, these take on varying roles, depending on the situation at hand, as we interact with our surroundings, perform tasks, reason, plan, and solve problems. We can describe these views, or some of them, at some level, as pictorial and imagistic.

Given that we tend to think of "pictorial representations" in the mind as graphical in nature (e.g., images, maps, diagrams), a more generalized term for these representations would be location-based views. In contrast, because propositional representations are more language-like, involving tokens that represent entities or objects, their properties, and interrelationships, these can be called object-based views. We can therefore refer to these more generally as the *where* and *what* representational systems, respectively.

The interdependency of pictorial and propositional types of representation, or the *what* and *where* representational systems of knowledge, was first described by Marr (1982), specifically with respect to vision. He asserted that processing of visual information must begin with the perceived image of the real world (i.e., it is directly location-based). Directly observed phenomena must first be selected and abstracted into key characteristics of the image. The resulting representation is the primal sketch, and then the 2½-D sketch, when perspective and surface orientation information have been added. These characteristics are subsequently interpreted using preexisting knowledge; objects are eventually associated with locations and groups of locations in the image. Objects, as concepts, are thus always higher order (and thereby, also *generalized*) information. This information can then be placed (or confirmed) within the *what* system; it also becomes knowledge that can be used to interpret subsequent images. This mechanism allows the dual *what* and *where* systems to grow and evolve over time in a controlled fashion, with built-in consistency checks to identify and deal with novel, contradictory, or unusual occurrences.

This is consistent with how we learn about our environment, discussed in earlier chapters, and lends further credence to the notion that the *what* system (our knowledge about objects in space) is indeed less fine-grained than the *where* system relative to stored spatial relationships. This also provides insight as to why a picture is worth a thousand words: For communication of information, the picture coincides with the first and most basic means of acquiring environmental information—through sensory perception. Thus, even though the picture may be stylized and the equivalent concept can be expressed in precise algebraic terms, the picture is usually quicker and easier for a person to understand. Figure 8.2 illustrates an example of this phenomenon.

The difficulty we often encounter in trying to give clear and precise verbal directions to someone who is not familiar with the area—and how we ourselves tend to recall the requested route—provides some insight into the role of imagery in how we view our environment and solve spatial problems. Precise spatial–metric expressions relative to distance and direction, are available in English. The real difficulty lies in our cognitive model of space, where spatial information is not usually stored as metrics in propositional form, but rather as a complex of representational forms, including imagery. This includes both perceptual images that we interact with and react to, and stored mental imagery. Without the usual sensory feedback providing metric information, most of us find ourselves visualizing the route—in effect, traveling the route within our minds by "playing back" a sequence of

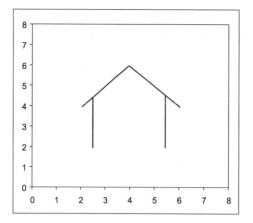

line from (2.5,2.0) to (2.5,4.5)

line from (5.5,2.0) to (5.5,4.5)

line from (2.0,4.0) to (4.0,6.0)

line from (4.0,6.0) to (6.0,4.0)

FIGURE 8.2. Generalized representation of a house in pictorial and in linguistic form.

stored images—in order to reconstruct the needed information as best we can.

SEPARATION OF "WHAT," "WHERE," AND "WHEN" KNOWLEDGE

There is a considerable tradition within geography and urban planning that explores how humans represent geographic-scale environments. Some of the literature relating to this was described in Chapter 7, within the context of how geographic knowledge is acquired (Couclelis, Golledge, Gale, and Tobler 1987; Downs and Stea 1973; Golledge 1988; Lynch 1960). For nonabstract objects, knowing *what* an object is, and being able to identify a specific type of object when encountered, involves detailed and precise geometric properties, particularly shape and the relative orientation of component parts, in addition to other attributes such as color and size. On the other hand, knowing *where* something is seems to be embedded in a knowledge of environmental scenes, including relative locations of objects, but these relative locations, such as "containment," "distance," and "relative direction," tend to be very generalized. For example, farms tend to be located *outside of* cities. Knowing where something is involves a general understanding of types of regions, or scenes, often characterized as being occupied by characteristic *features* or *objects*.

Knowing *where* something is within the context of a specific environmental scene also may involve paths between specific reference objects. These reference objects include recognizable landmarks, as well as oneself, and are recognizable as specific types of objects, such as a church, a school, or a gas station. Many landmarks may also have meaning as known and familiar objects within the environment (e.g., my home, the gas station I regularly use, city hall). In addition, an integrated form of *where* knowledge provides an overall spatial perspective of a well-known environment. This was described as configurational knowledge in Chapter 3.

There is a variety of neural, linguistic, and computational evidence to support this separation of *what* and *where* knowledge representation, which was first proposed by Ungerleider and Mishkin (1982). Some evidence derived from neurological studies indicates that this separation is at least partly a result of how the brain is functionally organized. A major division of labor is achieved by areas in the temporal and parietal lobes, whereby the former encodes object properties, such as color and shape, and the latter encodes spatial properties, such as location and distance (Anderson 1987; Farah, Hammond, Levine, and Calvanio 1988; Levine, Warach, and Farah 1985; Sergent 1991; Ungerleider and Mishkin 1982).

Evidence from formal modeling of simple learning systems supports a *what* versus *where* distinction in how knowledge is stored and used. In a PDP model of a simple visual system, Rueckl, Cave, and Kosslyn (1989) found that in their simulation, an additional level of efficiency was gained when *what* versus *where* functions were encoded separately. They also noted a disparity in the amount of information needed in the encoding of objects compared to locations, with over three times as many storage units necessary to store *what* information versus *where* information. By building a computer model of how people plan and reason about geographic space, Chown, Kaplan, and Kortenkamp (1995) later demonstrated how the *what* and *where* types of knowledge work together, resulting in an enhanced level of adaptiveness for dealing with novel situations.

Kosslyn and his associates built upon this neurological and computational evidence and investigated how the *what* versus *where* distinction plays out in the cognitive representation of categorical (i.e., general relations such as "next to") versus coordinate (i.e., metric) spatial relationships (Kosslyn 1987, 1992; Kosslyn et al. 1989). They noted that different kinds of representations for spatial relationships are useful for different purposes, corresponding to specific contexts that are qualitatively very different: First, people use spatial information to guide actions, ranging from eye movements to reaching and navigating; second, people must rely at least in part on spatial relations to identify scenes and objects within scenes by the interrelationships of multiple parts (Kosslyn, Chabris, Marsolek, and Koenig 1992).

An important means of identifying objects is through the interrelation-ships of component parts. For example, identifying a garden (or a scene with-in a garden) is based on the clustering of a significant number of flowering plants in reasonably close proximity (i.e., "nearness"). The same objects can also be encountered within the environment in many different positions and spatial orientations relative to ourselves. A cat, for example, can be seen sit-ting, facing the viewer, walking past us, or curled up asleep. We know that cats have fairly short, pointed ears, whiskers, a tail, fur, and so on. The fact that the cat may be curled up does not add or take away any of these compo-nents. Similarly, the fact that we may see a church from the side instead of the front does not add or take away any of its constituent parts. Even curled up, the cat's ears remain "on top of" its head. Such generalized, categorical spatial relations are necessary, because they remain constant in a qualitative sense, relative to the object in question.

In contrast, fairly specific metric information must be drawn upon for guiding action. Simply knowing that a table is "next to" a chair is not enough to avoid bumping into the table or to position oneself appropriately in order to sit in the chair. Such precise information is not necessary, or even relevant, for identifying scenes or objects. Specific, quantitative spatial relationships can also be stored. It is this kind of stored information that allows us to walk around our house in the dark without bumping into the furniture. We can nav-igate along any of a number of pathways, judging distance and position relative to our own bodies, and updating that information as we move along.

Because categorical relations are used for identification of objects, scenes, or events by referencing a cognitive model, these can be seen as re-lating to the *what* system. Coordinate (i.e., metric) relations, on the other hand, are used for navigating or otherwise precisely locating things in space. They can thus be seen as relating to the *where* system of the mind. Neisser (1989, 1993) provided a similar view. He argues that the *what* system oper-ates by recognition, comparing sensory information with stored representa-tion. The *where* system operates by direct perception, using the input of sen-sory information for the current scene to fill in any needed specifics dealing with spatial position.

As seen in the work of the Swedish psychologist Gunnar Johansson, mentioned in the discussion of schemata, temporal relationships are also an important element of knowledge. How something moves often provides im-portant information about both object identity and locational relationships. Determining temporal relationships, according to Kosslyn and Koenig (1992), is qualitatively different from knowledge relating to identity and location.

Thus, there is a third system, termed a motion relations subsystem, corresponding to what I call the *when* knowledge representation. For example, we may be looking directly at a deer at the edge of a wood but cannot see or identify what it is until it moves. As we move through our environment, we also experience landmarks and other elements in specific temporal sequences, depending on the direction we are traveling in space. How quickly we encounter these sequences of elements also varies according to how quickly or slowly we are moving.

Time relates to grouping information in two ways. First, properties of the patterns of movement per se define how moving objects will be grouped. For example, if we see a flock of geese flying past, we see them as a single unit, not as individual geese. This is called the Gestalt Law of Common Fate (Kosslyn and Koenig 1992). Second, a specific pattern of movement that tends to occur repeatedly in sequence will also be viewed as a unit. This is how we can recognize the beginnings of "urban sprawl" within our own community, even though the individual components in such a change in land use over time are complex.

Kosslyn and his associates argue that because of the fundamental difference in how they are used, the *where, what,* and *when* systems are qualitatively distinct from each other and, as such, each is encoded by separate processing systems within the brain. Kosslyn and Koenig (1992) emphasize that this does not mean that the portions of the brain in which these systems physically reside are inherently different; rather, this is a pragmatic separation dictated by the way the brain, as a whole, functions. Kosslyn et al. (1992) offer five principles to explain the functional separation of *what, where,* and *when* knowledge:

1. *Division of labor*—It is more efficient for separate components to perform different types of functions. This principle lies at the heart of the separation of knowledge into separate subsystems.

2. *Constraint satisfaction*—Precise information can be computed by satisfying a set of more generallydefined constraints simultaneously. This allows stored fuzzy relations, such as "next to" stored in conjunction with other types of information, particularly the identity of objects, to be used successfully in more exacting interpretation of new information.

3. *Concurrent processing*—Solving cognitive tasks in parallel is not only more efficient but also differing partial solutions are derived as an inherent part of the process. These partial solutions are compared and resolved for a better final solution.

4. *Weak modularity*—The subsystems of the brain are not entirely separate; rather, there is some overlap in terms of both function and the brain location where the function is physically carried out.

5. *Opportunism*—Physical portions of the brain that have been trained for one type of task can later be recruited for another, if needed.

Besides the neurological and simulation evidence just cited, linguistic evidence supports cognitive separation of *what* and *where* knowledge, although the distinction of temporal (*when*) knowledge has not been explicitly addressed in the literature. The marked commonalities of language across widely varying cultural contexts also indicate an underlying corresponding commonality in conceptual structure. Landau and Jackendoff (1993), in particular, focus on how language is structured. As stated in Chapter 7, they based their experimental studies on the hypothesis that the structure of language is influenced at least in part by how the brain stores and processes spatial information. They note that, in English, a rich store of nouns for naming objects incorporates a fairly specific encoding for object characteristics, including a substantial repertoire of names for coherent parts (eyelid, eyelashes, stems, leaves, etc.). Also included are terms to describe surfaces, shapes, and volumes (edge, slice, cube), and even negative parts (groove, notch, hole), which are used in a rich combinatorial system to describe objects. In contrast, locational expressions that refer to objects tend to be inexact and nonmetric. The key element in English to express place is the preposition, such as "above," "forward," and "upstairs." They also note that spatial prepositions in English are quite limited in number, a total of 79, including compound forms such as "in front of."

Landau and Jackendoff conclude that the language for describing object identity is qualitatively different from that describing location. The sheer difference in the number of nouns versus the number of spatial prepositions, which seems to be a common feature of language, lends credence to this conclusion. They also conclude that the *what* system of language is much more fine-grained than the *where* system, basing this on the assertion that the lexicon for describing objects is more precise than that for describing spatial relationships. However, they fail to demonstrate such a difference. How are object-related words such as "leaf", "chair," or even "cube" more precise than locational expressions such as "left of"? They note that the cognitive system of *what* versus *where* is generally acknowledged to be quite fine-grained and capable of capturing metric details of both objects and configurations, and they attribute this seeming mismatch between language and cognition to a tendency for language to generalize spatial relationships (implying that stored spatial knowledge in regard to spatial relationships are not generalized?). This mismatch is considered to be the reason for the difficulty often encountered in trying to give precise verbal directions.

Landau and Jackendoff, as well as Kosslyn and his associates, caution that *what* and *where* knowledge cannot be completely separate and independent. For example, explaining recognition of an object based on recognition of the spatial interrelationships of its component parts would be problematic if the two representations were entirely separate. As discussed earlier, it seems, rather, that there is a considerable amount of "cross talk" between the two representational systems. For example, knowing that *what* we see are a gas station and a car is not enough. We also need to know *where* the car is in relation to the gas station—as well to our own car and the road upon which we are traveling, if we want to enter the gas station. Obviously, we need to keep track of *what* is *where*, both in our immediate visual field and in our cognitive store of knowledge in general, in order to function in our environment. This also relates to the earlier discussion of how recognition of specific objects in our environment forms patterns, and that it is through the formation of patterns—through the *schemata*—that we are able to interpret and make sense of the perceived world. In order to do this, the two representations must be interlinked.

Lynch (1960) and, more recently, Golledge and Timmermans (1990), within a developmental context, provide discussions that show these interlinkages specifically for geographic-scale knowledge. Each individual has knowledge relating to objects that are "important" landmarks within the landscape. Some landmarks are simply navigational cues or anchor points along a path. These specific objects within the landscape act as signals telling us where to go next—to turn left, right, or go straight ahead. Specific types of objects can also tell us where we are within the landscape in a more general sense: The tall buildings on the horizon must mean that we are near a city. Such navigational landmarks may simply be recognized on the basis of their physical configuration and identification as a particular class of object (e.g., a skyscraper, or a large white house with red shutters). Other, familiar objects may have specific meaning to us, such as our house. Besides acting as navigational cues, such objects become associated with specific environments. Some of the most familiar objects, either from personal familiarity or cultural fame, become icons that are intimately identified with a particular geographic scene or location to such an extent that the location is often identified *by* that object, evoking an entire mental scene (e.g., the Eiffel Tower and Paris, and the Golden Gate Bridge and San Francisco).

HOW "WHAT," "WHERE," AND "WHEN" ARE ENCODED

Given the discussion so far concerning category theory, schemata, imagery, and how we learn and use information, a number of things can be said about how environmental information is cognitively stored. Although the mind is capable of remembering precise, quantitative detail concerning both objects and locations, actual numerical values are not usually stored, except where there is a specific need for them. An example of such an exception might include the length of the drive to work, simply because this often-repeated task is something most people deliberately try to minimize. I may know the exact dimensions of my living room—because I bought new carpet last year. Nevertheless, I know more or less the size of every room in my house.

Information about objects is stored in a richly complex and hierarchical system with multiple interlinked, taxonomic, parts-within-parts, and occurrence relationships. Similarly, locational information is stored in a hierarchy of spatial scales—views within views (some at the same scale possibly disconnected), with a complex of overall plan and route views, and landmark information. Some of this includes visual imagery of both objects and locations. This visual imagery is as remembered or constructed. Remembered images are themselves not normally exact visual memories, analogous to photographs, but are also composed to varying degrees. Thus, our *what, where,* and *when* knowledge systems are stored in generally the same way, forming three distinct but parallel and interdependent systems.

So it seems that our knowledge of objects, scenes, and motion and process comprise three cognitive subsystems within the mind, with significant differences in the way that spatial information about each is encoded. As noted in the knowledge acquisition perspective explored in the previous chapters, and the development of the schemata discussed earlier in this chapter, the *what* system of knowledge operates by recognition, comparing observed evidence with a gradually accumulating store of known objects. The *where* system operates primarily by direct perception of scenes within the environment, picking up invariants from the rich flow of sensory information. The *when* system operates through the detection of change over time in both stored object and place knowledge, as well as sensory information.

Nevertheless, these three systems cannot function independently. They are highly interlinked and work in parallel. Properties of specific objects in a hierarchy include where such objects are likely to be found in relation to other objects, and properties of places include what kinds of objects are likely to be found there. All utilize the same fundamental organizational principles of propositional structure, image–schematic structure, and metaphoric and metonymic mapping. All three systems, or views, also incorporate pictorial and linguistic representations.

Each of these views is also hierarchical in organization, which relates to the distinction between episodic and semantic memory, as described in Chapter 4. Episodic memory, as originally defined by Tulving (1972), contains memories of events and experiences, including images (or what we interpret as images) of specific places and things. Semantic memories deal with concepts and general principles. As such, specific places, things, or events become associated with generalized concepts (e.g., categories) as a part of learning. The learning process is how we gradually acquire understanding, forming generalizations and groupings that become progressively abstract categories.

Distinct category hierarchies are based on objects, places, and processes. Not only is the store of objects within the *what* system highly interrelated, with component part membership hierarchies and other associations, but the *what* system is also highly interrelated with the *where* system, with sensory information from perceived and remembered scenes providing the direct evidence from the experienced environment. The *what* system guides the *where* system in interpreting what things are in a perceived scene and determining appropriate action. Shown schematically in Figure 8.3, this highly schematized, generalized diagram of how knowledge is cognitively arranged in memory is distinctly similar to the process of knowledge acquisition shown in Figure 4.3. It certainly should be expected that the *form* of how knowledge is stored reflect the *process* of knowledge acquisition. The key difference is that

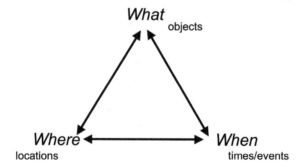

FIGURE 8.3. *What, where,* and *when* systems of representation are cognitively distinct, but highly interrelated. Each incorporates elements, relationships among those elements and to elements within the other two representational systems, and attributes of elements and relationships within a hierarchical structure containing categories and specific instances. These instances may be as remembered, or they may be cognitively constructed.

all three corners of the diagram in Figure 8.3 are composed of category hierarchies, as stored concepts. In Figure 8.4, knowledge storage is integrated with the process of knowledge acquisition and how the two relate—recalling that our knowledge of geographic space is a highly dynamic, open, multirepresentational and hierarchical system. The system is hierarchical in that spatial concepts are stored at many levels of abstraction, from our immediate perceived reality to totally abstract ideas. The system is dynamic in that not only are we continuously updating our knowledge of observed occurrences and abstract ideas, and the interconnections among them, but that we are also continuously changing cognitive representations of our knowledge, in order to deal with the specific situations and problems we encounter in our daily lives. The issue of whether our knowledge also is kept in some permanent and less changing form is still open to debate.

Given the discussion in this and previous chapters, we also quickly see that although there is some evidence for a degree of correspondence between quantitative–qualitative and the *what–where* systems of representation, there is no such correspondence between the pictorial–linguistic and the *what–*

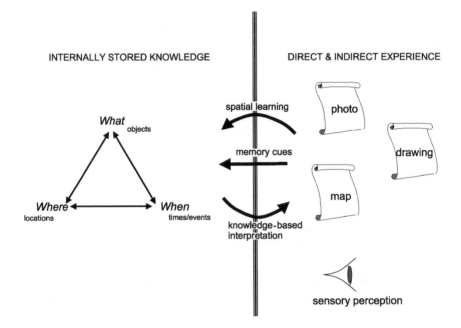

FIGURE 8.4. The interlinkages between internal knowledge and external knowledge, and experience.

where systems. In the most general terms, we can say that the pictorial–linguistic dichotomy exists within each of these systems—in terms of the characteristics of pictorial and linguistic forms of representation, as previously described. It is important to note, however, that our cognitive representation of geographic knowledge may use neither of these representational forms, as such. These are external forms that we may use as a matter of convenience to describe cognitive representation, given a commonality of properties. And as has already been pointed out, one can construe a range of intermediate cases.

How can our discussion so far about human representation and learning be reflected in computer-based representation? This is the topic of Part II.

PART II

THE COMPUTER AS A TOOL FOR STORING AND ACQUIRING SPATIAL KNOWLEDGE

Since the first satellites had orbited, almost fifty years
earlier, trillions and quadrillions of pulses of information had
been pouring down from space, to be stored against the day
when they might contribute to the advance of knowledge.
Only a minute fraction of all this raw material would ever
be processed; but there was no way of telling what
observation some scientist might wish to consult, ten or
fifty, or a hundred years from now. So everything had to be
kept on file, stacked in endless air-conditioned galleries,
triplicated at the three centers against the possibility of
accidental loss. It was part of the real treasure of mankind,
more valuable than all the gold locked uselessly away in
bank vaults.

—ARTHUR C. CLARKE, *2001*, p. 84

CHAPTER 9

NEW TOOLS,
NEW OPPORTUNITIES

So far in this book, I have examined human cognition of geographic space and time from the perspective of what has been derived through observation of human behavior and examination of traditional external knowledge representations. In Part II, I examine the formal representation of spatial and temporal knowledge. The intent here is not to review what can be called "traditional" formal representations as known in mathematics, for example. This is something that is not directly related to the main focus of this book. Rather, I focus on how cognitively informed representations can be advanced within a computing context.

The underlying purpose of a more human-centered approach to representation than is currently practiced in a computing environment is threefold: (1) to gain a better understanding of the implications of human cognition within the context of GIS as a means of indirect experience, of "exploring virtual worlds"; (2) to gain a better understanding of how computers, as tools, can aid people in complex problem-solving tasks related to the environment; and (3) to gain insight into how what we know about how people represent space, learn, and solve ill-defined problems can be used to improve the autonomous problem-solving capabilities of computer-based technologies in today's data-rich environment.

EXPLORING VIRTUAL WORLDS

There are a number of issues related to the use of GIS for learning through indirect experience. A desktop computer connected to a network can provide almost instant access to a tremendous wealth of information available throughout the world in digital form and, indeed, can convey whole virtual worlds. Information is portrayed on the computer screen in deceptively familiar forms: as text, pictures, diagrams, and cartographic images. Nevertheless, some obvious and not so obvious differences in these computer displays can affect the way that information is interpreted.

As discussed in Chapter 7, paper maps and other forms of traditional external knowledge representation are specifically designed by one person, or group of people, to convey a specific message. The representation utilized is selected by the designer(s) to aid in the conveyance of that message. Drawing a map and writing instructions for directing guests to a party at one's house, for example, is often done intuitively, and it usually works effectively. A published map is designed by a professional trained either in the semiotics and other conventions of cartography, or a graphic designer trained in the use of various graphic devices.

Now, with the widespread use of GIS and other forms of computer-based software systems with mapping capability, the end user—or reader of the map—is no longer a passive recipient. The user becomes the map designer. Although the results of a GIS query (e.g., "Show me the spatial distribution of people over 100 years old in the United States" or "Show me all state-owned parcels of land within the county") can often be displayed in a variety of forms (e.g., table, graph, or map), a map is often the user's preferred form of response. Such maps are usually drawn using colors and other cartographic design elements set as defaults within the software. Because these general defaults are set up with no way of knowing in advance what type of data or application the maps will be used to display, they can at best only account for aesthetics by, for example, providing visually pleasing color combinations and standardized progressions for monochromatic color schemes to represent a sequence of values. Even worse, the mapping software is often designed and implemented by individuals with no background in cartography or graphic design. The user has the capability of changing these defaults, but most often, the end user is also not a trained cartographer. As the possible result of an inadvertently inappropriate use of color or other symbology, the user may misinterpret the information displayed on the map. The symbology may also mask important patterns or other features in the data.

The other complexity of the use of cartography and other graphic forms in a modern, interactive computing environment is that much of the use of GIS and on-line geographic databases is exploratory in nature. The map designers/users have no "message" they are trying to portray in a static and single display. Rather, they are using the tool in a highly interactive fashion, without a preconceived hypothesis, exploring the data and applying general domain knowledge in looking for associations and patterns. He or she may change the colors in the original display to accentuate certain values, and then others. The user may then choose to display other data for the same area, or the same type of data for a different geographic area, compare a map with the same data for the same area displayed in a histogram, and so on. This means that cognitive principles about how people learn, and how they use maps, images, and other forms of graphics, as well as text, to gain knowledge from indirect experience need to be considered within a basic theoretical, representational framework for dealing with space–time information. In this way, computing technology can be used to facilitate and enhance the discovery and learning process, and to enable people to deal better with what would otherwise be overwhelming volumes of observational data.

It is these issues, and others, that derive from the differing nature of computers as a representational medium and tool for spatial exploration, as compared to traditional approaches. Other central issues include the following: How should traditional cartographic and other forms of external knowledge representation be rethought for this highly interactive, user-as-designer environment? What aspects of graphic symbology can be considered "natural" or intuitive for untrained users? Which aspects of current methods should be made available only to experienced users? How can the untrained user be guided through the process of discovering information within graphical displays?

COMPUTERS AS GEOGRAPHIC PROBLEM-SOLVING TOOLS

Strongly linked with how people learn and explore is the issue of how computers can most effectively be used to assist in geographic problem solving. The basic challenge is how to capitalize on the unique capabilities of computing technology in order to complement the unique capacities of human cognition, to utilize the speed and tireless computational accuracy of computers to aid people's imagination and ability to see visual patterns, using each capability to best advantage in helping people to solve complex problems.

How the data are represented within the database is as important as any visual representation for solving problems, because how the data are represented is central to how a problem can be solved, and to the ease or difficulty of arriving at a solution. Indeed, in an automated context, the database representation drives the visual representation. Geographic, or any other type of locational data, can be formatted digitally either in an array of grid cells, where the grid cells represent contiguous, small areas of space (also often called raster format), or as vectors of x–y coordinates. A series of coordinate vectors can delineate the boundary of some polygonal object, such as a lake or political district, or the spatial configuration of a linear object, such as a road or a river.

Attributes are associated with individual locations or objects. From a cognitive perspective, the grid format corresponds to a survey view. In a mathematical sense, this is generally equivalent to a field-based view. A vector format that represents boundaries around individual objects, or for individual linear objects, corresponds to a discrete view. This discrete view of objects can thus also be called an object-based view. At the same time, spatial location in either gridded or vector format can be expressed in either absolute or relative coordinates. Specifically, a vector format, emphasizing how objects are related to each other, corresponds in cognitive terms to a route-based view. These objects can include the links themselves as "edges" and their end points as "nodes." In mathematical terms, a representation of multiple interconnections, or routes, is a network.

Similarly, data or information can be represented in either temporal order along a time line (according to temporal location), or as distinct events (temporal objects). Time lines can be either open-ended, representing the linear and experiential view of time, or cyclical. Both temporal location and temporal object representations can also utilize either absolute (May 2, 2000; Tuesday; spring; etc.) or relative (before, after, during, etc.) temporal units of measurement.

Thus, a variety of types of spatially and temporally based database representations developed over the past 30 years correspond to the various general types of cognitive representations. These were developed with a focus on efficiency and other technical issues in an evolutionary and frequently ad hoc manner. Representational approaches were borrowed and translated from mathematics, cartography, and other fields. It is interesting to note that what, in retrospect, is a seemingly comprehensive array of spatial database representations, corresponding to the basic types of cognitive representations, developed without any underlying theoretical framework. The answer of "all of the above" in resolution of the very active raster versus vector debate of the 1970s and 1980s came mostly as a result of accumulated experience with computer-based systems for geographic data handling and display. Placing these also within a cognitive context is coming very belatedly, almost as hindsight. Already having these digital representations, however, it is the development of a comprehensive framework based on human cognition that will allow better understanding of how these should be best employed in helping people learn via virtual worlds and solve problems.

Just as people may shift their cognitive knowledge representation to suit a particular task at hand, each of these types of computer representations is advantageous for solving various types of problems. Figuring out the shortest route from one place to another, for example, is most easily achieved using a vector-type representation, whereas looking for overall spatial pattern is best accomplished using a grid format, or survey view. Changing cognitively from one view to another is automatic and fairly effortless for people, but it is a nonautomatic and time-consuming task for a computer.

There has also been a significant amount of research within the realm of what we now call geographic information science, toward the development of a formalized language for use in GIS and related computer-based systems. At the core of this effort has been the identification of a set of spatial relationships and how these interrelate. Historically, this research has taken three distinct paths. One utilizes algebraic and geometric principles directly, as perhaps best exemplified by Dana Tomlin's map algebra (1983, 1990). A second path has focused on how to extend the query language, SQL (standard, or structured, query language), used traditionally for database management systems (DBMS), for nonspatial applications in the spatial and spatial–temporal domains. SQL is itself built on principles of set theory from mathematics. This latter path is exemplified in work by Herring, Larsen, and Shivakumar (1988) and by Egenhofer (1989). A third path within the GIS research community has focused on natural language in the context of spatial expressions, but little in the way of implementation has been based on this approach. Another path that has arisen more recently is the application of human–computer interaction (HCI) principles. This shifts the focus from formalized language and relationships toward a form of interaction that relies more on the use of the senses (vision, touch, sound, etc.), in a manner more related to how we interact with the real world (Nyerges, Mark, Laurini, and Egenhofer, 1995).

All of this leads to additional questions: How much do these current representations coincide, at least on a conceptual level, with cognitive knowledge representations? Can new types of representations be developed for DBMS and GIS that better reflect human cognition? Can a robust and universal spatially based query language be built to represent a finite and elemental set of cognitive relationships? Need there be a mixture of textual and pictorial protocols for optimal human–machine interaction? If so, for what kinds of tasks is each best suited?

AUTONOMOUS PROBLEM SOLVING:
MACHINES THAT "THINK"

The idea of creating a machine that can think has been around since Charles Babbage (1792–1871) invented his "analytical engine," his follow-up to the "difference engine," which was successfully built and demonstrated as a numerical calculating machine. Although never built, the "analytical engine" was envisioned as having a "store" (memory), a "mill" (calculating and decision-making unit), and a control. This control was to dictate the sequence of operations, input by means of Jaquard loom punched cards. Commonly cited as the first computer because all of the basic components of modern programmable computers were anticipated in Babbage's design, Babbage and others at the time also recognized the potential for mechanized intelligence. It was the first design for a machine capable of performing something other than a fixed set of operations. Babbage himself described his analytical engine as capable of "eating its own tail," that is, altering its own program (Hofstadter 1979).

The field of artificial intelligence (AI) came into being in concert with the advent of modern computers in the 1940s and 1950s. The focus of AI is the development of computer systems that exhibit behaviors that we normally associate with intelligence in human behavior (Barr and Feigenbaum 1981). Winston (1992) has defined AI as "the study of the computations that make it possible to perceive, reason, and act" (p. 5). The field of AI incorporates study in four distinct areas.

The first, *robotics*, is concerned with programs capable of manipulating mechanical devices in complex ways. This area has in large part been driven by the practical requirements of industry for machines that can perform tasks with greater speed and repetitive accuracy than humans can achieve, or in environments that would be too hazardous. Perhaps the best known application of robotic technology is the mechanical welding arm used in auto manufacturing.

A second area, *computer vision*, is closely associated with robotics in that the utility of such devices is greatly increased if they can be equipped with sensory inputs; particularly the ability to "see." Computer vision is also important in a number of specific application areas, which historically have included medical and military uses. Computer vision has also become increasingly important in interpreting the massive amounts of imagery data—from satellites, areal photography and a variety of other sources—currently being collected. The main goal, image understanding, means that computers identify objects within a scene, interpret a scene as a whole and react accordingly, and detect new information from images.

In a third area of AI, *expert systems*, the focus is on encoding knowledge within a specific domain of expertise. Applications are almost infinite: from medical diagnoses to playing chess, to computer system design, to wine selection. The basic idea is to imitate how a human expert would reason about a particular, potentially unanticipated problem and come up with a solution, as well as to learn new knowledge that in turn can be used to increase decision-making abilities. Key issues in expert systems are how to represent domain knowledge, how that knowledge is used in deriving conclusions, and how new knowledge is acquired and integrated into the knowledge base.

A fourth area of AI is *language*. Here, the emphasis is to understand either spoken or textual statements in natural language form for the purpose of either interpretation of statements as computer commands or translation into another natural language. The nuances of accurate translation rely on a vast amount of highly interrelated, contextual knowledge, as in expert systems. So again, a main issue in automated natural language understanding is how to represent and utilize stored contextual knowledge.

Certainly one motivation for work in AI, historically, has been to realize the full potential of computers—to make them more flexible and independent of human intervention by imitating human decision making and learning. A second motivation has been to understand better the nature of intelligence and the thinking process. The former is primarily the domain of computer scientists and engineers, whereas the latter is that of psychologists, linguists, and philosophers. In using computers to better understand intelligence, the hope is that by constructing models of how people (or animals) represent information and use that information to think, we may learn more about the inner workings of the thinking process. The computer can in this way be used both to verify (or cast doubt on) results from human experiments and to derive new insights that in turn can be verified via human experiments. Ultimately, the hope is that AI can provide insight into what the *mind* is, in relation to the body and the environment (Wagman 1991).

Although the idea of "machines that think" has inspired an entire genre of science fiction stories dealing with both technological possibilities and the sociological implications, the current state of the art is still far away from replicating human capabilities. Robots that can navigate through natural (i.e., nonconstrained) environmental space are still mostly blind, relying on simple turning algorithms when they bump into something. Recognizing groupings and patterns, and matching patterns and identifying objects within an image by computer are still relatively time-consuming processes that often require human intervention.

Perhaps the biggest success story to date is in the area of expert systems. Computer programs that play chess are well-known examples, as well as programs for medical diagnosis. The numerous commercially available expert systems software packages available today have improved over time and are increasingly used for instruction, but they still cannot outperform a human expert. These programs, and their associated knowledge bases, are also always limited to a specific domain. For example, a chess-playing program cannot be used to make medical diagnoses.

One obvious element that is missing is the ability to imitate the human capability to derive and use metaphor and other forms of imaginative thought for connecting different knowledge domains. Because of the necessary reliance on much contextual information in generating correct interpretations of natural language that are very often nonliteral, again relying on imaginative thought, the interpretation of natural language has been recognized as a problem compounded. Voice and text recognition, as well as translation programs that have, however, recently become available are usable for practical applications and are also modest in cost. These still tend to be somewhat limited in what they can do, in that they have the ability to recognize the correct word from the surrounding words, but the voice and text recognition programs commonly available at present have no ability to place phrases within a larger cultural or situational context.

Thus, of the many questions relating to AI, there is the overriding question: Can a machine be built that replicates human capabilities? Many of the questions in the previous section arise again: How is knowledge represented? How is information processed? How is new information acquired and integrated into the preexisting knowledge base?

KNOWLEDGE DISCOVERY
IN DATA-RICH ENVIRONMENTS

Data relating to human and natural processes in the environment are now being collected in digital form from diverse sources and from very local (e.g., a single farm or urban property) to global scales. Collection occurs at a rate far beyond what we can see and interpret manually, either in graphical or textual form, in the sense of "exploring virtual worlds," without significant assistance from computers. This problem is made more acute by the accumulation of data through time regarding inherently dynamic processes. In the interest of observing change, data are not replaced as "obsolete" but are rather added to, in an ever-increasing and complex data store. As in the case of human cognition and learning, the raw data first needs to be filtered, with only relevant data selected and the rest ignored. Then, the relevant, or "interesting" elements need to be identified, interpreted, and either discarded as redundant, stored for future reference, or used to modify the appropriate knowledge structure. This multistep process has become known in a computing context as knowledge discovery in databases (KDD). These steps have been defined in various ways for KDD, but in general terms include (Fayyad, Piatetsky-Shapiro, Smyth, and Uthurusamy 1996):

- Selecting and compiling the data set.
- Cleaning the data, including recoding elements in a consistent form, and dealing with missing and erroneous elements.
- Analyzing the data to find patterns and associations.
- Interpreting and evaluating the derived information.
- Verifying the derived information through subsequent KDD as well as other methodologies.

"Data mining" is the term used to refer to the actual discovery portion of the process. Although these steps imply a linear process, as in human cognition, the process is very much iterative in nature. As such, MacEachren, Wachowicz, Haug, Edsall, and Masters (1999) have noted that the term "knowledge discovery" is somewhat a misnomer and would more appropriately be called "knowledge construction." The steps involved in the process of KDD conform also to the observation, analysis, theory development, and prediction–observation cycle institutionalized as the methodology of science.

KDD developed as a collection of various techniques in the fields of database management systems, statistics, visualization, and, not surprisingly, AI. Although KDD and data mining are viewed as highly interactive processes, the sheer volume of data makes adaptation and autonomous learning within the machine a much needed element. Techniques for the analysis, or data mining component, include Baysian classification, maps and scatterplot displays, decision trees, association rules, neural networks, and genetic algorithms.

The basic idea in KDD, as discussed by Adriaans and Azntinga (1996), is to reduce individual observations to points within a multidimensional data space. This makes the use of visualization techniques particularly important and useful in KDD when dealing with geographic data. A number of authors have recently discussed this and how visualization can and should be used as an important tool in each step of the process (Derthick, Kolojejchick, and Roth 1997; Lee, Ong, and Quek 1995; MacEachren, Wachowicz, et al. 1999). These authors have demonstrated how specific visual tools, such as maps and scatterplots, can be used to derive insights concerning patterns and interrelationships. But how do these relate to the cognitive process of perceiving and learning? How can what is already known about the role of vision and other sensory perception be used to improve the KDD process?

TOWARD A NEW PERSPECTIVE

Unlike the real-world environment, which has always acted as an extended knowledge store, with the individual reacting to environmental inputs and potentially changing the environment, the computer can become more. The computer can be an active participant in the thinking and problem-solving process both performing tasks that are components of a larger process, and communicating these results back and forth. This also means that ease of translation back and forth between internal and external representation becomes a consideration. The more that internal and external forms correspond to each other, the easier and thereby more efficient the translation between these forms becomes. This, however, can be viewed purely as a technical issue. The more important issue is that the more internal and external forms correspond to cognitive representations, the easier efficient—and effective—man–machine communication becomes. This is the central thesis of this book. As explored in Part I, sensory perceptions lead to cognitive information as interpreted observations that have been imbued with meaning. An accumulation of information in turn leads to knowledge via linkages, as integrated understanding, that may also be applied to various contexts. Data are observations collected via direct perception, or via mechanical or electronic sensing devices, that are then externally recorded. This can be viewed as part of the same process of gaining information and, eventually, of attaining knowledge.

The correspondences between the cycle of knowledge acquisition in cognitive and scientific contexts are shown in Figure 9.1. In the scientific context, theories are formulated via filtering the data to select those observations considered relevant or "interesting." This selection tends to reveal patterns or consistencies in an observed phenomenon. Theories represent formulations of apparent generalized characteristics, relationships or underlying rules of behavior for specific types of entities: This equates to the formulation of information as generalized characteristics and contextualized rules. The key difference between the cognitive and scientific perspectives is that data are externally recorded, and as such, are not subject to the fading of human memory. Data can be reexamined repeatedly without loss of detail and, perhaps more importantly, can be shared with others. Data are the raw material of formal and scientific analyses.

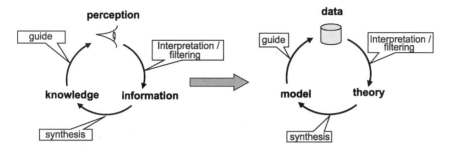

FIGURE 9.1. The cycle of knowledge acquisition from the cognitive perspective, on the left, and from the scientific perspective, on the right.

The similarities between the cognitive and scientific perspectives on learning are thus readily apparent. The scientific perspective, as a formalization of this cognitive process, has to a large extent already been translated into the computing context. Computers and computing technology, after all, began as a scientific tool, and they remain an essential tool for scientific learning and discovery.

So what can be said about computer representation of data and information? A fundamental tenet of representation, whether cognitively or computer-based, is that any representation is more effective for performing some tasks than for other tasks. This is why our cognitive knowledge representation is a multirepresentational and highly dynamic system. This is also why any computing system intended for any range of uses, including modern DBMS and GIS, incorporates multiple representations. Computer representations are also devised by humans, so it follows that computer representations must at least to some degree reflect human representation of spatial information.

This gives rise to subsequent questions. In particular, if cognitive knowledge representation is a highly interrelated, multirepresentational and dynamic system, then what are the practical implications concerning the difficulty of implementing such complex representations in a computing context, if these are much more than current computer representations? Would these also entail a major compromise in machine efficiency for performing various tasks and translating information from one form into another more suitable for a specific task?

These questions and the others mentioned in the preceding sections of this introductory chapter are among the issues that are explored in Part II.

CHAPTER 10

THE COMPUTER AS MEDIUM

THE CAPABILITIES OF COMPUTERS

The computer has become much more than a data storage and calculating device used mostly by scientists and businesses. Rather, it has become a multifaceted tool that is an essential part of our everyday lives. A computer may now serve as a mail system, typewriter, drafting system, appointment organizer, movie screen, game machine, and electronic atlas and encyclopedia (Andersen, Holmqvist, and Jensen 1993). As such, it is an extremely flexible, multimedia device. For these uses the computer has many similarities with its noncomputer counterparts, at least from the user's standpoint. These similarities are built in intentionally by software providers to give a familiar "look and feel." Many of the fundamental differences between computers and other means of representing information are not immediately obvious, but nonetheless are differences with profound implications. One fundamental difference is that the computer incorporates both an internal and an external representation that must work in concert: the screen display and the database.

The computer screen can be viewed as a representational medium that, like paper, can contain a map, a photograph, or a painting. Like a movie screen, it can also contain a dynamic image. Each of these forms presents a "window" onto another space that can be a portion of previously observed reality or some created space. The boundaries of this window define the boundaries of a virtual space situated within our normal experiential space. Like maps and photographs, this representation of virtual space is at a different scale than experiential space. Like the movie screen, it is also a dynamic medium and can produce multisensory output. A good example of this is the virtual campus tours now available on many university Web sites designed to attract prospective students. A modern computer display also typically includes several windows comprising various graphical and textual components at the same time.

Using virtual reality or, more specifically, immersive technology, the boundaries of a screen functioning as a window on some represented space are no longer consciously present. The resulting sensation is one of being transported into the space represented. The intent is to allow the user to rely on the perceptual mechanisms normally used for exploring experiential space in exploring these virtual worlds, such as turning one's head to see a different perspective. This is seen as one obvious means of letting the user explore the information with a minimum of preinterpretation.

As described by Manovich (2000), traditional representational forms (paper, film, videotape) are also static in the sense that the information is imprinted holistically. A photograph represents the reality in front of the camera lens given the particular direction in which it was pointing at a particular moment in time. Certainly, a new photo can be taken. A paper map can also be updated and a new edition published. But these represent complete replacements of earlier versions. In contrast, information stored in a computer database can be revised or otherwise selectively changed without the need to replace the physical medium in order to obtain a "clean" copy.

The computer database possesses a number of the characteristics of a screen display as a representational medium, in that any given spatial database represents a window onto another space, usually at a different scale, and can be selectively updated. It can also contain data in more than one representational format. Unlike a computer display, however, the space represented within a database cannot be viewed or otherwise perceived directly. Selected contents of the database are rerepresented on the computer screen for that purpose. It is also in this sense that a database can produce multisensory outputs. Like a computer display, a database can be randomly accessed. But although we can directly explore portions of a display in any order we wish by simply diverting our visual field, any access to the information contained within a computer database (for display or any other form of output) requires the use of an algorithm implemented as computer software (Manovich 2000).

The greatest potential power of the computer as a representational medium derives from its dynamic, conceptually multidimensional nature, and its ability to produce multisensory output. With recent computer technology, realistic and naturalistic simulations of possible—and impossible—worlds can be created for people to enter, interact with, and learn from. Such capabilities have been quickly adopted by the entertainment industry, as anyone who has played Nintendo, or watched someone else play, can attest. As the name implies, the virtual reality, or virtual worlds, that can be created by computer blur the distinction between what was described in Part I of this book as direct and indirect experience.

What we now call virtual reality is what the science fiction writer William Gibson had in mind when he coined the term "cyberspace" in his 1984 novel *Neuromancer*. That term has now become more commonly synonymous with the World Wide Web. We view this as a space unto itself that must be navigated. This is reflected in how the term "information superhighway," coined by then senator Al Gore to refer to a shared and interactive system for providing rapid communication and access to information, has also come into common usage (Everett-Green 1995). With virtual environments, users traveling the Web can cognitively place themselves within spaces created by the computer to one degree or another. We can thus update a familiar phrase and say that the *computer* becomes the territory.

Advances in hardware miniaturization are also making computers increasingly portable. Mobile phones and hand-held computing devices are rapidly being adopted for a wide range of uses, and distinctions between various types of devices are blurring. Web browsers are integrated within personal digital assistants (PDAs) and mobile telephones that allow on-line mapping capability. This, along with increasing reliance on the Internet, has fostered the demand and, indeed, the need for wearable computers that allow continuous access to information in multiple sensory modes. These multiple modes currently concentrate on sight and sound but can also include haptics (touch) and kinesthetics (muscular feedback). This is rapidly interweaving indirect experience through computers with direct sensory experience of reality, reflected most in what has become known as *augmented reality*. Such a blending has many potential benefits. In a geographic context, heads-up displays mounted in eyeglasses, perhaps with an earphone and a microphone, can guide hikers along unknown trails in response to vocal commands, aid people with visual or auditory impairment, or provide comparative imagery and other information for scientific fieldwork (Golledge 2001).

Computers that were once exclusively office-based tools have become almost constant companions, and we increasingly rely on them to access the wealth of information available via the Internet. Correspondingly, the users of GIS and digital geographic data generally have spread, from a fairly limited group of professional analysts and researchers 20 years ago, to the public at large, including schoolchildren. The availability of the free mapping and route planning program MAPQUEST via the Web is a case in point. Although the range of experience and sophistication of users of GIS and related software has broadened greatly, and we have entered the age of wearable computers and virtual reality, the user interfaces for geographic software systems have not fundamentally changed. GIS interfaces are still designed for one user at a time, utilizing keyboard, mouse, and display screen. They also do not take varying user capabilities or contexts into account.

In the early days of GIS and related software, the user interface was built, first and foremost, to provide a broad range of capabilities. In pushing the envelope of a nascent technology, ease of use was at best a secondary concern, thus requiring a prolonged learning process or intensive formal instruction to become a proficient GIS user. Current GIS user interfaces have been adapted to the modern, windows-oriented, graphical user interface (GUI) approach. Underneath the GUI facade, however, they remain a jumble of what in the larger commercial systems can be thousands of commands that map directly to software function calls. Although these commands are grouped into some broad categories (e.g., input, display, measurement), little or no thought has gone into how the overall structuring, or the look and feel of the interface, coincides with human cognition until very recently. This is now changing, as I discuss in Chapter 13.

THE CHANGING NATURE
OF ACCESSING GEOGRAPHIC INFORMATION

The wide availability of personal computers in homes and offices, their highly interactive nature, and the instant access to information they provide through the World Wide Web, have quickly made this technology the first-resort source of information for anyone with access to it. Given the properties of this technology as a representational medium, the nature of the information, as well as the process of gaining and disseminating information, has changed fundamentally. Aside from the speed and ability to present perceptually realistic virtual worlds, the information retrieved is no longer generated only by appropriate experts and then subjected to a process of selection and editing, as is traditionally the case for the information contained in any library, encyclopedia, or other publication. Anyone can publish as well as retrieve information on the Web. This means that the dissemination of information becomes a shared and cooperative process. The World Wide Web is therefore more like a communal bulletin board than a library or encyclopedia. While this is a more democratic arrangement, it also places the burden on the recipient of judging the quality, as well as the appropriate interpretation, of information retrieved.

This is also true of cartographic and other forms of computer displays generated by the user through the use of GIS or other forms of software. GIS are, in part, forms of external representation used to understand the world and convey geographic information, using computers as their basic medium. They not only combine a number of functions, including complex database retrieval, analysis, and mapping, but also put those functions for the first time completely in the hands of the end *user*. Certainly, the ability to access large amounts of stored geographic observational data and to instantly generate cartographic displays "to order" is a significant advance. The computer as a new representational medium presents new dangers as well as opportunity.

The major danger is that people view the computer map display as equivalent to the traditional paper map. Popular emphasis on the seductive, outward aesthetics of modern computer technology has reinforced a natural identification with what, on the surface, appears to be a familiar world representation with some extra features, such as dynamics. The traditional (paper) map is drawn with the intent of conveying a specific message or to present specific information. The data and map symbology are selectively organized by the person or persons drawing the map for that purpose. With GIS, however, cartographic displays are mechanically produced, literally with the push of a button. As discussed in the previous chapter, the unthinking use of traditional cartographic tools in this very different mode can easily lead the user either to misinterpret the information portrayed, or not recognize important information or patterns in the data.

The map has become much more than a final presentation of the results of queries or calculations performed by the computer. Similarly, the user has become much more than a passive receiver. Maps have become an intermediary representation as part of a highly interactive user interface. The computer display becomes a user-generated representation that intermediates between the human mental representation and the computer database representation, and is at the same time a representation of information that in its own right directly aids the thinking process.

The user needs to remain aware of the fact that the visual interface is indeed just that—an interface (i.e., a means of communication) between person and machine, and between internal representations on either side, and that, moreover, the form of the communication is much more driven by the recipient. The problem of equating GIS display with the traditional map has at least in part been compounded by much past GIS research focused on the translation of traditional forms of cartographic display into a database model—thereby creating a model of a model! This has the potential to make appropriate interpretation even more problematic.

While such models facilitated the rendering of (traditional) map graphics via the computer, and provided an obvious and expedient representational methodology for development of the earliest GIS, this approach quickly became institutionalized within the GIS community. The result is an approach to modeling geographic data that dominates the field several decades later and is based neither directly on human conceptualization of geographic space nor on the optimization of computing capabilities in representing some objective "reality." The translation of what is conceptually a map-based model into traditional database management representational forms, such as the relational model, or application of modern computer-based representational techniques, such as the object-oriented approach, have not resulted in any significant improvements in representational power or computational performance.

As discussed in Part I, human conceptualization of geographic space may include map-like representations, as well as other pictorial and nonpictorial representational forms. Through experience, the generators and users of computer-based geographic data have now discovered that, similar to (and indeed, in large part because of) human conceptualization, a multiplicity of forms is required in order to deal with the range and variety of types of information that currently needs to be stored and analyzed in various ways. Because of the growing difficulties with performance and flexibility of traditional representation methods in existing GIS, the need for increased attention on a theoretical and abstract level that focuses on human cognition has become generally recognized.

CHALLENGES OF MODERN COMPUTER TECHNOLOGY

Libraries have long been used as repositories of large collections of information. They use standard indexing and cataloging systems, meaning that once a person understands the system (i.e., the physical arrangement and the procedure needed for looking up the relevant index), the desired book can be retrieved fairly directly even from extremely large collections. Only one or two systems are used among most libraries for organizing their collections. Thus, knowing one library system facilitates access to the information contained in many libraries. Encyclopedias, known for over 2,000 years as summaries of a complete range of knowledge, are also organized to facilitate information retrieval. Today, topics are most often ordered alphabetically, but some are ordered according to the classification of the branches of science.

From their inception, computer databases have similarly served to facilitate fast search and retrieval of large, shared collections of information. Indeed, this is essentially the definition of the term "database" (Date 2000). The representational forms devised early on to handle these databases in the business world (e.g., payroll, inventory, and accounting) utilized explicit conceptual orderings and direct stored indexes to specific data items. Unlike libraries or encyclopedias, however, the information contained within computer databases had initially been fairly uniform.

As collections of data became larger, more integrated, and diverse, a more flexible method called the relational database was devised. This representational form does not rely on specific orderings or stored indexes, but rather on a system of relationships built into a software language for retrieval. A similar progression from individual and fairly uniform databases to increasingly large and diverse collections of shared information has occurred in parallel within the GIS realm. However, the representational approach utilized to cope with increasingly large and diverse collections of geographic data has relied on the use of multiple models rather than deriving a new approach based on a cohesive theory, due, in large part, to the additional complexity imposed by the multidimensionality of geographic data.

Modern networking technology has provided both new challenges and obvious opportunities in gaining access to information from a wide range of sources. The Internet allows almost instant access to databases and other forms of stored digital information from all over the world. The primary problem is not so much the explosion in the amount of data accessible at the same time, but rather the heterogeneity of these data and how they are organized. All forms of databases, as well as ad hoc collections of text and imagery, are available. Thus, the term "data warehouse" is now used to describe such heterogeneous and often spatially distributed data collections.

To access data through the Internet requires processing on at least three levels. The first requires the use of any one of a number of available search engines, which rely on sophisticated search algorithms that operate on entries in what functions as a standardized and physically distributed catalog. Once the desired information is selected and downloaded as a file, the user must then have the appropriate software for decoding the file. This is processing on a second level. Only then can the user actually access the data by rendering an image on the screen, browsing numerical data records. The problem of not only decoding the data once it has been downloaded but also of utilizing it within the user's local suite of available software, or integrating it within the database, has prompted much discussion about database interoperability and standardization of data formats, particularly in the GIS realm.

Besides these technical challenges, a number of cognitive challenges need to be addressed. As previously discussed in Part I, people's conceptual knowledge representations change as they accumulate an increasing level of expertise within any given knowledge domain. Correspondingly, people are able to apply additional methods of problem-solving techniques within the domain that brings already learned knowledge, as a whole, to bear. This, in a very general way, parallels the overall process of changing cognitive representation and problem solving as children develop, of being able to employ increasingly abstract and integrated knowledge from varying domains.

Also, different types of cognitive representation allow more effective problem solving for different types of problems. As a result, we cognitively rerepresent relevant portions of our knowledge to suit various situations. This implies that the information on a computer in a given topic should be presented differently to a novice than to an expert. For example, concrete, selected examples shown pictorially would be preferred for novices, whereas data summaries shown in abstract diagrams would be preferred for experts. Similarly, different representations within a local government GIS are used by the professional planners to project likely areas of future commercial land use expansion and future traffic patterns, and to present a new 15-year plan to the City Council to manage growth. People make some of these choices for themselves intuitively, when using software that provides a range of capabilities. Nevertheless, users can unknowingly make mistakes, causing confusion, frustration, or misinterpretation. As I have already discussed, ease of use for a particular task is central for any representation and, in terms of computer representation, includes external (i.e., interface) and internal (i.e., database) components.

From the human perspective, both the database and the screen representations are at least as important as content in any collection of computer-based information. How much do current space–time representational techniques coincide with or contrast with the way we cognitively represent space–time knowledge? How can space–time representational techniques used in current computer-based systems better reflect human cognition in representing certain types of data for certain types of tasks, without significant sacrifice of computing efficiency? Should certain capabilities be constrained for specific types of uses or users to facilitate man–machine interaction and thereby also enhance learning and problem-solving capabilities of such interaction? For example, should the number and/or type of display windows a domain novice can have displayed on the screen simultaneously be limited in order to avoid confusion? These remain active research questions within the field of study known as human–computer interaction (HCI). This field of study has only recently been applied to the GIS context, and has focused on external representation. Current research in this area, as it relates to geographic systems, is discussed in Chapter 13. Before dealing with computer interface issues and how people interact with stored digital data, we must first examine previous and potential approaches for storing digital geographic data.

SUMMARY

To summarize, an optimal computer representation requires a balance of computing and cognitive considerations, as well as how internal and external computer representations interact. In a cooperative man–machine environment, the machine becomes part of the highly interrelated, multirepresentational, distributed network of knowledge representation, as described in Part I.

In this current chapter, I have briefly examined opportunities, dangers, and challenges presented by modern computing technology for interacting with, exploring, and learning from digital geographic data as a representation of the real-world environment. Before the questions in the preceding section, and related issues concerning a better, more human-centered approach to geographic representation in a computer context, can be addressed directly, the context of current approaches used for representing digital data, particularly geographic data, need to be examined. Chapter 11, therefore, provides a historical overview.

STORING
GEOGRAPHIC DATA

STORING INFORMATION IN A COMPUTER CONTEXT

Storing information in a computer is usually referred to as storing and displaying *data*. It is therefore important to remember first that the term "data" is plural for datum. A datum is an observed fact. The word "datum" derives from the neuter form (*datum*) of the Latin verb *dare*, which means to give, or to convey. Data, as the plural of *datum,* are a collection of observed or inferred facts. But as the Latin root implies, data are used as the basis of calculations or analyses. In this sense, we must always keep in mind that data are means to something else—to provide information.

The term "spatial data" applies to any data concerning phenomena composed of elements that are distributed in two, three, or more dimensions. The locations of these elements can be expressed implicitly or explicitly using any system of coordinates. Describing spatial dynamics (i.e., the distribution of phenomena over space and time) presents some special problems because, as described in detail in Part I, the temporal and spatial dimensions have characteristics that are alike in some respects but different in others.

Spatial data include such things as bubble chamber tracks in physics and schematics in engineering. Geographic data, more specifically, are spatial data that normally pertain to the Earth.[1] Geographic data are also distinguished in relation to spatial scope. As described in Part I, geographic data pertain to observations beyond the scope of direct personal manipulation and interaction. Thus, geographic space is beyond the scope of tabletop space, and of a single room. It is often beyond what we can even see all at once.

A number of characteristics significantly differentiate geographic phenomena from other types of phenomena. The same characteristics make geographic data representation in general significantly more complex and difficult to implement in a computing environment than most other types of data.

First, natural geographic boundaries tend to be very convoluted and irregular. They subsequently do not lend themselves to compact definition or mathematical prediction. Geographic databases tend quickly to become large as a result. Geographic locations are commonly recorded utilizing specialized Cartesian, spherical, or other types of coordinate systems, including latitude and longitude, Universal Transverse Mercator grids (UTMs), township and range, or street addresses. Geographic location expressed in some of these coordinate systems cannot be converted algorithmically into other systems with any degree of consistency, such as the conversion of street addresses into latitude and longitude. This often forces the storage of more than one locational coordinate for geographic entities within the same database, which not only makes the required storage volumes even larger, but it also presents an additional level of complexity in maintaining the integrity of the database as data are added and updated.

Second, locational definitions of geographic entities are often inexact and can be scale- or context-dependent. For example, the boundaries between specific vegetation types in any given area, and the location of shorelines when considering a very local, human-scale area, are conceptual transition zones and not sharp boundaries. If viewing the same information from a statewide or national scale, these boundaries would most often be viewed as discrete. City boundaries are also fuzzy transition zones, if seen from the point of view of an economist, but cities do have sharply defined boundaries for the purpose of political jurisdiction. As discussed in Part I, our cognitive view of whether something is spatially discrete or continuously blended into neighboring elements is a function of the level of knowledge. Entities can often be seen as changing gradually through time as well, through the same factors. This means that gradual change through space *and* time must somehow be represented. For example, vegetation types change gradually through time as well as space. Abrupt changes, either spatially or temporally, tend to reflect man-made disruption or some unusual event. Thus, noting abrupt versus gradual change can be important for many applications.

Third, the spatial relationships between geographic entities that can be potentially described cognitively are very numerous. As such, it is a practical impossibility to store all of them explicitly. They are also often inexact and context-dependent. This is true of even very basic, relative spatial relationships, such as *near* and *far*, or *left* and *right*.

Fourth, geographic phenomena tend to be scale-specific, with phenomena at different scales interrelated. For example, global weather patterns affect the occurrence of excessive rain in California, which affects the risk of local landslides.

The combination of these properties make computer database representation of geographic data particularly difficult. Boundaries are represented as sharp demarcations in currently used data models, in part because of the discrete nature of computing hardware. An additional problem arises in the transformation of a space that is inherently multidimensional into computer memory, which is normally one-dimensional in nature. Moreover, *varying* views of the data, depending on application and level of knowledge, have only recently been recognized as a representational issue in geographic databases, since the development of shared, multipurpose databases.

The fundamental issues in the development of data models that can best capture the intrinsic characteristics of geographic data (and, ultimately, also knowledge) thus become, first, how to overcome the difference in the nature of the phenomena being represented and that of the representation medium, and second, how best to combine the strengths of the human and the computer for geographic learning and problem solving in a complementary fashion—the human's capacity for insight and intuition, and the computer's capacity for speed, untiring accuracy, lack of forgetting, and nonbias.

This first question has, until now, been the primary challenge of computer database representation. Past and current approaches for computer representation of nonspatial and spatial data are therefore reviewed as a context for addressing the second question.

DATABASES, DATA MODELS, AND LEVELS OF REPRESENTATION

Before reviewing specific representations, however, some additional terms need to be defined explicitly. Whereas "data" are a collection of observed or inferred facts, a "data model" describes the structure in which those data are arranged in user-oriented terms (Date 2000).

A "database" is a collection of interrelated data specifically designed to be shared by multiple users. Data redundancy is controlled, and a uniform approach is used for accessing and modifying data within the database. Database management systems (DBMS) incorporate a database, as well as the computing software and hardware, the users, and the management staff to run the system. DBMS, essentially computerized record-keeping systems, are used by virtually every enterprise today as a fundamental business tool for maintaining personnel, payroll, inventory, and other information. Such enterprises include corporations, government agencies, and scientific organizations. DBMS allow data to be kept secure, yet quickly accessed and updated by multiple users.

The field of DBMS borrowed the term "schema" (today called, more precisely, the "conceptual schema" in the context of DBMS) to denote the conceptual description of an entire database.[2] The schema includes not only a description of all data elements but also the set of *constraints* specifying how the data can be accessed and combined. No single individual, presumably, is aware of the entire database schema, except perhaps the database administrator. Each user or application has its own view of the database, which is called a "subschema." Each subschema includes only a subset of all elements in the database and may specify its own set of constraints derived from those in the schema. Schema and subschema are purely conceptual, and are separate from how the data are physically stored in computer memory. Although there must be a systematic and consistent mapping from one level to the next, the actual representation utilized may be expressed in very different terms. The various levels of representation from a DBMS perspective are shown diagrammatically in Figure 11.1.

The internal view is concerned with how the data are physically stored in the computer. The external view, the user-level or application-focused view of the database, comprises specific data elements and how they are interrelated with regard to a specific application or class of tasks. The conceptual view corresponds to views of the entire database (i.e., the schema), and portions of the database (the subschema). Translation software provided within the DBMS maps the data between the physical storage form and the schema, and between the schema and the various individual subschemas.

Since the concept of levels in views of the data was first developed by Codd (1970), a number of different authors have offered similar descriptions of representational levels, although the number of levels varies. Remembering that all of these descriptions are themselves conceptual models, the important aspect is that all describe a necessary continuum of mappings from the human-oriented to the machine-oriented perspective. The description given above is similar to that given by Chen (1976) in the DBMS context. He described four levels. With some modification for the purpose of the present discussion, I define these as follows:

1. The view (or views) of the data that exists in the mind of the user.
2. The data model, or information structure (i.e., a more formalized, logical structure of entities, attributes, and relationships).

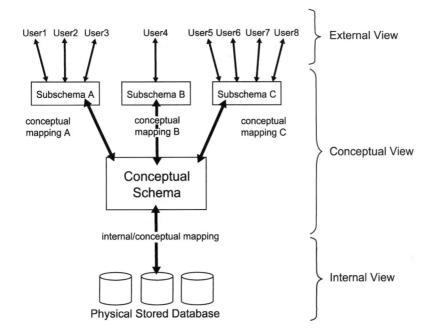

FIGURE 11.1. Representational levels within a database.

3. The data structure (i.e., the structure expressed in some implementation-oriented computer file format with a general conceptual scheme of how the data are linked, such as pointers).
4. The physical file structure (i.e., the view of the data that considers how data elements are stored in memory.

I will subsequently refer to these as the *cognitive level*, the *logical level*, the *implementation level*, and the *physical level*, respectively. These multiple data views (and the distinction among them) and corresponding levels, are a central element of DBMS software architecture and database design. More importantly, the idea of *levels* is central to the design and understanding of computer-based representations, generally. These were described in a geographic data modeling context by Peuquet (1984). From a data modeling and database design perspective, they provide a conceptual framework for deriving computer-based implementations from the top-down, or human-oriented perspective, rather than from a bottom-up, or computer-driven perspective, as had been done previously in DBMS. Each level represents an increasing degree of abstraction, from the conceptual to the physical level. Indeed, Chen built his entity–relationship approach as a formalized methodology for database design on the idea of varying views of the data.

From the perspective of software architecture, accommodation of multiple views of the data allows a number of functional characteristics that are critical in a shared data environment and have therefore long been fundamental to DBMS (Martin 1975):

- All computer programs that use the database can be independent of how the data are physically stored. This means that all computer programs, and the users themselves, can utilize the data model that is most appropriate for a specific task. Users also do not need to concern themselves with the physical location of the data in a distributed database.
- The database administrator can balance conflicting user needs while minimizing data redundancy.
- A standardized approach provided for accessing the database in turn simplifies data access, and makes security and data integrity measures easier to implement.
- Management of queries from multiple users, and access to distributed data over a network, can be optimized at a machine level.

Given this overall context, I examine methods for representing data, and particularly geographic data, in computer databases from a historical perspective in Chapter 12. It is this historical perspective that allows a better understanding of why current techniques are as they are, with their particular dominant representational approach and functional characteristics.

METHODS OF DATA REPRESENTATION
IN DATABASE MANAGEMENT SYSTEMS

DBMS and GIS, as such, both had their beginnings in the late 1950s and 1960s, developing as separate technologies in different contexts: DBMS for handling business records, such as inventory and personnel, and GIS for handling geographic data. GIS developed initially within federal- and state-level agencies. Some basic principles for representing data developed in DBMS, as well as the general history of development in that related field, offer a number of insights with regard to geographic data representation. Where the data happened to fit a particular DBMS model, a number of applications utilized DBMS technology for storing geographic data quite successfully.

DBMS are classified according to the data model, at the logical level, they present to the user. There are four main types: simple table, hierarchic, network, and relational. The schemas and subschemas specific to a particular database are defined according to one of these basic models. The ordering of the four types also represents a chronological progression within the DBMS field, prompted by the demands of handling progressively larger and more complex collections of data. Most dominant today, by far, is the relational type, and a significant number of commercial relational DBMS are currently available.

As the volume, complexity, and diversity of information contained within databases, as well as the diversity of applications, continued to increase, new representational approaches were developed. Date (2000) noted that the simple table, hierarchic, and network models were all defined after the fact. In other words, the DBMSs of these various types were developed first, as a response to an immediate practical need, and the corresponding data models, as formal information structures, were subsequently defined by induction. The relational DBMS was the first type to have the data model upon which it is based formally defined prior to implementation. The relational model was first defined by Codd in a paper published in 1970 (the same paper in which he introduced the idea of "levels" for database design). Commercial relational DBMS first began to appear in the latter part of that decade.

Although the simple table, hierarchic, and network models are considered obsolete in a database context, I review them very briefly here, along with the relational model, in order to show how the representational forms used within DBMS have progressed.

In a computing context, the most fundamental data model takes the form of a table containing rows, called *records*, and columns, called "data items" or *fields*. This model goes back to the earliest days of machine-based data processing, when information was physically stored on decks of punched cards. A data item is a single unit of information recorded about some entity, and corresponds to a physically contiguous set of columns (i.e., a field) in the file. A "record," the recorded set of attributes that pertain to a single entity, corresponds to the data stored in a single row of a table.

The next advancement was the approach introduced with the IBM Information Management System (IMS), first released in 1968 (Date 1995), and subsequently defined as the hierarchical model. The basic idea of the hierarchical model is to store data within *nodes* that are explicitly linked through a series of *parent–child* relationships within a branching, tree-like structure. The uppermost level has a single node, defined as the *root*, and is the only node that has no parent link, but has one or more child links. The lowermost level has multiple entity sets, defined as *leaves* or leaf nodes. Each of these can have only one parent node, as is the case for all nonroot nodes, but no child links.

The hierarchical database model allows rapid search and retrieval of information only when querying the database in a systematic top-down manner in conformance with the stored links, or pointers. This model can be very efficient for problem domains in which the information conceptually fits this model in a natural way; for example, a hierarchical model corresponds to geographic levels of governmental administration (national, state, county, township, or city), following a geographic subdivision. The Ground Water Site Inventory (GWSI) database maintained by the U.S. Geological Survey is another specific example of a hierarchical model that has served well for many years.

Nevertheless, the strict top-down structure was found to be overly restrictive for most real-world applications, even within the business world. No direct links between entities at the same level in the hierarchy (e.g., between cities or counties) are allowed. This would change the form to that of a many-to-many relationship, which by definition is not part of a hierarchical structure. Thus, information, such as space–time location, that may be common to many entities within the database would require many-to-many linkages to be explicitly represented. This gave rise to the network data model.

The essential difference is that unlike the hierarchical model, the network model allows any entity set to be linked to any other entity set within the database by any form of relationship (i.e., one-to-one, one-to-many, or many-to-many) (Martin 1977). This has two important advantages. First, it allows total flexibility in constructing a database schema that logically conforms to the nature of the data. Second, retrieval is simpler and faster, because data can be retrieved more directly, without the need to follow multiple links.

The pointers, or direct links, used in the hierarchical and network models consist of the storage locations of the linked records. These links are stored along with the data. As such, they represent overhead that increases the total volume of data that must be stored in return for added representational power and retrieval efficiency. The major disadvantage of the links, particularly in the network model, is not so much the amount of additional storage they require (although this certainly is a disadvantage), but rather their complexity. Maintaining the integrity of numerous interdependent linkages as the data are modified and updated is difficult, and correcting lost or corrupted links may require regeneration of the entire database. Even though the network model would seem a natural "fit" for many geographic applications in which there is a direct correspondence between the user view, or logical level, and the conceptual schema, there are few examples of current applications within the spatial domain. One example is Intergraph's Interactive

Graphics Design System (IGDS)/DMRS used for applications such as managing road and pipeline networks. The disadvantages of the network model in practice seem to outweigh the advantages in light of other alternatives for most applications.

The relational model was first described in the general literature by Codd in 1970, but the first commercial DBMS based on this model did not begin to appear until late in that decade. The relational model is firmly grounded in logic and mathematical theory, and does not utilize stored links to aid retrieval. As such, it represented a breakthrough in database representation in allowing independence of the external and conceptual views of the data from how the data are physically stored.

All data are stored in the form of tables, called "relations," as shown in Figure 11.2. Each relation is referenced by name. A row, or record, within a relation is called a "tuple," and a column is called an "attribute." The set of valid values that may be contained within any given tuple for any attribute is called a *domain*. Data stored in relational form are accessed and manipulated through a query language called SQL (standard, or structured, query language) and is based on set theory. This places certain restrictions on the relations: (1) There can be no duplicate tuples within a relation; (2) each relation (i.e., a table) is viewed as an unordered entity set, meaning that tuples (top to bottom) and also attributes (left to right) are unordered within a relation; (3) each tuple must contain exactly one value for each attribute.[3]

The relational model offers a number of advantages over its historical predecessors, which explain its longevity and broad use. The first advantage is that there are no pointers to store and maintain. All retrieval is based upon the operators within the query language. Second, the conceptual view of the relational model is as tables, and only tables. This is deliberately kept distinct from how the data are physically stored, so that the user is also shielded from the need to be aware of this physical arrangement. The result is a data model that is highly flexible and powerful, as well as elegantly simple and robust.

Parcel Number	Zoning	Total Area (sq. ft)
23-115-001	R-2	24,001
23-115-002	R-2	28,778
23-115-004	R-2	19,623
23-115-006	C-1	35,322

FIGURE 11.2. "Parcel-zoning" relation within a municipal relational DBMS.

The logicomathematical foundations of the relational model allow it to be applied to a wide range of data types, with operators included in the query language extended to accommodate new application contexts. The implementation of relationships as operators within a logically consistent query language also allows a great deal of flexibility for ad hoc data retrieval. Retrieval paths do not rely on built-in links. Oracle, Informix, Sybase, and DB2 are some of the more widely used of the many relational systems currently being actively supported.

As the range of applications increased, several inherent shortcomings were discovered in this overall scheme as well. The most fundamental of these is that the relational model can adequately represent only information that is precise and homogeneous. Nevertheless, this has proved to be no significant problem for basic business applications, which tend to conform to these characteristics. This is why the relational model strongly dominates the commercial DBMS software marketplace.

Current commercial GIS commonly incorporate traditional relational DBMS capabilities for storing and manipulating nonspatial data. Perhaps the best known example is ArcInfo. The name itself denotes a hybrid system, with the Arc component based on a specialized representation tailored for spatial data, and the Info component being a relational DBMS. This system, as well as other commercial GIS products, also now provides a generic relational DBMS facility.

Regardless of the success and widespread use of the relational model, and of software systems based entirely or in part on it, the model has a number of disadvantages for geographic data, as have been found in other applications.

Although the view of data as tables is a simple one, as mentioned earlier, it causes difficulties for representing multidimensional data. Because of the linear nature of time, extensions of the relational model to include temporal information for traditional business applications (banking, etc.) have been developed. Geographic representation entirely within a relational DBMS context has proven more problematic. There have been a number of experimental implementations of geographic representations using relational DBMS (e.g., Güting 1989; Maguire, Stickler, and Browning 1992; Shapiro and Haralick 1982; van Roessel 1987; Waugh and Healey 1987). As van Roessel's (1987) detailed analysis, in particular, has shown, it is indeed possible to store the complex coordinate and topological data commonly used for defining the spatial configuration of geographic objects within the nonrepeating requirements of this model. Nevertheless, the separation of the many individual coordinate pairs needed to appropriately represent characteristically irregular geographic boundaries within the relational model imposes an inherent and significant performance handicap.

Another difficulty is that intrinsically spatial operations (distance and direction calculation, drawing a map on the computer screen, spatial overlay, etc.) are not part of SQL. Numerous extensions to deal with geographic data have been proposed (Egenhofer 1991; Herring et al. 1988; Ingram and Phillips 1987; Ooi, Sacks-Davis, and McDonell 1989; Roussopoulos 1984), but with much variation in (1) the level of completeness they attempt, (2) how the operations are defined semantically, (3) their syntactical implementations, and (4) the degree of adherence to the SQL standard syntax structure. Egenhofer (1992) has suggested that this is the result of attempting to incorporate spatial concepts into a conceptual data model framework that is fundamentally ill-suited for representing spatial information.

Because of the inherent shortcomings for applications with more complex data, there was the development of extended relational systems during the 1980s. Extended-relational systems utilize a number of extensions suggested by Codd (1979). While still adhering to the basic theory of the relational model, these extensions include abstraction mechanisms for combining entities, properties, and associations into higher order units. Codd grouped these into two types: generalization and aggregation. Precedence also is introduced as a successor mechanism. These extensions represent relational operators of a different type than the set-oriented relations of the basic relational model. They were designed as a means to allow a hierarchical or heterarchical organization that could more naturally represent a wide variety of information types and allow the generation of higher level, user-oriented views. Postgress is probably the best-known DBMS based on the extended relational model, and at least one prototype implementation was tailored for very large geographic databases (Guptill and Stonebraker 1992).

More recently, object/relational and object-oriented database management systems utilizing the object-oriented[4] approach have been developed either as attempts to superimpose the object-oriented approach onto the relational model or as completely new systems, respectively. These systems have been criticized for a lack of a firm theoretical foundation, and as such, for taking a step backward into ad hoc database system implementation (Darwen and Date 1998; Stonebraker et al. 1990).

SPECIALIZED REPRESENTATIONS
FOR GEOGRAPHIC DATA: THE BEGINNINGS

As mentioned earlier, the specific characteristics of geographic data spurred a separate but broadly parallel sequence in the development of representational forms designed specifically to represent geographic data, and in the software systems to handle and manipulate those data, known as GIS. Geographic data have historically most often been two-dimensional, representing the surface of the earth as a plane. Other forms, known as 2½-dimensional, represent a surface as a warped plane and are most commonly used for displaying perspective views of terrain. Three-dimensional forms are found in some specialized systems for dealing with subsurface (geologic or oceanic) or atmospheric phenomena. Although current representational techniques can handle temporal, as well as spatial information, these have major limitations. Current techniques view the world as static, derived from the simple fact that the first computer representations of geographic data were really intended as digital versions of the traditional static (and flat) paper map.

The first representation of geographic space for use with a computer is generally attributed to Waldo Tobler (1959), who created what he called "computer maps" in the late 1950s. The first computer maps were usually gridded, in what became known as the raster format. This type of data model was used for the earliest developments in this area simply because the only computer output device commonly available at the time was the line printer. Thus, the map was quantized into a matrix of cells, so that it could be printed a row at a time, in order to build up the variations in shading that formed an entire image. The grid is a familiar representational technique used for quantitative geographic data, as well as other forms of spatial data, used since the time of the ancient Greeks. The level of general familiarity and seemingly overall comfort with this representational form was undoubtedly a major factor in the popularity and longevity of one of the first generally available computer mapping programs, SYMAP (SYnagraphic MAPping), a program developed by Howard Fisher at Northwestern University in 1964. The other major factor that contributed to its rapid adoption was that the facility to draw maps quickly and "automatically" via computer was suddenly available to anyone who had access to the computing technology. Because, by this time, most larger companies, governmental offices at the federal and state level, and universities had computers, this was the beginning of the democratization of cartography. Although maps produced via SYMAP were visually crude and did not compare to the sophistication and accuracy of manually produced professional products (see Figure 11.3), this method provided speed, flexibility, and also eliminated the need for a cartographic draftsman.

The earliest geographic information systems either utilized this same grid representational format or the points, lines, and polygons on the map were copied, line for line and point for point, using what became known as the vector data model. The development of the vector model seemed a natural extension of traditional, manual, drafting techniques that cartography shared

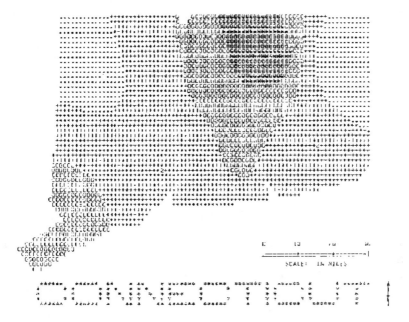

FIGURE 11.3. A computer map drawn on a line printer via SYMAP.

with other areas of the graphic arts and engineering fields. As such, basic forms of this model were developed in parallel. The very first developments in computer representation of geographic and other forms of spatial data in vector form were motivated by the desire to utilize the mechanical accuracy and exact repeatability that computer-rendered drawings could provide. These developments also drove, and were in turn driven by, development of computer output devices (both hardcopy plotters and cathode ray tube [CRT] displays) that could draw in vector mode. When these devices became more generally available, vector representation became a viable alternative to the raster model as the representational form upon which to base GIS. What is generally acknowledged to be the first GIS, the Canada Geographic Information System (CGIS), became functional in 1968 and was based upon a vector data model.

Even with the higher resolution graphics that vector-based computer graphics devices provided, at least in the beginning, raster and vector data models both continued to be used and developed in parallel. The use of grid-based mapping continued for a long time, in large part because of cost. Every computing installation had a line printer; thus, everyone who had access to a computing facility had access to a line printer. This was not the case for vector-mode plotting devices. Installations that did not support graphics as a major activity could not justify the extra cost. Certainly, the first exposure to computer mapping for many university students, until well into the 1980s, was through the use of SYMAP-like systems.

For over two decades, developments in models specifically designed to represent geographic data continued to occur primarily in academic and government settings. These efforts were focused on physical-level concerns of database efficiency, and data models specifically used for geographic data were classified into vector and raster types. More recently, these have more generically been called entity- and location-based models, respectively.

VECTOR MODELS

The first vector and simplest data model used in the earliest vector-mode GIS and drawing software became known as the spaghetti model. This is a direct, line-for-line translation of the paper map. The basic logical entities are points, lines, and polygons. As shown in Figure 11.4, each entity on the map, as a data model, becomes one record in the data structure and is defined as strings of x–y coordinates. The two-dimensional data structure is translated into a list,

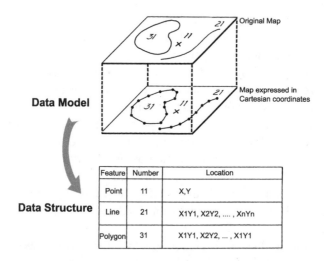

Feature	Number	Location
Point	11	X,Y
Line	21	X1Y1, X2Y2, ... , XnYn
Polygon	31	X1Y1, X2Y2, ... , X1Y1

FIGURE 11.4. The spaghetti model. From Peuquet (1984), originally adapted from Dangermond (1982). Graphic image courtesy of Environmental Systems Research Institute, Inc. Copyright © 1984, Environmental Systems Research Institute, Inc.

or one dimensional model at the physical level. Although all entities are spatially defined, no spatial relationships are retained. This is why this representation became known as the "spaghetti model" (i.e., a collection of coordinate strings heaped one on top of another, with no internal structure). A polygon recorded in this manner is represented in the data structure by a closed string of x–y coordinates that define its boundary. For adjacent polygon data, this results in recording the x–y coordinates of shared boundary segments twice— once for each polygon.

The spaghetti model is not very effective for most types of spatial analyses, because any spatial relationships must be calculated. Nevertheless, the lack of stored spatial relationships, which are extraneous to the plotting process, makes the spaghetti model efficient for reproducing the original graphic image. It has thus often been used for applications that are limited to the simpler forms of computer-assisted cartographic production. Corrections and updates of the line data must rely on visual checks of graphic output.

The U.S. Census Bureau developed the GBF/DIME (Geographic Base File/Dual Independent Map Encoding) format in 1968, for digitally storing addresses and urban maps to aid in the gathering and tabulation of census data (U.S. Census Bureau 1969). This model was a major advancement, in that for the first time, topology was explicitly incorporated as an integral component of a vector representation. In the GBF/DIME format, each street, river, railroad line, municipal boundary, and all other linear features, is represented as a series of straight line segments. A straight line segment is defined between two nodes. A "node" is defined as the location where the line changes direction or intersects with another line (Figure 11.5). In addition, the identifier, or name of the census tract and census block polygons, on either side of the line segment is recorded. In this way, some relative spatial relationships are explicitly stored and can be used for analysis, although in the original GBF/DIME model, the primary purpose was for automated error checking. The GBF/DIME model assigns a direction to each straight line segment by recording a from-node (i.e., a low-address node) and a to-node (a high-address node). This constitutes a directed graph that can be used to check automatically for missing segments and other errors in the resulting file, by following the line segments that comprise the boundary of each census block (i.e., polygon) named in the file. This topological information allows the spatial definitions of map vectors to be stored in a nonredundant manner, which is particularly advantageous for defining the boundaries of adjacent polygons, as shown in the example of a generalized topological model in Figure 11.6. Another feature

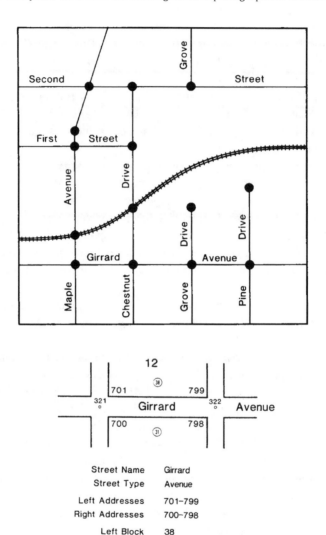

FIGURE 11.5. Graphic elements of the GBF/DIME model and corresponding contents of a GBF/DIME file record.

worth noting is that each line segment is spatially defined according to the definition of the model, using both street addresses and UTM coordinates, in recognition of the fact that street addresses cannot be directly derived from conventional Cartesian or polar coordinate systems.

The problem with GBF/DIME or any similar topological model is that individual line segments do not occur in any particular order. To retrieve any specific line segment, a sequential, exhaustive search must be performed on the entire stored file. To retrieve all line segments that define the boundary of a polygon, an exhaustive search must be done as many times as there are line segments in the polygon boundary!

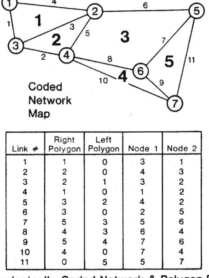

Coded Network Map

Link #	Right Polygon	Left Polygon	Node 1	Node 2
1	1	0	3	1
2	2	0	4	3
3	2	1	3	2
4	1	0	1	2
5	3	2	4	2
6	3	0	2	5
7	5	3	5	6
8	4	3	6	4
9	5	4	7	6
10	4	0	7	4
11	0	5	5	7

Topologically Coded Network & Polygon File

Node #	X Coordinate	Y Coordinate
1	23	8
2	17	17
3	29	15
4	26	21
5	8	26
6	22	30
7	24	38

X,Y Coordinate Node File

FIGURE 11.6. A generalized topological model. From Peuquet (1984). Graphic image courtesy of Environmental Systems Research Institute, Inc. Copyright © 1984, Environmental Systems Research Institute, Inc.

HIERARCHICAL VECTOR MODELS

To overcome the very major retrieval inefficiencies seen in the simpler types of vector models described earlier, each type of data entity is separately stored in a hierarchical fashion. Specifically, data describing polygonal, linear, and point features are organized and stored separately from each other in the data structure (i.e., at the implementation level). Because polygons are composed of the linear features that comprise their boundaries, and linear features are comprised of strings of point locations, links built into these models relate one type of feature to another, as shown in Figure 11.7. It is also important to observe that hierarchical vector models normally (but not necessarily) also include topological information.

One of the earliest hierarchical vector models, POLYVRT (POLYgon conVERTer), was developed by Peucker and Chrisman (1975) and implemented at the Harvard Laboratory for computer graphics in the late 1970s. The general data model and data structure forms of this specific representation are illustrated schematically in Figure 11.7. In POLYVRT, the term

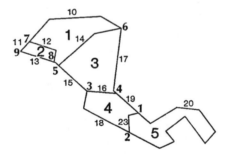

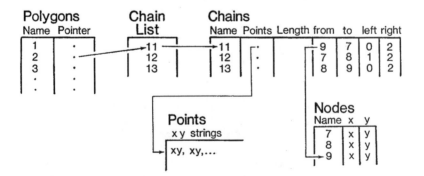

FIGURE 11.7. The POLYVRT data model.

"chain" denotes the basic line entity. A "chain" is defined as a sequence of straight line segments that begin and end as a "node," the intersection point between two chains. The coordinate information defining the spatial configuration of each chain is not stored as part of the chain record within the data structure. Instead, a pointer to the beginning of this information within the separate Points file is recorded. Similarly, pointers within the Polygons file lead to the individual chains that comprise it. Note that the individual chain records contain the same explicit direction and topology information used within GBF/DIME: from- and to-nodes, as well as the left and right adjacent polygons. If the chain defines an outer boundary of the entire area, such as for the Chain 13 in Figure 11.7, this outer area is denoted as polygon "0."

This hierarchical model also has considerable versatility, particularly if it includes topological information. The model can be augmented for the representation of more complex data. It does not violate the basic concept of the model to add additional levels to the hierarchy, such as an additional level of polygonal features. Using such a modified POLYVRT structure to represent a map of the United States, for example, the higher level polygons could be states and the lower level ones could represent counties.

The hierarchical vector model provides a significant functional advantage in a computing environment over nonhierarchical vector models for retrieval and manipulation. First, the segregation of the different classes of elements (i.e., polygons, lines, nodes, and points) in storage allows for more efficient queries. For example, a query to identify the set of polygons adjacent to a given polygon need only deal with the polygon and chain portion of the data. The actual coordinate definitions are not retrieved until explicitly needed for operations such as plotting or distance calculations. At the physical level, this separation of entity types allows much greater efficiency in both storage space and retrieval time. Nevertheless, this separation also creates the need for stored links, or pointers, to other entity types. These nondata elements significantly add to the total volume of stored information. Pointers also represent a potential problem with ensuring and maintaining data integrity, because incorrect pointers can be extremely difficult to detect or correct. The initial generation of this type of representation can also be labor-intensive and time-consuming.

TESSELLATION MODELS

As mentioned earlier, the grid representation was used for the first maps represented by a computer. This is really the simplest form of an entire class of models that have become known most widely by practitioners as "raster" models. Nevertheless, this class of models is best termed the tessellation-type model, for reasons that are explained below. This was also commonly called the raster, or scan line, format, because the maps were printed out one raster or scan line at a time to produce the entire rectangular grid. The name became widely used in the 1970s, when data captured directly in digital form via satellite imagery, as well as computer CRT displays capable of graphics, became available. These operated, and still operate, in scan line fashion. As a result, the raster model also became the predominant representation for applications requiring image interpretation, including photographic and satellite imagery (Jensen 1996; Rosenfeld and Kak 1976). Light reflectance or other raw observational values are employed to identify and delineate objects represented in an image and other attributes.

Although the raster model was the major alternative in practice to vector models for storing geographic data in earlier years of computer-based geographic data representation and use, this general class of model encompasses much more than representations based on a rectangular or square mesh. This class includes any infinitely repeatable pattern of a single polygon or polyhedron with no gaps or overlaps that form a mesh. The term for this in geometry is "regular tessellation." A tessellation in two dimensions is analogous to a mosaic, and in three dimensions, to a honeycomb (Coxeter 1973). These other tessellations come into use when a tiling is particularly beneficial for a specific application.

Tessellation, or polygonal mesh, models represent the logical dual of the vector approach. Individual entities become the basic units for which spatial data are explicitly recorded in vector models, and locations are stored as attributes. With tessellation models, however, a discrete location, or spatial area, is the basic unit for which information is explicitly recorded.

As known to the ancient Greeks, only three regular tessellations are possible in two-dimensional space (square, triangular, and hexagonal meshes), and only five are possible in three dimensions (tetrahedron, cube, octahedron, dodecahedron, and icosahedron). These are shown graphically, left to right, respectively, in the upper and lower rows of Figure 11.8. All three possible types of regular planar tessellations have been used as the basis of geographic representations. Each has different functional characteristics based on the different geometries of the elemental polygon (Ahuja 1983). Irregular tessellations, in which the size of the polygons within the mesh is variable, have also been employed.

Of these various tessellation models, the regular square or rectangular mesh has remained the most widely used for a number of reasons. On a logi-

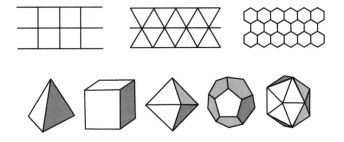

FIGURE 11.8. The regular tessellations in two and three dimensions.

cal level, the grid is a very familiar structure that is also compatible with Cartesian coordinates. On an implementation level, it has two major advantages: (1) it is compatible with the array data structure built into standard computer programming languages, and (2) it is compatible with a number of different types of hardware devices used for spatial data capture and output. Although a number of modern computer languages provide a great deal of flexibility in representing data through both additional intrinsic and user-defined structures, the conceptual familiarity and simplicity of the grid structure and its use in computer languages allow a straightforward and direct translation of the grid data model into a data structure.

The primary advantage of the regular hexagonal mesh is that all neighboring cells of a given cell are equidistant from its center point. Radial symmetry makes this model more advantageous than the rectangular mesh for radial search and relative location. It has also been used in at least one operational GIS, AGIS/GRAM, originally developed for the U.S. military to track ships in the open sea and later released as a commercial product. Although this product is no longer offered commercially, the regular hexagonal mesh is used as a conceptual model for a range of applications, such as watershed management and habitat analysis.

A characteristic unique to all triangular tessellations, regular or irregular, is that the triangles do not all have the same orientation. This makes many procedures involving single-cell comparison operations, which are simple to perform on the other two tessellations, much more complex. Nevertheless, this same characteristic gives triangular tessellations a unique advantage in representing terrain and other types of surface data by assigning a z value to each vertex point in the regular triangular mesh. The triangular faces themselves can represent the same data via the assignment of slope and direction values. Regular triangular meshes, however, are rarely used for representation of this type of data. Irregular triangular meshes are used because the varying size of the elemental triangles allows the density of elevation points to vary directly with the variability of the surface, with a denser distribution of points (and thus larger triangular facets) where the elevation is changing, and a more sparse distribution of points over flat areas. This not only corresponds to the way that measurements of elevation are traditionally taken in the field, but it also translates to a very storage-efficient surface representation.

In terms of processing efficiency of general procedures to compute spatial properties, such as area and centroid calculations, or to perform spatial manipulations such as overlay and windowing, the algorithms initially devised for operation on square grids can easily be modified to work in the case of a triangular or hexagonal mesh. These, in fact, have the same order of computational complexity (Ahuja 1983).

Hierarchical Tessellation Models

Hierarchical tessellation models, like hierarchical vector models, provide significant performance efficiencies. Square and triangular meshes can each be subdivided into smaller cells of the same shape, as shown in Figure 11.9. The critical difference between square, triangular, and hexagonal tessellations in the plane is that only the square grid can be recursively subdivided with the areas of both the same shape and orientation. Triangles can be subdivided into other triangles, but the orientation problem remains. Hexagons cannot be subdivided into other hexagons, although the basic shape is approximated. These hexagonal "rosettes" have ragged edges (cf. Figure 11.9). Ahuja (1983) described these geometrical differences in detail. Each of these tessellations

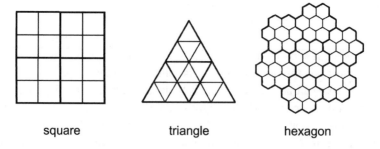

square   triangle   hexagon

FIGURE 11.9. The three regular tessellations in recursively subdivided form.

thus acquires specific characteristics, in addition to retaining the characteristics derived from its basic (i.e., single-level) cellular geometry.

There are several very important advantages of a regular, recursive tessellation of the plane as a representational approach. As a result, this particular type of model received a great deal of attention within the computer science community for a range of spatial data applications, particularly during the 1980s (Samet 1990a, 1990b). The most studied and utilized of these models is the quadtree, based on the recursive decomposition of a grid (cf. Figure 11.10). More recently, hierarchical tessellations have generated interest among researchers in cartography and GIS as a means of dealing with global databases, as I discuss later in this section.

Besides the general model just described, the term "quadtree" has also acquired a generic meaning, to signify the entire class of hierarchical representations that are based on the principle of recursive decomposition, many of which were developed in parallel. The "true," or region quadtree, was first described within the context of a spatial data model by Klinger (1971; Klinger

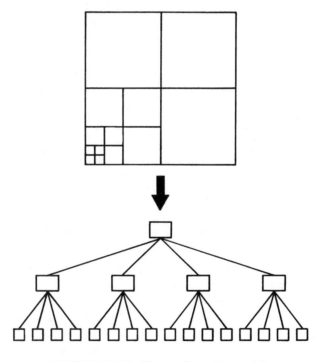

FIGURE 11.10. The quadtree data model.

and Dyer 1976), who used the term "Q tree." Hunter (1978) was the first to use the term quadtree in this context. Finkel and Bentley (1974) used a similar partition of space into rectangular quadrants. This model divides space based on the location of points within the space rather than regular spatial decomposition. Although this was also originally termed a "quadtree," it has become known as a "point quadtree" to avoid confusion. It is an adaptation of the binary search tree for two dimensions (Knuth 1998).

From a logical level, a quadtree (or more specifically, a region quadtree), as defined earlier, translates in cartographic terms to a multiple-scale scheme based on powers of two. This means that data can be explicitly stored at multiple scales within a single model. Although all recursive tessellations can be viewed as having the variable scale property, the triangular and hexagonal versions do not have direct compatibility with Cartesian coordinate systems. Because of these advantages of the region quadtree, this representation is the basis for at least one commercial GIS, SPANS (SPatial ANalysis System).

This model, like others, also has its disadvantages. A major one is that real-world geographic phenomena usually do not lend themselves to the regular scale progression intrinsic to any regular, recursive tessellation. The progression of scales that best fit interrelated phenomena of, for example, the hydrologic cycle from global (e.g., climate patterns) to very local (soil absorption properties) scales, may skip or fall in between the geometric progression of scales dictated by the data model. Similarly, the progression of scales used in standard cartographic documents that may be (and have been) used as archival data sources for digital databases do not fit these geometric progressions.

From an implementation viewpoint, the advantages of all hierarchical tessellations for geographic phenomena include

1. Recursive subdivision of space in this manner functionally results in a regular, balanced tree structure. Implementation of tree structures is straightforward, and efficient retrieval techniques are well known.
2. The recursive subdivision facilitates physically distributed storage, and greatly facilitates browsing operations.

The cost of all of these advantages, however, is increased storage volume.

All of the regular hierarchical tessellations, including all derivative forms, may be extended into multiple dimensions (Jackins and Tanimoto 1980, 1983; Reddy and Rubin 1978). The oct-tree, or three-dimensional quadtree, is probably the best known of these. Individual quadtrees representing different classes of data can also be spatially registered to form multiple layers, as can be done in a gridded database. A comprehensive discussion of quadtrees and their variant forms, as well as an extensive bibliography, has been provided by Samet (1990a, 1990b).

Hierarchical tessellation models have been recognized as a promising approach for dealing with global databases at multiple scales. It has been documented that a global tessellation is best based on triangles (Dutton 1983; Tobler and Chen 1986). Geoffrey Dutton's (Dutton 1996, 1999, 2000) octahedral quaternary triangular mesh (O-QTM) is a global hierarchical data model based on an octahedron—one of the five Platonic solids—inscribed within a sphere (see Figure 11.8). The key to such a data model is an associated referencing system based on some Peano curve ordering that allows direct retrieval at any scale. A number of alternative schemes have been suggested. Goodchild (1989; Goodchild and Shiren 1992) and Otoo and Zhu (1993) also suggest octahedron-based schemes. Fekete (1990; Fekete and Davis 1984) developed an applicable model based on the icosahedron, and Bartholdi and Goldsman (2001) provided a generalized triangular tessellation addressing scheme. However, none of these schemes get around the problem of the mismatch between the predetermined scale progression of such models and the scales at which archival data have already been recorded.

Irregular Tessellations

Irregular tessellations are not suited for general-purpose use, but in a number of applications, irregular tessellation are uniquely advantageous in representing some types of data, and for some specific types of tasks. The four types most commonly used for geographic data applications are based on square, triangular, and variable (i.e., Thiessen) polygon meshes. The basic advantage of an irregular mesh is that it eliminates the need for redundant data, and the structure of the mesh itself can be tailored to the areal distribution of the data. This scheme is a variable resolution model in the sense that the size and density of the elemental polygons vary over space.

An irregular mesh can be adjusted to reflect the density of data occurrences within each area of space. Thus, each cell can be defined as containing the same number of occurrences. The result is that cells become larger where data are sparse, and smaller where data are dense. The fact that the size, shape, and orientation of the cells reflect the size, shape, and orientation of the data elements themselves is also very useful for visual inspection and various types of analyses when displayed graphically. This is the basic idea behind cartograms, in which map entities are adjusted to a size proportional to the number of occurrences.

Perhaps the irregular tessellation most frequently used for spatial data representation is the triangulated irregular network (TIN), or Delunay triangles, where each vertex of the triangulated mesh has an elevation value (cf. Figure 11.11). TINs are a standard method of representing terrain data for landform analysis, hill shading, and hydrological applications for three primary reasons. First, it avoids the "saddle point problem" that sometimes arises when drawing isopleths based on a square grid (Mark 1975). Second, it facilitates the calculation of slope/aspect and other terrain-specific parameters. Slope/aspect values can also be used in determining shading values for the triangular facets that result in an effective 3-D effect for visual display. Third, the data are normally recorded at points distributed irregularly in space.

A problem associated with TINs is that many different, possible triangu-

FIGURE 11.11. Left: Triangulated irregular network (TIN) derived from U.S. Geological Survey 7½-minute quadrangle Digital Elevation Model data for State College, Pennsylvania. Right: The shaded relief map derived from this TIN.

lations can be generated from the same point set. Thus, there are also many different triangulation algorithms. However, one unique triangulation, the Delunay triangulation, is the dual of a Thiessen polygon mesh. The problem of differing triangulations can thus be solved by first constructing Thiessen polygons from the point array and then triangulating on the basis of these polygons. This will always produce the same triangular array from a given set of points. The analytical derivation of a Thiessen polygon mesh (also called a Voronoi diagram) has been studied by a number of people (Fortune 1986; Guibas and Stofoli 1985; Kopec 1963; Rhynsburger 1973; Shamos and Bently 1978).

From a conceptual standpoint, Thiessen polygons can be thought of as constructed by bisecting the side of each triangle at a 90° angle; the result, as shown in Figure 11.12, is an irregular polygonal mesh in which the polygons are convex and have a variable number of sides.

Although visually straightforward, this approach is difficult to implement in a computer context. Therefore, other approaches have been developed for computer implementation (see Guibas and Stofoli 1985 for an effective analytical solution). Once the Thiessen mesh is derived, it is a simple task to derive its unique triangular mesh dual by bisecting each Thiessen polygon boundary and extending the line to the point on either side. These points become the vertices of the triangulation.

The first direct, documented practical application of Thiessen polygons was in the determination of precipitation averages over drainage basins by Thiessen (1911), for whom Thiessen polygons were later named. Thiessen polygons are useful for efficient calculation in a range of adjacency, proximity, and reachabililty analyses, including closest point problems, smallest enclosing circle (Shamos and Bently 1978), the "post office" problem (Knuth 1997), and others.

Two extensions of the basic concept have also been developed. The first of these is to assign a positive weight to each of the points, which represents the point's power to influence its surrounding area, to produce a weighting for each polygon. Described by Boots (1979), this has particular advantages for marketing and facility location siting problems. Drysdale and Lee (1978) generalized the Voronoi diagram to handle disjoint and intersection line segments, circles, polygons, and other geometric figures.

Although these various, irregular polygonal tessellations are each uniquely suited to a particular type of data and set of analytical procedures, they are very ill-suited for most other spatial manipulations and analytical tasks. For example, overlaying two irregular meshes is extremely difficult, at best. As a result, irregular tessellations are unsuitable as database models. But because the tasks for which TINs and Thiessen polygons are particularly

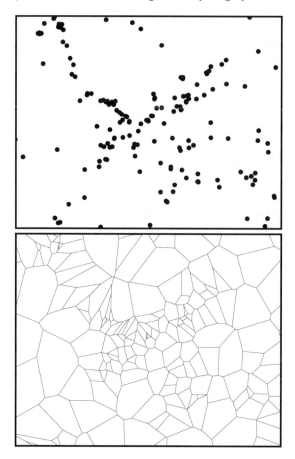

FIGURE 11.12. Top: U.S. Environmental Protection Agency Toxic Chemical Release Inventory (TRI) sites for Pittsburgh, Pennsylvania, 1994. Bottom: Thiessen polygons created from these point locations.

advantageous frequently are needed in a variety of geographic analysis contexts, these irregular tesselation models are included in most commercial GIS as the representational basis for specialized subsystems.

Because these data models are, by definition, irregular, they cannot be implemented simply as arrays. TINs, in particular, are implemented by storing the edges of the triangular facets as vector connections between points in the irregular point distributions that comprise the vertices of the triangles. Seemingly, because of this, TINs have sometimes been regarded as a vector-type model. This could be viewed as a simple matter of confusing the logical-level data model with the implementation-level data structure. However, it is

not that simple. The edges of the triangular facets become the primary element of interest—at a logical level—for some specific applications, such as
streamflow modeling.

Scan Line Models

A special case exists if the square mesh is the parallel scan line, or raster,
model. The critical difference with the parallel scan line model is that the cells
are organized into single, contiguous, rows across the data surface, usually in
the x direction, but do not necessarily have coherence in the other direction.
This is often the result of some form of compaction, such as raster run-length
encoding. This is a format commonly used by graphical scanning devices.

Although this model is more compact than the square grid, it has many
limitations for processing. Algorithms that are linear or parallel in nature (i.e.,
input to a process to be performed on individual cells does not include results
of the same process for neighboring cells) can be performed on data in scan
line form with no extra computational burden, in contrast to gridded data, because null cells (i.e., cells containing no data) must also be processed in the
uncompacted, gridded form. Many procedures used in image processing fall
into this category. Other processes that do depend on neighborhood effects
require that scan line data be converted into grid form.

Another form of scan line model is a family of curves that generates a
track through space in such a way that n-dimensional space is transformed
into a line and vice versa. These space-filling curves were discovered in 1890
by the mathematician Giuseppe Peano (1973). Peano curves, as they are also
known, preserve some of the spatial associativity of the 2-D dataspace and the
single dimension formed by the single line trace. An example of a simple 2-D
Peano curve is shown on the left side of Figure 11.13. With this particular
version, called a Hillbert curve, all changes of direction are right angles. The
graphic on the right side of the figure shows the same Peano scan at a smaller
scale.

Peano curves possess several useful properties as a database indexing
scheme. These properties were originally summarized by Stevens, Lehar,
and Preston (1983):

1. The unbroken curve passes once through every locational element in
   the dataspace.
2. Points close to each other in the curve are close to each other in
   space and vice versa.

FIGURE 11.13. A right-angle Peano curve at two scales.

3. The curve acts as a transform to and from itself and n-dimensional space.

Because of these properties, Peano curves have proven useful as indexing schemes that allow efficient mappings between two (or multiple)-dimensional Cartesian coordinate space and the linear address space of physical computer memory. The first known practical application of Peano curves for digital geographic databases was as the areal indexing scheme for the Canada geographic information system (CGIS) (Tomlinson 1973), and they have since been used for spatial indexing in various software systems. This scheme divided gridded areal data into "unit frames." The frame size was determined for convenience of retrieval and processing. Each unit frame in the system was assigned a unique number starting at the origin of local coordinate system. From that point, frames are sequenced so that they fan out from the origin as the frame number increases. This arrangement is shown in Figure 11.14 for the first 84 unit frames. This numbering scheme, named the Morton (1966) matrix after its designer, represents the trace of a Z-shaped (or N-shaped) Peano scan (cf. Figure 11.15). This spatial indexing scheme utilizes the property that areas close together on the earth will likely have a minimum separation in a sequential digital file. This has the effect of reducing search time, especially for small areas. Further enhancing the operational efficiency of the Morton matrix specifically as a database indexing scheme is that addresses can be directly computed by interleaving the binary representation of the geographic x and y coordinates (cf. Figure 11.16). This addressing scheme was examined more fully by White (1981) and Torpf and Herzog (1981). Variations on this scheme have also been discussed by a number of researchers, including Abel and Smith (1983) and Lauzon, Mark, Kituchi, and Guevara (1985). The basic idea relies on the recursive properties of Peano curves and

	0	1	2	3	x ⟶			
0	0 / 0000	2 / 0010	8 / 1000	10 / 1010	32	34	40	42
1	1 / 0001	3 / 0011	9 / 1001	11 / 1011	33	35	41	43
2	4 / 0100	6 / 0110	12 / 1100	14 / 1110	36	38	44	46
3	5 / 0101	7 / 0111	13 / 1101	15 / 1111	37	39	45	47
y	16	18	24	26	48	50	56	58
↓	17	19	25	27	49	51	57	59
	20	22	28	30	52	54	60	62
	21	23	29	31	53	55	61	63

FIGURE 11.14. The Morton matrix indexing scheme.

the direct correspondence between the Z-shaped Peano curve and the area quadtree structure.

The properties of Peano curves also have significant utility for image-processing applications (Stevens et al. 1983), including data compression in the spatial and spectral domain, histogram equalization, adaptive thresholding, and multispectral image display. These and many other techniques for manipulating and analyzing imagery data are sequential operations (i.e., they are linear in nature). The property of preserving some of the spatial relationships in the one-dimensional Peano scan data allows improved interpretation, and thus improved results, from these procedures.

SUMMARY

Computer representations in DBMS and GIS have followed a broadly parallel yet separate history of development. All of the models just discussed were developed in response to a practical need and focused on implementation. Nevertheless, the first model in DBMS to be based on a generalized theoretical framework was introduced over thirty years ago, whereas representation in GIS and other types of geographic data handling software remains ad hoc— at least in practice. Although these geographic data models were initially developed or first applied in a computing context in the 1960s and 1970s, placing existing models within an overall taxonomic framework did not appear until

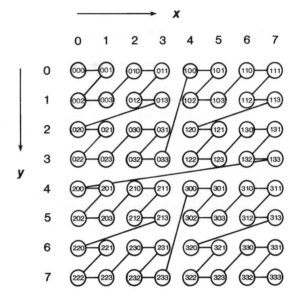

FIGURE 11.15. The relationship between the Morton matrix indexing scheme and the Z-shaped Peano curve.

the 1980s (see Peuquet 1984) except for acknowledgment of two broad classes. And unlike the case in the DBMS community, where a new and implementable perspective for representation with a firm theoretical foundation was seemingly derived with one stroke, development of such a theoretical basis for the representation of geographic data has remained a research priority within the wider GIScience community for over 10 years.

The fundamental reason for this lag, alluded to in this chapter, is the greater problems posed by the nature of geographical data. The relative suc-

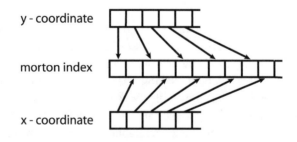

FIGURE 11.16. Conversion of Cartesian x–y coordinates into a quadtree address via bitwise interleaving.

cess of DBMS at handling business-oriented, list-type data with single-model systems, combined with the lack of a comprehensive theoretical foundation, resulted in a debate in the GIS community about whether the raster- or vector-type representation was better. In Chapter 12, I discuss geographic representation in GIScience, within the broader context of scientific development. This begins with a more detailed "postmortem" of this debate, then continues with a discussion of the current dialogue relating to geographic database representation and recent research in developing advanced representations.

NOTES

1. In the modern context of space exploration, however, the term "geographic" may also apply to data pertaining to other planets and objects in outer space.

2. The terms "scheme" and "subscheme" are also used in this context.

3. A relation that satisfies this last property is referred to as *normalized*. Why this particular property is given a specific name (and, indeed, added significance) is based on the idea of nonredundancy of data within the database, thereby minimizing the total volume of data to be stored. A relation that satisfies only the basic requirement of only one value for each attribute is, more precisely, said to be in *first normal form*. A relation that satisfies additional requirements relating to independence among attributes may be in second or third normal form. These degrees of normalization were included in Codd's original proposal for the relational model (1970). Higher relational forms, including Boyce–Codd normal form and beyond, have been introduced more recently in order to provide even greater levels of data independence (Date 2000).

4. The term "object" in this context has a more specific meaning, which I explain in the next chapter. Until then, the meaning of the term "object" as it has been used until now will suffice.

CHAPTER 12

A NEW PERSPECTIVE FOR
GEOGRAPHIC DATABASE
REPRESENTATION

THE BIG DEBATE: DEEPER THAN SOFTWARE

A well-known debate that occurred in the GIS field in the late 1970s and throughout most of the 1980s was commonly called the raster versus vector debate. In general terms, the debate was about which was the better representational approach. Spatial data handling systems of the time could generally be described as being monolithic in their database design in that they were based on either a vector or a raster representation. This monolithic design was due mostly to the relatively early state of development of spatial software. Another factor, however, was that builders of GIS were looking to other types of data processing software as design models. In particular, the related technology of DBMS at the time utilized a single-model design very successfully, as it still does. As an example of the degree of separation in the use of raster and vector data models, ESRI (Environmental Systems Research Institute), one of the longtime major providers of GIS software, marketed two separate GIS (PIOS, the Polygon Information Overlay System, and GRID) that were based on vector and grid models, respectively.

With the advent of vector graphic devices in the 1970s, the topological vector format, and particularly the hierarchical topological vector format, became the dominant spatial data model for geographic software focused on urban and regional planning. Essential data included linear and polygonal types of geographic objects, such as streets and census blocks, for tasks such as traffic and land use planning. The grid or raster format was found to be more effective for applications in which a particular location and locational pattern were the major emphasis, such as forestry and other natural resource management applications. This format was also used for image processing applications in which data provided in raster format needed to be interpreted.

As mentioned in Chapter 11, the development of representational techniques within the GIS community progressed during this time, with a general emphasis on the short-term and implementation-oriented solution, and an overall bottom-up approach to database design (Goodchild 1992; Peuquet 1988a). This was partly because GIS, in the beginning, were in large part developed internally by government agencies, or under contract to those agencies, to take advantage of the newly available capabilities of computers to fill an urgent need for handling large amounts of geographic data. While in retrospect some have pointed to this as being shortsighted, this is the usual way any new methodology begins, with both push and pull factors. Push factors stem from a need that traditional methodology can no longer satisfactorily fill and provide the motivating force needed to justify the expenditure of time and resources. Pull factors stem from the availability of an enabling technology that makes a solution possible.

In the specific case of GIS, the obvious pull factor was the availability of computers, but this was also a constraint due to the lack of maturity of computing and computer database modeling techniques. Undeniably, hardware capacity and software tools and techniques were limited and primitive relative to the tasks to be accomplished. It can be argued that this alone would force an emphasis on implementation. As I mentioned, early developments in geographic databases were driven by public agencies, with immediate and specific needs for computer-based geographic databases. Only through the substantial budgets made available by governments at federal and state levels were initial advancements in geographic databases made possible, by acquiring what was an infant technology at the time.

In this initial phase of any new methodology, preexisting and familiar approaches are adapted to the new, enabling technology. Besides minimizing the time needed for achieving useful results in the short term (thereby helping to fill the need and gaining a return on investment), this is also learning by doing. Perhaps more importantly, "jumping in" and using the existing and familiar paradigm for getting started in the development of any new field is a very rational approach. It allows things to be tried out in a new context as quickly as possible, in order to see what works and what does not; it also allows previously unseen problems to reveal themselves. The automobile was originally called the "horseless carriage" and used a chassis very similar to that of the buggy—with the same wheels and suspension. It is also the usual progression in science to acquire knowledge in any new area by beginning at a practical level, trying things out empirically, and gathering observational data about various approaches. From these learn-by-doing experiences, some generalities can be made inductively. Nevertheless, theoretical advancements at this stage tend to occur only when necessary to accomplish a specific task. The integration of topology in the development of DIME by the U.S. Census Bureau in 1968–1969 is a prime example in the case of geographic database representation.

There were articles on the larger theoretical context of GIS, some on a philosophical level very early-on, but these were only occasional contributions that did not gain much attention at the time. Sinton (1978) interpreted Berry's decade-old (1964) geographic matrix framework—linking space, time, and attribute—within a GIS context. Chrisman (1978) discussed the implications of absolute versus relative space for geographic database representation. Dacey (1970) anticipated the need for a spatial query and manipulation language based on fundamental geometric principles.

As the software and hardware, and data in digital form became more available, the number of users and uses of GIS methodology grew. Users began to address an increasingly broad range of questions and applications within the same software. As a result, the relative functional limitations of each of these approaches to geographic database representation became more apparent, and the issue of how to broaden the functional capabilities of GIS arose. Because of a monolithic GIS database design philosophy and the speed/capacity constraints of computing technology at that time, the solution was interpreted as determining which of the two types of representation could best handle the complete potential range of functions. This was also seen as a means to avoid the overhead of converting data from one data model to another (Peuquet 1981a, 1981b).

Formation of an overall taxonomy of existing representational approaches (i.e., putting known approaches into a cohesive framework) was the next phase in the development of geographic representation. These models also gained much attention as practical guides (Kainz 1987; Peuquet 1984). At this stage, leaps still tend to not be made until something is noticed that no longer works or does not fit. The development of helicentrism after the discovery of retrograde motion is one of the most famous examples in the development of science in general.

During the 1980s, some hybrid models, which have characteristics of both raster and vector representations, were also developed for specialized purposes. But these were also attempts to test the extant framework and potentially extend it. The Strip-Tree (Ballard 1981) and the Vaster (Peuquet

1983) models are perhaps the best known of these hybrids. In the early 1980s, commercial GIS increasingly acquired the capability of handling both raster and vector data, simply out of functional necessity. Not until the late 1980s was the dual nature of raster and vector models, and the intrinsic need for both, generally acknowledged and described in the literature from a theoretical perspective (Peuquet 1988a).

In terms of the dual framework, the basic logical component of a vector model is seen as a spatial entity, which may be identifiable on the ground or created within the context of a particular application. These may thus include lakes, rivers, roads, and entities such as "the 20-foot contour level." The spatial organization of these entities is explicitly stored as attributes. Conversely, the basic logical component of a tessellation model is a location in space. The existence of a given object at that location is explicitly stored as a locational attribute. From this perspective, one can clearly see that neither type of data model is intrinsically a better representation of space. The representational and algorithmic advantages of each are data- and application-dependent, even though both theoretically have the capability to accommodate any type of data or procedure.

From the current perspective, each of the basic vector and tessellation representational approaches is seen as object- (or entity-) and location-based, respectively. Each is therefore also intrinsically more effective for answering one of two fundamental spatial queries, which themselves can be viewed as logical duals of each other:

- The object-based view—Given a specific entity or entities, what are its associated properties (one of these properties may be its location)?
- The location-based view—What entity or entities are present at a given location?

Most current commercial GIS utilize a multirepresentational database design and incorporate both raster- and vector-type representations for coordinate data, as well as links to a DBMS for storing attribute data. This allows procedures to be linked to the most suitable representation for performing a given task.

Also, during the late 1980s and early 1990s, Goodchild (1987), Anselin and Getis (1992), and others claimed that the real power of GIS lies not so much in being able to answer such direct queries, as in dealing with complex analytical issues of "why" and "what if." These are explanatory and predictive modeling questions that require the combination of location- and object-based information, and their attributes, in complex ways. Such analysis was viewed not as using direct queries and the GIS database representation per se, but as mostly utilizing traditional statistical tools.

The issue often discussed at the time was how to integrate existing statistical techniques, often with specialized representations of their own, into the GIS software. More recently, it has been recognized, however, that traditional methods for spatial analysis are often inappropriate for the modern context. Traditional statistical techniques assume certain distributional characteristics about the data that can no longer be said to be true in current large, shared, heterogeneous databases. Moreover, the very large and exhaustive nature of these databases requires development of more inductive procedures that utilize computers as exploratory aids for people in simply finding meaningful patterns and extracting other information from the data. Openshaw, Charlton, Wymer, and Craft (1987) recognized this at the time, and Openshaw's Geographical Analysis Machine (GAM) drew much attention, but the importance and scope of a new approach to geographic analysis in a computing context has only been recently acknowledged generally. This new, exploratory approach requires drawing from what we know about cognition, perhaps, in part, simulating how people learn from direct and indirect experience.

From the perspective of representation and capturing observed information, researchers were recognizing the need to model *human concepts* of reality rather than cartographic representation (Burrough 1992; Nyerges 1991; Peuquet 1988a; Raper and Maguire 1992; Usery 1993). Corresponding to human problem solving, each type of model, or view of reality, is a particularly useful form of representation for addressing certain types of problems and answering certain types of questions. Helen Couclelis (1993) couched the duality of GIS data models in cognitive terms, renaming the two basic approaches of geographic data representation as objects and fields, instead of vectors and rasters, respectively.

This coincided with the more general recognition that a coherent theoretical foundation for geographic representation was lacking, and that such a theory was needed to interrelate all forms of representation: cognitive, graphical, and database. This was highlighted in the call for development of GIScience as an interdisciplinary field to address the issue of geographic representation and other related issues on a theoretical level (Goodchild 1992). This general recognition was an indication that GIS research generally, as well as geographic database representation specifically, had reached a particular state of maturation within the normal progression in the development of any methodology or science. The field that had been evolving for over two decades became *GIScience*, with the name emphasizing an important shift from an implementation focus to that of developing a coherent theoretical framework. Such a theoretical framework in GIScience would in turn do four things with regard to representation:

1. Allow for effective representations to be derived in a systematic and predictable fashion for *specific* application contexts.
2. Allow new and potentially better solutions to be discovered within the interstices, or as extensions, of that framework.
3. Provide formalization of concepts of space that previously had been only intuitive or informal in nature in a number of disciplines, including geography.
4. Through implementation of these formalisms, and their use by people, provide additional insights into human cognition and our understanding about how people store and use geographic knowledge.

WHAT ARE OBJECTS, ANYWAY?

Even though some researchers have recognized the need to model human concepts of reality in geographic database models directly instead of modeling the map, what this actually means seems to be confined for the most part to those actually working in the area of geographic cognition. Besides the relational database management systems (RDBMS) approach, the object-oriented programming approach is being used with increasing frequency by GIS researchers, as exemplified in the efforts of Herring (1992) and Worboys (1994). The conceptual basis of approaches for geographic data representation has not changed, however.

The object-oriented programming approach was originally developed to provide programming tools at the implementation level, so that representations of entities as understood in a normal, conceptual sense could more easily be built. The standard data structure elements available at the programming level (e.g., lists, tables, and pointers) are augmented, so that they can be combined in higher level units used to describe objects, classes of objects, and their interrelationships. The basic principles of object-oriented data modeling —particularly *encapsulation* and *inheritance*—represent an important advance and significant departure from traditional data modeling approaches. Through *encapsulation*, data elements are defined on the basis of their behavior as well as their measured attribute values as modular units. These are distinct and unique to each object. The principle of *inheritance* of properties and behaviors from an object's parent object type or component objects facilitates construction of taxonomic and partonomic hierarchies. The combination of these principles provides a powerful and natural mechanism for translating a cognitive-level representation eventually into the implementation level.

The object-oriented programming approach has been used only as an implementational tool, and the results are thereby called object-based models. In the models developed in this way, geographic data elements are still conceptually divided into the same, point, line, area, or grid cell units, but these are now considered "objects." Certainly, a pixel- or a vector-based view of the world does not match our everyday perception of it (Gatrell 1991). By taking a bottom-up approach in geographic database design, many researchers have not gone back to examine the cognitive level, or the external view, as shown in Figure 11.1. Thus, "objects" programmed utilizing object-oriented techniques do not reflect objects or entities in any everyday sense.

While these modern programming techniques are intended to allow modeling of truer and more natural representations of entities, the schema and subsequent subschema imposed by raster and vector models changes the GIS user's (and indeed the GIS researcher's) view of reality on an ontological level. Thus, for example, a forester who is an experienced user of a GIS for analysis such as ArcInfo or ArcView based on a "traditional" vector geographic model will begin to think of forest stands in an analytical context more as "polygons" than as complex areal entities with often ill-defined boundaries. Our perception and, indeed, our understanding of cognitive entities in a normal sense (e.g., lakes, roads, counties, etc.) is in a very real sense filtered by and interpreted through the ontology of points, lines, polygons, and pixels used by current spatial data models.

So why is the GIS field stuck in a primitive and artificial ontology of points, lines, polygons, and pixels in computer representation? Why does the data model used for computer representation drive the user view and the kinds of analyses that can be performed, instead of the other way around? The problems can be seen as threefold: First, changing a long-used conceptual paradigm is difficult. Second, points, lines, polygons, and pixels are convenient for dividing space to fit into the discrete units of computer memory, providing sharp (albeit frequently artificial) boundaries. Third, learning how to separate the logical-level data model from the implementation-level data structure as computer programs is, again, very difficult. Understanding the differences between these levels and being able to translate a model from one level to the next is not an everyday distinction, and requires training and practice frequently not emphasized in computing courses or curricula. Certainly, the long-used names for the basic representation types—vector and raster—do not help.

So the inertia—and indeed, the comfort—of familiar forms as data models was reinforced as these same familiar forms lent themselves to computer implementation in a fairly direct way. Thus, conceptual- and implementation-level views have become intermingled. By not differentiating levels in a database design and the disconnect of that design from conceptualization of everyday reality, those working directly with GIS technology develop a sort of recursive ontology (forest stands *become* polygons). As already mentioned, it is the researchers who have worked primarily in the areas of cognitive psychology and geographic cognition, as well as philosophy, who have remained outside of the recursive ontology trap in applying their perspective to geographic representation in GIS.

Drawing upon philosophy, Couclelis (1993) linked the field-based versus object-based dichotomy of geographic database models to the cognitive distinction of location- and object- (or entity-) based views of the world. From a database design standpoint, as discussed in Chapter 11, renaming these basic model types also serves to emphasize the focus on the cognitive level. Couclelis derived her observation from the philosophy of Hooker (1973b), who rectified the notable differences in various mathematical and logical structures specifically with regard to quantum theory.

Hooker analyzed the seemingly contradictory components of quantum theory: field theory and particle theory. He concluded that the mathematical structure of each of these components reveals two "great and highly disparate ontological structures" (Hooker 1973a, p. xiv), and furthermore, that the cause of the interpretational difficulties encountered with quantum theory in the scientific community is the existence of this duality in the ontologies and, correspondingly, in the mathematical language used to describe each of these dual aspects.

In Hooker's analysis of field theory and particle theory, he distills out three distinct components in the structure of these (and indeed any) physical theory: the ontological type, the conceptual structure, and the mathematical structure. These correspond directly to the top 3 levels used in the database design scheme previously described in Chapter 11. In a lengthy exploration of the structures contained within quantum theory, he managed to rectify the relativistic and nonrelativistic aspects as two dual and complementary views of the same phenomenon, corresponding to the discrete versus continuous views of space–time, respectively. As Couclelis stated, these two differing ontologies hold, that (1993, 1999):

- Entities exist in time and space that have attributes and relationships.
- Locations in space–time are the entities, and these have attributes and relationships.

In the first ontology, phenomena exist *in* continuous space and time. In the second, space is divided into discrete elements that we interpret as things. The first is an ontology of objects, and the second is an ontology of fields. Both also directly correspond to the object- and location-based views used in GIS data models.

THE SPACE–TIME TYPOLOGY COMES FULL CIRCLE

Seen from this perspective (and the perspective used throughout this book),[1] the conceptual paradigm change required in geographic database representation is not one of adopting something that is entirely new—something foreign or unfamiliar. As we have already seen, there is a rich literature stretching back to the ancient Greeks to fall back upon. The only change that is really needed regarding the basic conceptual paradigm we use is to put more emphasis on our own natural, everyday, intuitive notions of space and time, and *revert*—in a geographic database–modeling sense—back to the pure notions of what Hooker (1973b) called "atomic" and "plenum" ontologies.

Relating back to the philosophical discussion in Chapter 2, there is a fundamental theoretical framework that can serve as a robust basis for moving forward: the notion of the discrete versus the continuous view. These can be briefly defined as follows:

• In the "discrete" (or entity-based) view, distinct entities, such as a lake, a road, or a parcel of land, are the basis of the representation. Their spatial and temporal extents are denoted as attributes attached to these entities.

• In the "continuous" (or field-based) view, the basis of the representation is space and/or time. Individual objects are denoted as attributes attached to a given location in space–time. Using land ownership information, for example, the particular parcel number would be an attribute of the entire space it occupies, with locations denoted in some continuous coordinate field.

For both views, there may attributes that are either absolute (e.g., a lake may have associated with it measured values of specific pollutants, etc.) or relative in nature (e.g., entities adjacent to the lake), or both. In addition, individual entities or locations—as the fundamental element of these views, respectively—may have various interrelationships stored as attributes. These may include is-a and part-of relationships for the object-based view that in turn form conceptual taxonomic and partonomic hierarchies, respectively. For the field-based view, part-of relationships would form a scale hierarchy. Given these characteristics, object-oriented data modeling techniques seem particularly well-suited for specific implementations of this representational framework.

Moreover, just as the discrete view can be applied to entities in space (static entities) or in space–time (dynamic entities), it can also be applied to events, as entities that have a temporal but perhaps no spatial extent. An example of such a purely temporal event would be a bankruptcy or an election. An example of an event that occurs in space and time would include an earthquake or a storm. Whether the temporal (or spatial) extent of any object is a point or some interval is dependent on the temporal (or spatial) scale used to record the data.

Because these two views are duals of each other, the same spatio-temporal phenomenon can be described as either discrete or continuous. From a cognitive standpoint, as described in Part I, which of these views is preferred depends on a number of factors, including the amount of total knowledge in a given domain and the particular task at hand. Therefore, computer data models used by humans should provide the choice between the two views, as dual models within the database design.

THE DISCRETE VIEW

In this view, a spatial value, or set of values, is associated with each spatial entity, which describes its spatial extent. Similarly, a temporal value, or set of values, usually called the lifespan of the entity, is associated with each temporal entity that describes its temporal extent. A set of associated specific operations is also defined for these objects, some of which are associated with its spatiotemporal configuration type, such as point, area, instant, or time interval.

In a discrete view, spatiotemporal entities are perceived as linked via spatiotemporal relationships. For instance, roads go through towns, two cities are 50 km from each other, one harvest yield in a given agricultural district succeeds a previous one. The topological and metric relationships, now well known thanks to 30 years of experience with GIS, are the spatial relationships needed for database operations. Likewise, a robust system of "topological" temporal relationships for time intervals was defined by Allen (1983). Temporal metrics are a simplified case of spatial metrics (distance is the only type of temporal metric, because time is normally assumed to be one dimensional). In order to support these applications, data models should provide the concept of spatiotemporal relationships as operators.

Data models in this view can be either simple, incorporating the basic spatiotemporal types: point, line and area, instant, and time interval. More sophisticated models may include the following:

• Support for an extensible, derived hierarchy of various spatiotemporal entity categories (e.g., city, county, state). These derived categories can incorporate complex types made up of a heterogeneous set of configurational types, such as lines and areas.

• Support for structural constructs—object class, attribute and even relationship—to have a spatial extent/temporal lifespan, thus achieving orthogonality between the structure and the space–time dimensions.

THE CONTINUOUS VIEW

In this view, each continuous space–time extent is characterized by the domain and range of its defining function. Often, the domain is the whole spatial and temporal extent described by the spatiotemporal database. It can also be any contiguous subarea and/or time interval selected from within the database. The range is the entire set of values for some measured attribute (e.g., temperature, population density) associated with the given space–time domain. There can be any number of such attributes and corresponding attribute ranges. A set of associated functions is defined for a given attribute and may be specific to the nature of that attribute. Spatial and temporal relations are valid as operators, as in the discrete view.

Similar to models in the discrete view, data models can be simple, incorporating a set of attributes that can be treated conceptually as flat "layers." More sophisticated models may include the following:

• Support for a multiscale spatial and/or temporal hierarchy. These derived hierarchies can represent regular subdivisions (e.g., area quadtrees, octrees, etc.)

• Support for an irregular subdivision of space and/or time that is dictated by a particular application context (e.g., population density or elevation, implemented via TINs) or distribution of specific clusters of attribute values (e.g., R-trees).

• Support for structural constructs in which entities have identity as sets of space–time locations.

INTEGRATED IMPLEMENTATIONS

Most spatial applications need both discrete and continuous views of space and time: Some forms of information are "naturally" perceived as spatial entities, while others are perceived as continuous fields. For instance, weather forecast programs use discrete areas for counties, discrete points for cities, and continuous spatial fields for temperature and rainfall. And indeed, most current commercially available GIS software provides this capability. In the same way, temporal applications need both discrete and continuous views: Some temporal entities are "naturally" perceived as discrete and separate "events" (e.g., storms, volcanic eruptions, elections, etc.). Others are perceived as time functions, such as temperature and precipitation.

Moreover, discrete and continuous views for spatiotemporal databases need to be functionally interrelated, supporting a common language and a common interface, without requiring transformation of one view into the other. Many current GIS do not support such a common interface, yet such an integration would allow combinations of spatial and temporal relationships, as operators defined within a unified query language, to form complex queries that can operate on discrete and continuous views simultaneously. For example; find all houses built after the last flood within 100 meters of the river, and on land less than 500 meters' elevation.

GEOGRAPHIC DATA MODELS BASED ON COGNITIVE STRUCTURE

Increasing attention has been paid by researchers to ontological and high-level conceptual issues relating to geographic data modeling in the digital context. This is evidenced by the series of COSIT (International Conference on Spatial Information Theory) meetings, held biannually since 1993. As important as these efforts are, they also reveal a disconnection in the research on geographic representation. Research in the realm of high-level abstraction, which has grown directly out of the spatial cognition and philosophy traditions, has so far remained in the theoretical realm. In contrast, efforts at implementation remain largely focused on technical issues and are largely stuck in a world view of points, lines, polygons, and pixels. Very little has been done so far to extend a more natural, cognitive world view into a usable (i.e., implementable) framework for space–time database representation. This evidence of the third part of the problem mentioned earlier—the *connection* between theory and practice—is still lacking.

Raper and Livingstone (1995) recognized this and offered an example of specific representation for observational geomorphologic data based on a high-order, natural cognitive world view. Recently, Mennis, Peuquet, and Qian (2000) offered the first general-purpose, implementable framework in what is called the pyramid model. This model represents an extension of the triad model, previously developed by Peuquet (1994), and represents a fundamentally new approach to geographic database representation in three ways: (1) It is based on cognitive principles of how humans store everyday knowledge, and it carries that principle through in *all* representational levels; (2) it provides the means to create multiple interpretations of temporal and multiscale geographic observational data (i.e., it allows multiple world views); (3) it allows knowledge to be added through declarative expert knowledge (e.g., prompted by interactive visual analysis) as well as the application of data mining and other statistical techniques (i.e., it allows additional knowledge to be acquired and integrated into the preexisting structure). In addition, it has been implemented and demonstrated on real-world data.

As shown in Figure 12.1, the pyramid model incorporates two distinct yet interrelated components to represent data and knowledge. The data component provides the "raw material" from which objects are constructed and corresponds to the *raw image* of perception as described by Marr (1982). The knowledge component represents derived knowledge interpreted from the observational data (the recognition of visually displayed objects and their classification into categories) and a rule base for deriving objects and their categorical membership from the observational data. (This also corresponds to Tulving's 1972 episodic and semantic components of memory, as previously discussed in Chapter 4.) As such, the data and knowledge components represent levels of abstraction and a "cycle" of knowledge development (in conjunction with visualization) analogous to how humans use sensory perception and cognitive representations to learn about the world (cf. Figure 9.1). The

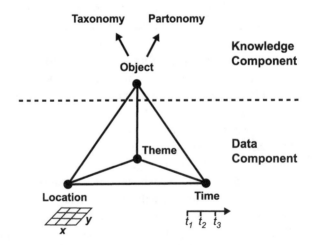

FIGURE 12.1. Overview of pyramid model data and knowledge components. From Mennis, Peuquet, and Qian (2000). Reprinted by permission of Taylor & Francis Ltd. *http://www.tandf.co.uk*.

pyramid model can therefore be thought of as both a knowledge development *process* and a representational *structure*.

The location-, time-, and object-based views shown in Figure 12.1 correspond to the *where*, *when*, and *what* perspectives of cognition, respectively, as shown in Figures 8.3 and 8.4. A "theme" is defined within this model as an attribute-based view of observational data and corresponds to the attribute layers used in current GIS. The knowledge component concerns the representation of derived geographical objects, their classification, and their interrelationships. Here, objects include both object classes and specific instances of objects identified from the observational data. It is important also to note that an *object* within the pyramid model represents a geographic entity having a unique and cohesive identity. It is related to a specific combination of observational data stored in the location, time, and theme perspectives, but it is stored separately. In order to link an object instance with its observed properties stored within the data space, an object is represented formally within the pyramid model by an *object template*, which defines the characteristics of any object. More specifically, such a template would specify basic space–time attributes (e.g., size, shape, duration) and behavioral rules (e.g., conditions conducive for occurrence, spatial and temporal associations) for that object, as well as alternative names, taxonomic/partonomic relationships, and space–time location. The information fields given in an object template will vary to suit the particular type of object being represented and are not intended to describe the object exhaustively. Rather, the object template provides, in part, a guide for finding the complete life history within the data component (when desired) and represents enough information to define its unique identity.

Object classes are represented in a similar manner, but the attributes of a class record range over values that define the *criteria* for membership in the given object class rather than values that record the observed behavior of a specific object. The locations, times, and themes described, with value ranges, in an object class template serve as a rule base for discovering specific instances of that object class from the data stored in the data component.

At the implementation level, the conceptual object templates and object class templates are represented as utilizing *frames*. As already mentioned in Chapter 4, the frame concept, originally proposed by Minsky (1975), has become a key means for knowledge representation in computer vision and artificial intelligence (AI) research as a formalized knowledge structure. A frame is a complex data structure designed to portray a particular type of object or process: it is a distinct unit of information. Frames may be permanent, or they may be created and discarded during the process of problem solving.

Each frame consists of a set of *slots*, each slot having a *name* and associated *values*. The values within a given slot can consist of relevant attributes (color, spatial location, etc.), as well as components and relationships to other objects or processes. Thus, a simple frame for the object class "kitchen" may look like the following:

Object name:	kitchen
Part-of:	house
Is-a:	room
Min. size:	81 sq. ft.
Contains:	stove, refrigerator, sink

As such, it can be thought of as a structure that emphasizes the nodes in a relational network, with linkages implemented through part-of, is-a, and other linkage forms.

In an application demonstration of the pyramid structure within the Apoala prototype system (Mennis, Peuquet, and Qian 2000), storms were retrospectively identified and tracked utilizing weather record data for a portion of the northeastern United States. These data were used to populate the data component. An example frame for an object class template is illustrated in Figure 12.2. This particular frame gives information representing a "typical"

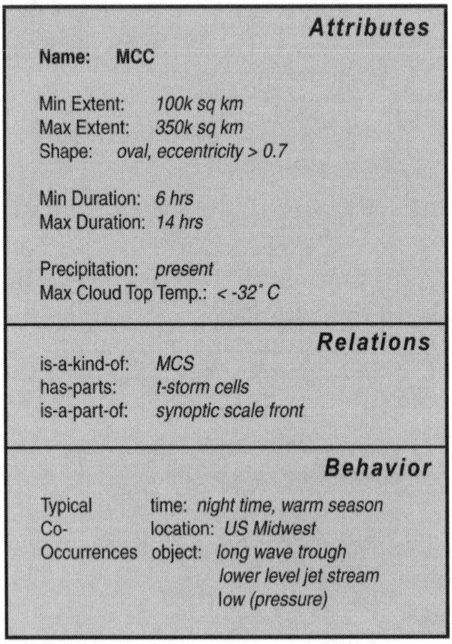

FIGURE 12.2. Frame representation for an object-class template of a mesoscale convective complex. From Mennis, Peuquet, and Qian (2000). Reprinted by permission of Taylor & Francis Ltd. *http://www.tandf.co.uk*.

mesoscale convective complex—a particular type of storm. Figure 12.3 shows the frame representation for a particular observed storm, noted as mesoscale convective complex (MCC) 9.

As shown in Figures 12.2 and 12.3, there are a number of critical differences between a frame representation and other types of other implementation-level data models. Unlike the relational model, slots can (and often do) have multiple values. Unlike most implementation-level models, they can be ranges or relative values (e.g., greater than, less than). It is also possible to have slots that are completely empty, or that use default values. All of these values directly correspond to how people characterize objects—with stored information that is often inexact and incomplete. Another important difference is that frames can contain not only values and value ranges as such (e.g., "blue," "32," or "between June and October"), but also behavioral rules or software procedure names. Behavioral rules (if–then–else) can designate default or "typical" behaviors. Procedure names stored within a frame are explicitly associated with the particular object or object class, and can in this way be used to denote behaviors as well. As procedure names within a particular software system, they can also be invoked directly by referencing the object.

Attributes	
Name:	MCC [9]
Location:	Pittsburgh to State College, PA
Avg Extent:	135k sq km
Max Extent	197k sq km
Avg. Shape:	oval, avg eccentricity = 0.84
Time:	August 29, 1998, 1 pm - 12:30 am
Duration:	11.5 hrs
Avg. Precipitation:	2 mm/hr
Min Cloud Top Temp.:	-64° C
Relations	
is-a-kind-of:	MCC
has-parts:	t-storm cells [1, 2, 3, ...n]
is-a-part-of:	synoptic scale front [4]
Behavior	
Co-Occurrences	time: afternoon, night, warm season
	location: Pennsylvania
	objects: lower level jet stream
	low (pressure)

FIGURE 12.3. Frame representation for an object template of an observed mesoscale convective complex. From Mennis, Peuquet, and Qian (2000). Reprinted by permission of Taylor & Francis Ltd. *http://www.tandf.co.uk*.

Frames can contain not only default rules and values, but specific named slots can also be used to note necessary or boundary conditions for classes. In addition, relationships expressed within the frame structure are not limited to is-a, kind-of taxonomic relationships, as shown earlier in the initial example and conventionally described in textbooks. They can contain other forms of taxonomic, partonomic, spatial, and temporal relationships (becomes, before, next to, etc.)—whatever forms of relationships are supported within the software.

From the experience with the Apoala prototype of the pyramid model, frames proved, at least in this case, to be a very flexible and "natural" means for implementation of a geographic data model based on cognitive principles. The primary restriction rested in the practical limitation of what procedures and functions could be implemented in the supporting software, because frames allow the inclusion of function names, as well as rules, as means to denote behaviors and relationships.

Mechanisms for utilizing stored rules have long been studied and implemented in AI research, particularly in expert systems. Software packages known as "inference engines," which perform the needed logical operations on rule chains and are designed to be independent of any particular knowledge domain, have been available for decades. The combined use of frames for a knowledge-based representation with an inference engine is a powerful mechanism for reasoning and learning in a geographic context. This was first demonstrated in an early application of AI techniques to the geographic context (Smith, Peuquet, Menon, and Agarwal 1987). Definitions for objects and object types can be modified, and definitions for new objects discovered via automated inference can be added to the knowledge base utilizing an inference engine.

While some relationships such as "becomes," "during," or "contained within," are straightforward to implement algorithmically, others that are fuzzy—being context- or scale-dependent in nature (e.g., "near")—are more problematic. Although solutions for a few specific relationships have been proposed (Jungert 1992; Peuquet 1988b, 1992; Peuquet and Xiang 1987), this area needs more research.

THE BIGGER PICTURE

Some characteristics of the storage of conceptual knowledge by people also coincide with a number of requirements for large geographic databases, including the following:

1. *Hierarchical organization*—The complete enumeration of all observations and possible relationships among these observations for all but relatively small subsets is both physically impossible and unnecessary.
2. *Abstraction*—As a direct result of item 1, the data should be structured in a way that implies much more information than is explicitly stored.
3. *Associative retrieval*—If the collection of data is to be both large and efficient, there must also be a way to retrieve specific information through association with related information.

Being able to structure data in such a way requires higher level knowledge concerning the nature of the phenomena represented and how component elements interact. It also requires techniques for representing and using that knowledge in a consistent and unified manner.

It must be remembered that the example of the pyramid format in the previous section is only that—an example. As discussed in Part I, cognitive representations of the world are highly dynamic and intertwined. Because some representations are intrinsically more useful for some tasks than for others, humans, as well as computers, need a multirepresentational model with the ability to switch among representations, as needed, for the task at hand. This also means that all three views in the data component may not be needed for a given application. The object-based view in the knowledge component, however, is always essential.

Cognitively, the level of abstraction accumulated in stored information within the knowledge component increases with increasing knowledge and experience of the environment portrayed. By extension, we should model the same process in the computer if we are to (1) utilize human cognition and development as both a motivation and a guide for designing and building more powerful and flexible computer data models—models that can effectively store interrelated observational information, learn from that information, and store the resultant derived knowledge; and (2) have truly user-friendly database/software systems that can adapt to users' levels of expertise and communicate in a natural and intuitive manner.

Implied in the first requirement, in traditional data model terminology, such a representational approach must be robust. What this means from a cognitive perspective is that the approach must be able to handle exceptions in observational data and marginal cases, as well as cases that conform to categorical rules. There must also be a means of sifting through observational data to decide what is "important" and needs to be integrated into the stored information by causing higher level abstractions or rules to be refined, re-membered as an observational occurrence, or perhaps discarded as redundant information.

The second requirement is not as much a direct requirement for the form of the stored information itself, but rather how it is compatible with lan-guage and graphical methods of presenting and receiving information from and to users with varying levels of knowledge and experience. This is analo-gous to internal versus external representation in cognition. This means that a space–time query language is needed, as well as a visual interaction proto-col. The issues then become, What form should these external representa-tions take? (e.g., algebraic or natural language), and, How are they mapped to the internal representation? Given that such a mapping presupposes knowl-edge of the representation being mapped into, the discussion of language and visual issues in subsequent chapters focus on this representation. There is also recognition here that because internal and external representations are closely interrelated, once some details of the representational approach for in-ternal representation are worked out, the representational approach for com-munication will naturally flow from this.

The issue of whether the internal representation changes with the level of knowledge raises the question of whether the computer representation changes to match either the user's level of knowledge or the level of knowl-edge contained within the database. The answer to the first would seem to be, no—no more than an adult would need to have a child's representation of the world in order to communicate with children, or an expert have a novice's representation of a given domain in order to communicate with a novice. All that is needed is a mutually understood system of linguistic, graphical, and/or other form of symbols. The Piagetian perspective—taken in the general sense of a representational progression—does, however, offer some insight as to how a general representational scheme, or model, would be progres-sively "populated" by information with increasing amounts of observational data and higher level knowledge.

As shown with the example of the pyramid model, we cannot ignore im-plementation issues. Whatever theoretical framework is derived for computer representation of geographic data must be carried through to the implementa-tion for database design level. This must also translate into models that are natural for humans to use, and that run at the physical, machine level with an acceptable degree of efficiency, which simply means that (1) queries and other tasks utilizing the physically stored database are accomplished quickly enough for an interactive environment, and (2) the required amount of physical storage overhead (e.g., pointers) does not overwhelm the amount of actual information stored.

SUMMARY

GIScience in general, and geographic representation specifically, currently seem poised to make the leap away from imitating previous, well-known methods and models. This is necessary in developing the means to accomplish tasks that were not possible, or were very limited, given previous methods and models. At the same time, the additional capabilities offered by the medium of computers are in part fueling developments. This is the push–pull phenomenon that I mentioned at the beginning of this chapter. The end result—I hope—is that we can take fuller advantage of computing technology and its unique capabilities for learning from the vast amounts of geographic data in digital form. Whatever the new framework, it will still be built on previously derived ideas and techniques but will begin at a much higher conceptual level that can be utilized in deriving a firm theoretical basis for going forward.

In Chapter 13, I examine approaches for interacting with shared information. In parallel with direct and indirect experience with the real world, this includes "seeing" through visualization of realistic scenes, particularly what is now called virtual reality, and through language and symbolized graphics.

NOTE

1. Material in this and the subsequent three sections was previously published in Peuquet (2001).

INTERACTING
WITH DATABASES

THE TRADITIONAL QUERY LANGUAGE APPROACH

Just as language is an essential means of communication between people, it is also an essential means of communication between the human user and a computer system. But in the man–machine context, this is usually much more one-way communication and takes the form of a command language (i.e., issuing commands or requesting that the machine to do something).

These commands can potentially take various forms. One class of user command constitutes a basic retrieval of information contained within the database. Historically, this is also the origin of the term "query language"—the user is querying the machine for information, as in the following examples:

1. List the records for all houses on East Main Street that contain a commercial enterprise.
2. Display all locations that have less than 10% slope.
3. What is this? (using the mouse to point to a location on a map display)
4. Display the location of the fire station nearest to the university.
5. List all state-owned parcels of land east of the city and within 2 miles.
6. Display areas that experienced a landslide within 1 week of a major storm event.

These examples all involve retrieval constraints that specify a value, or range of values, and perhaps relationships among multiple geographic phenomena. All imply a particular ontology and, at least in a general way, a particular kind of conceptual representation. Some queries (1, 2, 3) refer to attributes of particular entities or particular locations. Others refer to spatial (4, 5) and/or temporal (6) relationships. Although temporal relationships cannot be explicitly handled in current GIS, simply because current geographic data models do not support the representation of the temporal dimension, these, as in the case of some data models, can be handled implicitly to some extent. Thus, instead of being able to say "before" or "during," dates and/or times can be compared on the basis of standard mathematical relationships (e.g., "less than" or "equal to").

Most current GIS do not have a query "language" as such, but rather rely on pull-down menus and point-and-click interaction. While relatively easy to learn, this type of computer interaction is not as flexible or powerful as a language, with a syntax and vocabulary that can be combined in an effectively infinite number of combinations and strung together to make complex statements. Relational DBMS do use a language interface, however, as the primary mode of interface. A feature of this language interface is that it has a cohesive and theoretical basis, derived from principles of mathematics and logic.

The language interface for relational DBMS was defined at the inception of the relational model itself, in the very early 1970s, at a time when many GIS were using large collections of separate and often uncoordinated commands that in reality were function calls (or little more than direct function calls) made to individual software modules. The standard language for relational database systems is standard query language (SQL), which is supported by essentially every relational DBMS in the commercial marketplace. Although this language continues to evolve, its principal component is known as the *relational algebra*, which involves a set of operators that take relations (tables) as their operands and return a new relation as the result. The original algebra, as defined by Codd (1970, 1972), consisted of eight operators, including the four traditional set operators (*union, intersection, difference*, and *Cartesian product*) and what he called the special relational operators (*select, project, join*, and *divide*). Loosely defined, select extracts all tuples (rows) from a relation that satisfy the given constraint, and returns the result as a new relation. Similarly, *project* extracts all attributes (columns). *Join* forms a new relation by combining two relations on the basis of a common attribute. Figure 13.1 shows examples of these operations against the given input relations. It must be noted that the operations as shown in Figure 13.1 are generalized for the purpose of explanation and are not meant to reflect the exact syntax of SQL.

Besides Codd's original eight operators, SQL has been considerably extended to include row- and columnwise calculations or comparisons. Group (and ungroup) operators have also been added that create relations whose at-

QUADRANT

QUAD#	SLOPE	ASPECT
1	8	N-NW
2	27	N
3	5	NE
4	10	S
5	18	SW
6	7	NW

EVENT HISTORY

DATE	QUAD#	EVENT
Aug. 1987	1	wildfire
Aug. 1987	2	wildfire
Mar. 1988	6	control burn
May. 1988	2	replant
Sept. 1988	3	cut brush

Select   Result:
QUADRANT where QUAD# < 3

QUAD#	SLOPE	ASPECT
1	8	N-NW
2	27	N

Project: Result:
QUADRANT over QUAD#, SLOPE

QUAD#	SLOPE
1	8
2	27
3	5
4	10
5	18
6	7

Join:
QUADRANT and EVENT HISTORY over QUAD#

Result:

QUAD#	SLOPE	ASPECT	DATE	EVENT
1	8	N-NW	Aug. 1987	wildfire
2	27	N	Aug. 1987	wildfire
2	27	N	May.1988	replant
3	5	NE	Sept. 1988	cut brush
6	7	NW	Mar. 1988	control burn

FIGURE 13.1. Examples of Select, Project, and Join operations. Adapted from Date (2000).

tributes themselves are relations, thus allowing hierarchical structures to be generated. Additional operators allow data maintenance (e.g., create, update) and security restrictions, among other types of functions the user may want request the DBMS to perform. In other words, SQL is not simply a language for data retrieval, as the commonly used term "query language" implies. Rather, SQL, like natural language, has become a high-level, symbolic representation of information (Date 2000).

SQL has also been extended to include time. A significant amount of research in temporal reasoning, using the relational model following Allen's initial theoretical insights, investigated what an elemental set of temporal relational operators would include (Allen 1983; Al-Taha 1992; Freksa 1992; Montanari and Pernici 1993). Work on a standard temporal database query language resulted in TSQL2 (Snodgrass 1995). Within the DBMS community, efforts to establish the concepts developed within the context of this language as an integral part of the international SQL standard specification.

While SQL is based on something known as the "relational algebra," there is an alternative, theoretically equivalent form called the "relational calculus." The key difference between the two, as described by Date (2000), is that the relational algebra is *prescriptive*, whereas the relational calculus is *descriptive*. In other words, in the relational algebra, the procedure needed is expressed (thus, the relational algebra is also called *procedural*), whereas in the relational calculus, the desired result is described. Thus, in the style of the relational algebra, the following is a sequence of two loosely stated commands for "List the records for all houses on East Main Street that contain a commercial enterprise":

Join structure__inventory and land__use over street__address, then
Select from result where (struct = house) and (land__use = commercial)

The equivalent (again very loosely) expressed in the style of the relational calculus would be something like the following:

Get street__address for properties such that there exists a commercial enterprise in a house.

Relational calculus is based on a branch of mathematical logic called the "predicate calculus." The relational algebra itself is based on algebra, as a system of operators, and utilizes principles of combination such as associativity and commutativity. The equivalence of these two approaches was proven by Codd (1972), who demonstrated that any arbitrary expression in the calculus could be converted to a semantically equivalent expression or set of expressions in the relational algebra.

A query language called QUEL, based on the relational calculus, did compete with SQL (originally called SEQUEL) early on; the two languages were used in two early implementations of the relational model (Chamberlin 1976; Chamberlin and Boyce 1974; Held, Stonebraker, and Wong 1975; Stonebraker, Wong, Kreps, and Held 1976). The question to be raised, then, is why the standard database language is based on the relational algebra and not on the relational calculus. Does one have advantages over the other, or was this the result of some other circumstance, or even chance?

There is no clear answer. Even with a language based on the relational algebra being the long-accepted standard, the standard introductory texts on DBMS still go into some detail on the equivalence of these two approaches. However, there are a couple of speculations as to why the relational algebra came to dominate on the basis of the different nature of these two approaches. First, the procedural approach of the relational algebra is compatible with standard practice in business and science, where documenting the procedure used to achieve a given result is not only expected but also often required. Second, a more direct correspondence between the relational algebra and standard programming languages, which, historically, are also procedural languages, makes the relational algebra easier to implement.

While both of these issues could have impacted earlier development of interface languages, they are now greatly ameliorated by virtue of modern programming environments. Procedural programming languages are still the norm, but sophisticated modern programming tools greatly aid implementation tasks. Automatic generation of the procedural code in some user-readable form from nonprocedural language statements, then, also becomes an easier feature to implement. Nevertheless, it is interesting to note that the procedural form of programming language that has long dominated (e.g., Fortran, Pascal, C, and now Java) is commonly called third generation, while nonprocedural programming languages (e.g., Lisp) are called fourth generation. The implication is that nonprocedural programming languages are an advancement. The reason for this is that having the programmer (i.e., user of the language) describe the result and not be concerned with the details of how a particular task is done, is a much more natural and user-friendly form of expression. Indeed, the statements listed at the beginning of this section are not only quite normal-sounding but they also conform (very generally speaking) to the relational calculus form of expression. The only exception is item 3, which is a question rather than a declarative statement.

It must also be noted that in order to meet the needs of DBMS users, SQL has not adhered strictly to the relational algebra. As described by Date (2000), it has evolved somewhat into a hybrid, containing elements of *both* the relational algebra and the relational calculus. The overriding reason comes back to representation and the need for expressive power beyond flat tables and scalar variables: range variables (i.e., the ability to designate a *range* of values defined within the database itself) and the existential quantifier (there exists at least one value) of the relational calculus. As already mentioned, SQL has also been extended to allow relations that themselves contain relations, allowing hierarchies. In this way, metadata can also be referenced, if they are included within the hierarchy. These, in essence, extend the underlying data model.

Because most GIS do not have an interface "language" as such, GIS have incorporated standard relational DBMS in an effort to handle the nonspatial components of the database in a more effective fashion, as well as to provide a user interface language for spatial data manipulation. The adoption and adaptation of SQL would be an attractive prospect because of not only the obvious linkage with standard DBMS but also the advantages of the relational approach, particularly its conceptual simplicity and extensibility. Standard SQL, however, has severe limitations in that neither SQL nor relational DBMS are designed to handle the special characteristics and relationships inherent in geographic data. While there have been a number of attempts to extend SQL to include spatial operators (e.g., distance and direction), these have met with limited success (Egenhofer 1991; Herring et al. 1988; Ingram and Phillips 1987; Ooi et al. 1989).

A number of researchers have argued that extending SQL to accommodate geographic data is inappropriate as a general approach. The reasons they present also involve issues relating to the underlying representation. Some of the criticisms made in the early 1990s are now less valid given the current level of extensions in standard SQL, but there are still a number of problems. First, SQL has no way of dealing with graphics (Schenkelaars 1994). This includes the "What is this?" kind of query. Similarly, there is no mechanism for specifying how output should be presented, given that in relational DBMS, the results of all queries are presented in tabular form. Nevertheless, a number of solutions have been proposed over the past two decades (Egenhofer 1991; Frank 1982; Roussopoulos 1984). Second, while value ranges can be specified, there is no way to deal with inexactness (e.g., "near" as an operator or value; "all gas stations near the airport") (Burrough and Frank 1996). Third, SQL cannot identify objects, because it is value- and not entity-based (Egenhofer 1992). Given that the relational model is based on a tabular (i.e., row and column) view of the data containing observational attribute values, the notion of storing identity criteria is not part of that model (Egenhofer 1992). This becomes a critically needed feature in the case of temporal databases, in which attributes for the same entity can be expected to change over time, yet the entity remains the same. Related to this, nondeterministic behaviors over space–time cannot be expressed (e.g., "Towns are *usually* smaller than cities") (Mark, Frank, and Efenhofer 1989).

A notable exception to GIS not incorporating a language, aside from the adoption and adaptation of SQL, is MAP (the Map Analysis Package). This

package, developed by Tomlin in the early 1980s to implement what he originally termed the "map algebra" (Tomlin 1983, 1990), assumes a grid data
model. As implied by the name, the map algebra is also procedural. It manipu-
lates single-factor map layers through the use of spatial operators and generates new map layers as the result. Originally, this explicitly geographic user
interface language incorporated arithmetic and set-oriented operators that
computed a new value for every location within the given rectangular area,
based on the corresponding locations in the input layers. An example MAP algebra expression would loosely be something like

Add map1 to map2 for newmap

Some uniquely spatial operators, including distance, are also included. These
operate over an area of the input layer, or layers, to calculate the value for
each location in the resultant layer. The original language was subsequently
extended to distinguish three operator classes, each with arithmetic and set
operators that function on single locations, immediate neighborhoods, or
larger, input-defined zones.

Tomlin's MAP algebra has problems equivalent to those of SQL. While it
uses gridded map data as input and generates the same as output, including a
separate command for displaying that output, there is no way of dealing with
any other form of data, and also no way to deal with inexactness, or to identify
objects independent of explicitly stored variables denoting that identity.

COMPUTER GRAPHICS AND VISUALIZATION

How the use of maps and other forms of graphic display, as well as language,
are an important form of indirect experience has already been discussed in
Chapter 7. Recently, a growing but still nascent research effort has included
the areas of geographic visualization and virtual reality for geographic exploration and analysis. Both scientific visualization, or ViSC, and virtual reality
rely on sophisticated modern computer graphics and fast processors. Both
also attempt to integrate computer technology with human vision and cognition to best advantage. The aim of these efforts so far is to explore the possibilities of these various graphically oriented interface paradigms.

Geographic visualization, or geovisualization, derives from cartography,
with influence from ViSC. ViSC, as such, had its beginning in a widely cited
report by McCormick, DeFanti, and Brown (1987). On the one hand, ViSC focuses on the use of 3-D and dynamics in displays for the purpose of understanding physical objects by allowing visually realistic renderings of them to

be inspected in ways not possible with the real object. This includes zooming, panning, and rotating; making the opaque transparent; and interactively slicing through an object. ViSC has been used in applications such as aircraft design, weather analysis, and examining the human body (MacEachren, Edsall, et al. 1999). It also focuses on data exploration of high-dimensional data, through the use of highly abstract diagrammatic devices. In essence, this creates visual metaphors for nonvisual phenomena.

Work so far in geographic visualization has stayed close to its cartographic roots. Renderings tend to be either symbolized maps or realistic synthesized scenes but are usually dynamic and 3-D, as well as interactively manipulable. Animation is an obvious means to introduce the temporal dimension in maps and scenes. As a result, animated maps have generated a lot of interest within the context of geographic visualization, but as Peterson related, "What happens between each frame is more important than what exists on each frame" (1995, p. 48). This reveals one of the current limitations of representing space–time processes through visual means—but again, the limitation is in the underlying data representation. Dynamic maps and scenes utilize the snapshot data model, creating frame-by-frame "map movies." Nevertheless, the ability to control the dynamics of the display interactively provides a powerful tool for gaining insights into geographic processes through visual exploration to find patterns and relationships (DiBiase, MacEachren, Krygier, and Reeves 1992).

In connection with this, researchers in geographic visualization have endeavored to provide a formalized framework from which a set of rules for governing computer-generated displays can be formulated. The motivation is to provide the most appropriate and effective graphical renderings for a given use and type of user. This in part stems from current research on semiotics as applied to cartographic design, as described in Chapter 7.

Bertin's (1981) system of graphic variables (*location, size, value, grain, color, hue, orientation*, and *shape*) was the first attempt to formally define the elements of symbology that can be manipulated in order to convey information. Subsequent authors have extended his original variables to include *color saturation, pattern arrangement* (Morrison 1974), and *texture* (Caivano 1990). These have also been extended to deal with dynamic maps, even though Bertin explicitly did not consider the temporal dimension. DiBiase et al. (1992) found Bertin's original variables to be valid within the temporal context and added three explicitly temporal variables: *duration, rate of change*, and *order*. MacEachren (1995) subsequently added *frequency, display time*, and *synchronization* to this list. See MacEachren (1995) for further discussion of these variables.

As noted by Kraak (1999), cartographers have come to use the term "visualization" in its broadest sense of simply "to make visible," and in this way, it encompasses all of cartography. The implications of this are twofold and complementary. First, it widens the possibilities of investigating the computer as a new cartographic medium and the implications of new capabilities in a changing role for cartography. Second, it recognizes the rich heritage of cartography that can, and should, be brought to bear in advancing geographic visualization.

A note of caution needs to be interjected here regarding formalization. Semiotics (i.e., maps and other forms of non-photo-realistic or nonimmersive images as a symbolic system) was developed within the context of traditional cartography and graphics, where such graphics were designed by professionals to convey a specific message. The original goal of semiotics was therefore to provide a formal framework, including a set of rules for representing the intended information in a way that "gets the message across" most effectively.

But it is precisely because of how and why semiotics developed that the principles learned can be particularly useful in the current environment of user-generated maps and the "democratization" of cartography through the use of modern computing environments. Semiotic rules can at least to some extent be utilized in the design of mapping software to minimize the risk that the user will unknowingly produce a misleading map. However, it must also be remembered that large proportions of the images produced through geographic data handling software are used for data *exploration*, and not as final products for presentation. In this latter type of graphic, there is no "message." The exploration, rather than presentation role, will undoubtedly result in some semiotic principles being extended or recast for the characteristics of computer graphics.

Much more can, and should, be done to help the user produce graphics with qualities and characteristics that aid knowledge discovery, to facilitate seeing patterns and relationships that are real, and to avoid generating spurious ones. Besides applying the principles of semiotics in the design of specific types of computer-generated displays, much can be done through the application of some basic principles of cognition and learning in tailoring the type of display generated for a particular type of user and application. For example, cognitive research has shown that novices within a particular knowledge domain have a much harder time dealing with abstract representations relating to that domain than do experts; users identified as novices could be shown realistic renderings as a default, while displays for expert users might default to more abstract and symbolized displays.

But when and how should these rules be applied, so that the user's imagination can still have free reign? Should an electronic "guide" be provided instead? Which rules of semiotics should or should not apply? How should the visual representation be adapted to suit particular types of uses and data? When is one type of display better than another? How can the principles of visualization be extended to include other senses besides vision (e.g., haptic, auditory)? Much research is needed to study these issues within the specific context of computer interfaces for GIS and other forms of geographic data-handling software.[1]

VIRTUAL ENVIRONMENTS

The use of virtual reality (VR) as a form of geographic visualization is a current area of particular interest for both individual and collaborative data exploration and problem solving. This has more recently also become known as virtual environments (VE). Both VR and VE, as interchangeable terms, refer to navigable, 3-D, graphical rendering of a portion of a real or imagined environment. Because of the dominance of the term "virtual environments" in the recent literature relevant to the current discussion, this will be the term I use. There are two types of VE: *desktop VE*, which utilizes conventional computer hardware, and *immersive VE*, which utilizes specialized devices such as the CAVE or the Power Wall. The intent is to make users feel that they are physically "in" the simulated environment, instead of just looking at it, as if through the "window" of the computer screen. Although immersive VE engage multiple senses, including vision, hearing, balance, and touch, the head-mounted displays, tactile gloves, and other specialized equipment required are very expensive and currently unwieldy.

Both forms of VE have been used for a number of years with great success in landscape and urban planning (Verbree, van Maren, Jansen, and Kraak 1999), military applications (Darken, Allard, and Achille 1998), and flight simulators, where experienced reality is portrayed as faithfully as possible. A number of efforts in the geographic realm have utilized VE to render types of environment that could not be experienced in the real world or that were completely abstract (Cook 1997; Wheless, Lascara, Valle-Levinson, and Sherman 1996).

As is particularly the case for VE in military applications, rendering environments as realistically as possible is in part motivated by the expectation that such displays would minimize the "cognitive effort" required to explore and interact with the information presented in a 3-D, dynamic, simulated world (Cartwright et al. 2000; Dykes, Moore, and Wood 1999). A number of additional capabilities are very attractive in that they offer humans the potential for interacting with and learning from digital geographic data. These include overcoming the physical constraints of the real world, with barrier- and gravity-free movement. The ability also to directly control the scale and speed of the display allows the user to fly around in the environment at any desired speed and level of detail. However, research has already shown that simulated "natural" forms of movement (e.g., walking or flying) with such unnatural control can quickly result in disorientation (Fuhrmann and Kuhn 1999). A number of researchers have suggested showing conventional cartographic views along with that of VE to solve this problem (Darken and Cevik 1999; Fuhrmann and MacEachren 2001; Moore, Dykes, and Wood 1999). This provides an abstracted overview for orientation, along with the detailed view. Others have suggested ways to control the degree of detail (and thereby the amount of information) presented at a given time (Hoppe 1998; Koh and Chen 1999).

VE are seen as having great potential for providing a means of interacting with geographic data in a way that (1) is intuitive, "natural," and most appropriate for novices or people in stressful situations; and (2) allows a means of accessing a vast amount of information at a high level of detail.

DEVELOPMENTS IN HUMAN–COMPUTER INTERACTION

A large volume of literature on human–computer interaction and human-oriented interface design for general computing systems has been developed over a significant period of time. Consideration of cognition in human–computer interaction with geographic-based software, however, has been actively considered among researchers for only the past 5 or 6 years. An early prototype for exploring new directions and more intuitive, user-friendly interfaces for geospatial systems outside the VE context was *The Geographer's Desktop* (Egenhofer and Richards 1993). This utilized the metaphor of overlaying maps on a light table to replace the files and folders of standard interfaces. The idea was to allow the spatially aware environmental scientist to relate GIS operations to the more familiar process of manipulating paper maps as an integral part of the spatial analysis process. More recently, Blaser, Sester, and Egenhofer (2000) investigated the integration of sketching with speech in the early part of the problem-solving process.

This development parallels current efforts in the computer science community toward the universal usability of computer interfaces that has recently gained a considerable amount of attention (see Cartwright et al. 2000; Shneiderman 2000). As Shneiderman (2000) stated, the goal of universal computing, in part, is to accommodate a variety of users from different cultural, social, and economic contexts, with different skill levels, knowledge, and physical ability. Universal computing also focuses on how to bridge gaps between what users know and what they need to know in order to gain maximum benefit, and how, at the same time, to support a broad range of hardware, software, and network access.

Although the use of traditional maps by the visually impaired has been studied (Coulson 1991; Golledge 1991; Olson and Brewer 1997; Tatham 1991; Wiedel 1983), similar efforts need to be undertaken to ensure equity and ease of use in computer-generated displays. Golledge (2001) has noted that GIS rely exclusively on visualization as the prime mode of interface, which effectively excludes the blind or visually impaired.

Shepherd (1994) was among the first to propose the use of interfaces specifically for GIS that utilize multiple sensory modes for various types of users and situations, and suggests that research is needed in developing them. Oviatt (1996) conducted empirical research with a simulated mapping system. Her results indicated performance degradation with voice-only map interaction but performance improvement with multimodal interaction. This has obvious implications for the feasibility of hands-free mapping and navigation systems. Sharma and his colleagues (1999) developed a system based on an interactive, large-format map display that responds to voice commands and hand gestures. Blaser et al. (2000) suggested that sketching can be combined with speech for the early part of the problem-solving process. Neves et al. (1997) combined VE with more conventional displays to achieve a more natural integration of multiple sensory modes in a GIS setting. Some specific strategies beyond modality have also been suggested as potential improvements of human–computer interaction for GIS. Kraak and van Driel (1997), for example, emphasized the implementation of distinct thematic, spatial, and temporal links for navigating geographic databases via cartographic displays.

As noted by Cartwright et al. (2000), numerous efforts within the computer science community not directly related to geographic applications also have obvious relevance, particularly in the area of information visualization, including browsing large images (Plaisant, Carr, and Shneiderman 1995), and image libraries (North and Korn 1996), and linking multiple visualizations (North and Shneiderman 2000).

It is particularly interesting to note that although the use of VE seems to provide a vehicle for more "natural" and intuitive interfaces, it is likely that this approach will not supplant the need for more abstract types of display. As Mark and Freundschuh (1995) point out, the map provides a more concentrated and manipulable information representation. They cite Montello, who states:

> Maps represent environmental and geographic spaces, but are themselves instances of pictorial space! I therefore expect the psychological study of map use to draw directly on the psychology of pictorial space rather than on the psychology of environmental space. (1993, p. 315)

Thus, while realistic displays have their uses, abstract displays will remain a key element of computer interfaces for exploring and learning from digital geographic observational data.

A complicating factor in developing better interfaces is that there is a potential for GIS and related software utilizing the World Wide Web for collaborative decision making for individuals who are locationally distributed (MacEachren, Brewer, and Steiner 2001). The advantage of networked geographic software and databases is that displays (including VE), and the interaction with them, can be shared and coordinated. This allows a shared view of the problem space or geographic area being examined via visualization in real time. MacEachren et al. (2001) noted three new interface issues that have not previously been encountered because interfaces have historically been designed for use by only one individual at a time. First, how can the software switch between varying display orientations among multiple participants without the participants becoming disoriented? Second, how can differing conceptual views be traded and shared? Third, how should the software interface coordinate commands among multiple individuals?

Researchers have recognized that better interfaces can only be derived through the use of a cognitively oriented theoretical framework and intimately involve the functional linkages between cognitive and computer representations (both visual and database) (Davies and Medyckyj-Scott 1996). To this end, Hernández (1995) has proposed a theoretical framework based on that developed by Palmer (1978), which makes the role of the observer explicit. Various dichotomies of representation types, including declarative versus procedural, propositional versus pictorial, and quantitative versus qualitative, provide guidance in terms of mappings they allow to be established between various internal (cognitive) and external world views. These mappings are also dynamic, because processing operations are imposed upon the information. Hernández takes this a step further in adopting the view of

Winograd and Flores (1986), and is reminiscent of Bertin in a visual context, that the essence of computation involves the manipulation of symbols, and the attribution of the meaning of those symbols. Since human–computer interaction is, in essence, an information-processing activity, the study of this interaction must involve a holistic consideration of human cognition rather than a focus on individual actions or tasks.

BEYOND QUERIES

Realistic graphics, symbolized graphics, and language are all important forms of direct and indirect experience, each having unique aspects. These are thereby important tools for conveying information between human and computer, and are also in some sense limited in their expressive power and vary in the types of information they can best represent. In other words, there can never be a perfect single mode of communication that is both formally specified and allows for all possible complexities in the world. Like the human exploring his or her real environment, the task really requires a combination of modes working in concert. This has already been happening in an ad hoc manner with computer interfaces for geographic and nongeographic database systems.

A language-based interface was the only option in the early days of relational DBMS, and for GIS, for providing the capability to formulate a wide range of complex queries. And indeed, this by itself could better serve the nonspatial context. Because of the more difficult task of developing a language with the expressive capabilities required for the geographic context, GIS interfaces tended to consist of a collection of function calls to access the explicitly spatial capabilities, including graphic output, and a separate DBMS subsystem with its separate language. When graphic CRTs and windows-oriented computing environments became the norm in the late 1980s, interfaces for GIS quickly integrated the graphical capabilities provided by the windows environment. The emphasis was on making interfaces easier to learn; thus, point-and-click interfaces became commonplace. Nevertheless, the kinds of questions that users could ask through these interfaces did not change.

There is thus another important issue, beyond the issues discussed in this chapter so far, that relates to the paradigm used for current user interface facilities. Although there are current efforts within computer graphics and visualization to develop better tools for exploration, current interfaces for GIS are geared toward straightforward retrieval and manipulation operations—

queries. Such capabilities are basic and, indeed, essential for any type of database system. The true power and utility of database and information systems, however, lie in the ability to perform analysis. Such analysis can be conducted on any of a number of progressive levels of inquiry beyond straightforward retrieval and manipulation, specifically:

1. *Exploration*: What patterns of spatial associations are there between vegetation type and various animal species?
2. *Explanation*: What factors would account for the current distribution of the spotted owl?
3. *Prediction*: If a given plot is clearcut next year, what changes in species mix of flora/fauna on the plot will occur in 1 year? In 6 years?, In 30 years?
4. *Planning*: What are the optimal vegetation age/species mix and spatial distributions for maintaining a stable spotted owl population in an area, while simultaneously conforming to governmental regulations and specific economic goals?

This also relates to what Bertin (1981) called *information levels* contained within a map: the *elementary* level, the *intermediate* level, and the *overall information* level. These also coincide with the levels of knowledge within cognitive theory I discussed in Part I. The *elementary* level corresponds to straightforward retrieval and tabulation of stored information: "How many shoe stores are there on Main Street?" Bertin's *intermediate* level of information corresponds to the exploration level of inquiry just listed. His *overall information* level corresponds to both the prediction and planning modes.

So, how can questions be posed to geographic software that address tasks beyond the retrieval, or elementary, level? Within current GIS interfaces, all of these (1 through 4) must be derived in round-about fashion through series of retrieval-type queries and manipulations of the data. Moreover, this also requires the guidance of a human with a significant amount of knowledge concerning the given domain. The underlying manipulations required at the algorithmic level may themselves be at varying levels of sophistication, from simple calculations (e.g., distance) to statistical procedures (regression), to complex mathematical models (soil loss).

The exploratory level of inquiry has become increasingly important with the advent of exhaustive and extremely large, shared databases. Exploratory data analysis on large and essentially exhaustive databases provides an opportunity for acquiring knowledge about the earth and our environment never before possible. These databases represent phenomena beyond what can be perceived directly or in sufficient detail. The computer can be a tool—as an extension of human capabilities—that allows us to "see" the world vicariously. This process of finding patterns and relationships through observation and interaction is an essential component of human learning.

One recent project that involved developing various methods for exploratory analysis and other, higher level forms of inquiry involving space–time data was called the Apoala project (Peuquet 2000). This effort also resulted in the pyramid model, discussed in the previous chapter as part of the larger effort to build a complete prototype environment for advanced geographic data representation and analysis. A portion of this effort was devoted to the development of methods for exploratory data analysis that allow a high degree of interaction. One outcome of this work was the development of graphical interactors that support both the linear and cyclic conceptualizations of time (Edsall and Peuquet 1997; Edsall, Kraak, MacEachren, and Peuquet 1997). An example of such interactors is shown in Figure 13.2.

In providing a general interface capability that allows predictive modeling and planning, one assumption sometimes made is that a more advanced query language is required. Indeed, it would seem that being able to handle the kind of complex, natural language statements as phrased in the examples given earlier would significantly aid human–computer interaction. Natural language understanding has been an area of research since the 1950s, and the state of the art in this area still falls significantly short.

It is within the capability of modern language parsing software to deal with the syntactical structures of natural language; in other words, to be able to pick apart the separate phrases and determine their interrelationships. There is also no problem with storing the amount of vocabulary required to

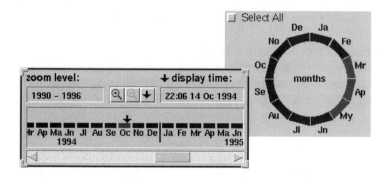

FIGURE 13.2. An example of a linear temporal legend (left) and a cyclical temporal legend (right).

deal with individual terms for any level of everyday or scientific discussion. The real problem is with semantics. As I described earlier in this book, natural language (as well as any symbolic graphic representation) and knowledge structure are closely intertwined. This shared knowledge structure is the key to enabling the wide range of nuances and variability in meanings to be appropriately understood depending on context. In Sowa's (2000) review and evaluation of the current state of language understanding by computer, he stated that the programs have yet to attain a level of understanding comparable to, say, average high-school students.

Even between humans, the precision and nonambiguity of mathematical language is preferable to natural language when precision and exactness are desired, so that there is no chance for misunderstanding. It also facilitates understanding through parsimony of expression, because complex concepts can be represented by a single symbol (e.g., $\int$ and ∞). This exactness and parsimony is indeed why mathematical notation was invented. But no single kind of language, nor graphics alone, is sufficient to deal with complex analytical tasks. For centuries, mathematical expressions have always been introduced, explained, and elaborated on by natural language. Graphics and diagrams also help to make the point clear.

The more complicated the concept or task, the more combined means of communication seem to be necessary. Diagrammatic symbology is particularly helpful in conceptually grasping multiple interrelationships in space, as anyone who has read a map can attest. This holds for grasping multiple interrelationships in time as well, and is why PERT (Program Evaluation and Review Technique) charts are so valuable in planning projects involving many steps.

This observation was utilized in another component of the Apoala effort, which investigated how to combine language and graphics in a way that promotes a high degree of human–machine interaction for computer-aided data exploration and higher level analysis (Qian and Peuquet 1998). This component utilized visual language built on the diagrammatic programming metaphor. More specifically, the work involved what is called a visual programming language, formally defined by Shu as "a language which uses some visual expressions (in addition to or in place of words and numbers) to accomplish what would otherwise have to be written in a traditional programming language" (1999, p. 203).

It is important to distinguish between a visual environment and a visual language. As distinguished by Shu (1999), a visual environment focuses on *showing* what is going on in terms of the system state (i.e., what the software is doing), the results of queries, and tutorial help. This is the entire environment that we commonly call the graphical user interface (GUI). In contrast, a graphical language focuses on telling the computer to do something. Thus, a visual programming language could be one element within a broader visual environment, or GUI, and indeed, this is how it was implemented in the Apoala prototype.

Visual programming languages have been implemented at a number of conceptual levels. Some deal with the intricacies of low-level control of the computer and are used as an aid for students learning computer programming techniques. Structures in a visual language at this level include mechanisms for specifying and manipulating the data format, as well as *while* and *for* loops. As such, these are procedural languages. Included at this level would also be visual programming versions of a database language, such as SQL. The visual programming language in the Apoala prototype, however, was aimed at a higher conceptual level—that of the user focusing on an analytical problem.

The visual programming language within Apoala was designed on a level comparable with the diagrams of various sorts that scientists and analysts have traditionally used (in notes, class discussions) to puzzle through an analytical process, and to describe the results. An example of the Apoala visual programming language is shown in Figure 13.3.

Part of the aim was to design the interface at a level of conceptual detail that would enable the user to deal with individual statements or commands that correspond to a natural level of detail for understanding the process or phenomenon being studied. This meant striking an intermediate level that would neither lose the user in extraneous details (e.g., retrieving individual data elements) nor be overly general.

The basic idea was to use this familiar and seemingly natural form of notation but to also make such diagrams executable by the computer in a highly interactive manner. This meant that the language primitives, which in this case consisted of lines, arrows, and annotated boxes and connectors, were utilized to define a formal syntax. Thus, diagrams composed according to this syntax via a drag-and-drop interface (including a menu of operations to pick from) could be parsed and interpreted in a manner equivalent to a purely textual language. The components could be executed and modified interactively and step-by-step as they are drawn and intermediate results shown, or the entire program could be saved and executed in "batch" mode.

The advantages of such an analysis-oriented and highly interactive graphical programming language include the following:

- The user is freed from the details of computer control, the format of the data, and the other intricacies of traditional computer programming and debugging. All data are simply treated as *sets*.

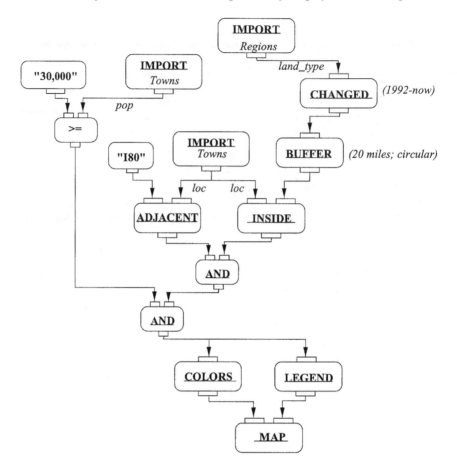

FIGURE 13.3. A simple query in the APOALA visual programming language. From Qian (2000).

- The user is freed from the usual linear language form and can instead pictorially see how all of the components of a potentially very complex process interconnect.
- Any errors in defining elements of the model or how they connect are immediately detected by the language interpreter.
- The model for the process or phenomenon represented in this manner can be interactively derived, tested, and subsequently run on various data sets.
- The human-oriented and executable descriptions of the model become merged.

The Apoala visual programming language is generally modeled after a data flow diagram—a popular device used as a modeling tool in software engineering to design and document the overall logical flow of processing in complex software systems. Today, such diagrams are also widely used in industrial applications for workflow analysis and process management. The notation of data flow diagrams was developed initially from graph theory and basically consists of *processes* and *flows* represented graphically as circles and lines. Processes take in one or more inputs, perform some operation, and produce output. The flows represent "pipelines" of data between "nodes" at which some operation or process is performed on the data. Flows have direction, and both processes and flows are labeled. These same notions have been utilized for the Apoala visual programming language, as seen in Figure 13.3, with some minor modification of the standard symbology (circles have become rounded rectangles).

Like data flow diagrams, the processes expressed in the Apoala visual programming language can also be "layered." This is particularly useful for complex processes that themselves have a number of subprocesses. Each of these can be converted into a "macro"—essentially a metaoperation that can be used like any of the built-in operations. The only difference is that for these "macros," an underlying diagram can also be displayed. In this way, simple, easy-to-follow diagrams can be maintained, and details can be uncovered by progressively and selectively "peeling away" the layers of macros.

SUMMARY

In Chapter 11, I discussed what have become "standard" or "traditional" data models, and in Chapter 12, I summarized emerging directions in geographic representation at the external representational level (i.e., the conceptual view). In this chapter, I examined the complementary question of how developing language and visual interface capabilities have the potential to allow these new types of database representations to be fully utilized.

Even though recent work on new types of interfaces for handling and analyzing geographic data is promising, there is much left to be done, particularly in addressing issues discussed earlier in this and preceding chapters. Derived information (or higher level knowledge) regarding phenomenological relationships, in addition to observational data, needs to be stored in order to use geographic databases in explanatory, predictive, or planning modes in a manner that does not require step-by-step specifications and instructions. Interfaces and the underlying representations must incorporate ways of handling inexactness. They must also provide multiple representations in a manner that enhances, rather than hinders, the user's understanding.

At the same time, this does not mean that the goal in digital data analysis should be to make this an autonomous, computer-based process. Again, the computer should extend human capabilities, ultimately achieving a cooperative problem-solving and analytic environment. What this means in terms of a user–computer interface is a unified facility that combines language with graphical capabilities in the form of both realistic, real-world imagery and symbolized graphic display. The purpose is to allow direct interaction of the human user with the computer in a way that simulates both the direct world experiences, through seeing and other sensory perceptions, and indirect experiences, through language, diagrams, and maps. This is the essence of "exploring virtual worlds."

With the context set, I examine representational requirements that represent challenges for realizing geographic databases that can both reflect and take advantage of human cognition to aid human–machine cooperative analysis and learning.

NOTE

1. These issues coincide, at various levels, with the research themes and issues delineated by the International Cartographic Association Commission on Visualization and Virtual Environments, documented in MacEachren and Kraak (2001).

CHAPTER 14

ISSUES FOR IMPLEMENTING ADVANCED GEOGRAPHIC DATABASES

EXTENDING CAPABILITIES IN REPRESENTATION, INTERACTION, AND LEARNING

As seen in Chapters 11 and 12, advances are being made at the conceptual level in developing database and knowledge representations based on cognitive principles, as well as new means of interacting with them. While deriving such an overall framework is a central component for moving forward, specific functional capabilities must be examined with a view toward practical implementation. Just from the standpoint of accommodating the characteristics of observational geographic data more fully, the representation of information in the temporal as well as spatial dimension is essential, as are ways to deal with vagueness, uncertainty, context-dependency, and incompleteness. With the need for knowledge discovery in very large and complex databases, algorithmic implementations that capture the functional characteristics of metaphor at least to some degree would be an invaluable tool.

Each of these specific capabilities has been addressed within the research literature in various application contexts. While none of these issues has been solved as yet, the real challenge will be to implement these capabilities in combination. Beyond these specific issues, there is also some insight to be gained from the experience of the relational model, specifically with regard to an overall theory.

In this chapter, I examine these issues and try to put each within the context of solutions proposed in the existing research literature.

ABOUT TIME

The development of temporal capabilities in geographic data models, and GIS in general, began with Langran's groundbreaking work in the late 1980s (Langran and Chrisman 1988), and was prompted at least in part by the work on temporal DBMS earlier in that decade.

It was during the 1980s that the advancing speed and storage capacity of computer hardware technology made it practical to store the significantly increased data volumes necessitated by the addition of the temporal dimension. Temporal capabilities in DBMS are seen as important for transaction-oriented applications, such as banking and medical systems, in which details of past history, as well a the preservation of individual changes through time, are of critical importance. Work in temporal DBMS focused from the beginning on extension of the relational model (Barbic and Percini 1985; Stonebraker 1986; Studer 1986), although there was work that also utilized hierarchical (Schiel 1983), object-oriented (Afsarmanesh, McLeod, and Parker 1985), knowledge-based (Dayal and Smith 1986), and semantic (Urban and Delcambre 1986) data modeling approaches. These efforts within the DBMS community were discussed in detail by Tansel et al. (1993) and summarized by Roddick and Patrick (1992).

More recently, there have been a significant number of studies within the DBMS community on database models and operators for applications involving moving objects (Erwig, Güting, Schneider, and Vazirgiannis 1998, 1999; Güting et al. 2000), where movement is usually stored as trajectory vectors in 3-D space for coherent objects. Current temporal DBMS can handle discrete change, but continuous change can only be implied from one recorded state to the next for any given entity that is tracked within the database (bank account, etc.). "Traditional" temporal DBMS applications (e.g., banking) are most often required to deal only with discrete change (transactions). There are, however, a number of specialized moving-objects database (MOD) systems available commercially, and these do not utilize the relational model. The Omnitracs system developed by Qualcomm for the transportation industry is one example (Qualcomm 2001). MOD systems are designed for applications such as the tracking of delivery vans, taxicabs, or military vehicles.

Although there is now a large body of literature on temporal database models and numerous temporal DBMS prototypes (Böhlen 1995), there are no commercially available, relational DBMS with more than rudimentary temporal capabilities. This may soon change, because there is a growing consensus on a unifying conceptual data model, known as the bitemporal[1] conceptual data model (BCDM) (Jensen and Snodgrass 1996). As Jensen and Snodgrass

(1999) have noted, the BCDM manages to retain the simplicity of the relational model, while also providing the ability to capture the unique aspects of temporal dynamics. Besides the firm foundation for building DBMS with full temporal capabilities from scratch that a uniform theoretical model provides, utilization of a viable extension of the relational model for data that includes the temporal dimension also allows the possibility of implementing temporal capabilities on top of an existing, nontemporal DBMS. Known as a *stratum* approach, this has a number of practical advantages for the shorter term, including compatibility with the familiar and much-used relational approach and ease of implementation (Torp, Jensen, and Snodgrass 1998).

It is interesting to note that although the BCDM model allows DBMS to be built from a unified, long-proven, and familiar theoretical foundation, it has the limitation of being able to represent only discrete change, in a spatial context, for rigid or point objects. This indeed fits the needs of the vast majority of business-oriented applications. As mentioned earlier, the representation of continuous change requires a different representational approach. Although MOD systems can represent continuous change in space, they also are limited to tracking rigid or point objects through space.

It is important to keep in mind that there are two kinds of change that are particularly important for geographic objects: They can change location in space (i.e., *move*, as well as change in spatial extent by *growing* or *shrinking* and *changing shape*). Depending on the application, change can consist of movement of point objects, which have no notable areal extent, or rigid objects. This type of movement-only change applies to the tracking of delivery vans, taxicabs, or military vehicles. Many geographic applications involve continuous change of objects in space as they both move and change shape (e.g., forest fires or urbanized areas). One commonly cited exception is the case of cadastral systems, in which property boundary changes are recorded in discrete transactions.

Most of the early research on the development of temporal capabilities in geographic data models, as was the case in DBMS, focused on the extension of traditional spatial and nonspatial data models to incorporate the temporal dimension (Hazelton 1991; Kelmelis 1991; Langran 1992; Pigot and Hazelton 1992), while some efforts utilized extended DBMS (Al-Taha 1992; Gardels 1994; Lolonis and Armstrong 1993; Rafaat, Yang, and Gauthier 1994). Recent attention has also been focused on the use of the object-oriented approach as a means of conceptually extending vector or raster approaches (Wachowicz 1999).

Extending the raster model to incorporate time can be accomplished using existing GIS by simply recording sequences of "snapshot" images in the same manner that a series of attribute "layers" are normally recorded. Each grid in the temporal sequence represents a "world state" at a particular moment. In terms of representation in the temporal dimension, this model is even more restrictive than the discrete changes of temporal, transaction-oriented DBMS. Not only is change limited to representation in discrete increments but also temporal increments apply to every location, whether or not change occurred at a given location.

The incorporation of time into this location-based model is often assumed to be the obvious solution for representation of spatial change. This stems from the modern mathematical tradition of the Minkowski framework of a combined space–time cube, and it does provide a convenient visual geometric. The combined treatment of space and time also seems to be supported by natural language (at least in the case of English), in which time tends to be expressed in spatial terms (e.g., He is *close* to graduation). Nevertheless, as discussed in Part I, the extension of locational concepts to include time is only a convenient metaphor. Treating time as space is sometimes useful for gaining insight into specific problem contexts. It does not allow the unique and specific properties of time to be explicitly represented, which is essential for a general-purpose representation.

This fundamental restriction causes several specific problems. Although the snapshot approach maintains the conceptual simplicity of the raster model and the world state for any stored time that can be retrieved directly, the *changes* between world states are not explicitly stored and can only be interpolated from two successive, stored world states. As already mentioned, changes also cannot be recorded selectively. This means that wherever a location has not changed from one snapshot to the next represents redundant data. In implementing the snapshot model by "fooling" existing GIS, an additional drawback is that there is no query support for temporal relationships ("before," "after," "during," etc.).

Extensions of the vector model to incorporate the temporal dimension rely on the concept of "amendment vectors" (Langran 1992); changes in an initial location and the time at which they occur are recorded as amendments (additions) to the original recorded map vectors and previous lines, or parts of lines, are marked as obsolete at that same time. This, then, is a transaction-oriented representation of time. Beyond the limitation of handling only discrete change, and only for boundary changes, the storage of these amendment vectors also becomes unwieldy as individual geographic objects evolve over time, because these changes alter the topology of connected map vectors. The problem is compounded when various new objects come into existence, then perhaps disappear and reappear (Peuquet and Qian 1996).

So far, efforts in developing space–time data models have provided pragmatic solutions for some application areas but have fallen short of the longer
term need. Extending the vector model to include the additional dimension of
time is simple conceptually, but becomes overly complex and unwieldy on the
implementation level. With the representational basis being object boundaries, there is also a focus in the temporal dimension of representing changes
in the locations of those boundaries. Extending the relational model has been
shown to be an effective solution for including temporal data when the other
data is also conceptually linear in form, as in the case of most business applications. Extending the relational model to include space as well as time, however, has led to forms of implementation that are both complex and voluminous. Experience has also shown that extending a spatial data model to
include temporal data, or vice versa, will also result in forms of implementation that are both complex and voluminous (Peuquet 1994; Sellis 1999).

Given the discussion in previous chapters concerning both human cognition and the past history of data model development, trying to incorporate all
information within a single, monolithic representation for multipurpose databases is bound to be less user-friendly and less flexible than a multirepresentational approach. The traditional (and statically oriented) object–field
dichotomy utilized in GIS data modeling, with object- and location-based
representations, has also resulted in a relative neglect of time-based representation for geographic data within the research community. This third
element of what/where/when perspectives is needed in order to represent
explicitly the evolution of entities and locations, and the interrelationships of
these through time. Certainly, it is possible to capture temporal information
within object- and location-based representations, but this can be done only at
the cost of significantly increased volume and complexity. Similarly, calculating specific temporal interrelationships, such as relative temporal coexistence
of specific entities or relative temporal configuration of various events that
are not explicitly stored, becomes computationally taxing.

Such a time-based, conceptual-level model was proposed by Peuquet and
Duan (1995) in ESTDM (the event-based spatiotemporal data model). This
model records the timestamp for any change and associated details describing
each specific change in temporal order. The sequence of timestamps can thus
be viewed as a time line, or a compacted temporal (1-D) vector corresponding
to the forward progression of time. New events that occur as time progresses
are added to the end of the list. The time line is compacted, in that temporal
locations where no change occurs are not recorded. It is assumed that each
event list and associated changes relate to a single, thematic domain (e.g.,
land use or population), although these stored events can relate to a change

occurring at either a location or set of locations, or an object or set of objects. A similar model, with associated query operators, was described by Claramunt and Thériault (Claramunt and Thériault 1995, 1996).

Hornsby and Egenhofer (2000) have also described a visual language for the explicit description of change relating to objects, called the change description language. This work focuses on the appearance–disappearance of entities and, particularly, of the identity transitions from one object to another that are possible. Identity is seen as a means of tracking and querying the existence of specific objects and types of objects, independent of specific attribute values.

Fortunately, ontological issues specific to time seem to be recognized and reasonably well understood for the purpose of database representation. This includes linear versus cyclic views of time, multiple times (world time, valid time, user-defined time), continuous or discrete change, branching time, and alternative time lines. The nature of concepts that are inherently both spatial and temporal (e.g., motion), and the interactions of the spatial and temporal dimensions, need to be better understood, however.

INEXACTNESS AND SCALE ISSUES

There is inherent inexactness built into all spatial, temporal, and spatio-temporal databases. *Fuzziness* and *uncertainty* are different kinds of inexactness that derive from both the inherent nature of the information and as a result of translating the information into digital form. Developing means for handling this complex issue in spatial databases has therefore long been an active area of research within the GIS community (Fisher 1999). Handling inexactness attracted much less attention in the DBMS community because historically this area had focused on application contexts that for the most part do not involve inexactness as a normal characteristic (e.g., banking and inventory). Only recently has any work on inexactness involving the combined effects of uncertainty in the spatial and temporal dimensions has appeared within the DBMS literature.

Fuzziness concerns inherent imprecision and derives directly from two inescapable characteristics of the real world and of human knowledge. First, entities in the real world often do not have sharp boundaries. For example, forests, shorelines, and urbanized areas all tend to be bounded in space by transition zones. Boundaries are also often fuzzy in the temporal dimension. In the example of vegetation types, under usual circumstances, there is no sharply-defined instant in time when an area changes from one vegetation type to another. This tends to be a gradual change. Of course, the sharpness of such distinctions also depends on scale. For example, viewing land use change from month to month would appear to be very gradual indeed, but viewed from decade to decade, it would appear discrete. There is also often fuzziness in categorization. The distinction between what may be classified as "urban" versus "rural," depends on context and is often a personal judgment call. Specific data classifications and object definitions used within any given spatial, temporal, or spatiotemporal database are therefore best defined within an application-dependent semantic framework.

In contrast to fuzziness, *uncertainty* concerns the lack of information— something that is not exactly known, for example, the exact location of a specific truck at a given moment (e.g., somewhere between New York and Philadelphia). As in this example, inexactness can also have combined spatial and temporal effects. Uncertainty can be introduced at the time the information is gathered and depend on the fineness of the measurement scale used, as well as the spatiotemporal resolution: Temperature measured on an hourly basis will have less uncertainty than temperature measured daily. Similarly, temperature measured in tenths of a degree will have less uncertainty than temperature measured in whole degrees.

Uncertainty can also be introduced through artificial discretization for computer representation of what are intrinsically or conceptually continuous phenomena. In the case of spatial data, such discretization was introduced historically through the cartographic tradition of drawing crisp lines to define geographical entities on paper maps. The development of the vector data model as one of the standard representational schemes for geographic data was derived directly from the idea of (crisp) lines on maps, so the vector model continues the introduction of uncertainty in entity boundaries. However, the level of uncertainty is often compounded when preexisting maps are digitized as sources of archival geographic data in building geographic databases (Burrough and Frank 1995; Molenaar 1998).

The snapshot form of space–time representation introduces a different kind of spatial and temporal uncertainty into the data. As mentioned earlier in this chapter, the actual changes that occur at locations between recorded temporal locations are not explicitly stored. These can only be derived by comparing the pixel value differences between successive snapshots. For discrete change, the exact time that the change occurred is not recorded. For continuous change, the trajectory of values between two snapshots, or the exact time that some change began, is not known. It may also be that some changes are completely missed because of the particular temporal resolution used. Even when change is recorded in an asynchronous manner and represented correspondingly, for example, in a time line, changes that actually occur may not be recorded.

Artificial discretization, a problem within geography and other fields that deal with spatial data, is known as the modifiable areal unit problem (MAUP; Fotheringham and Wong 1991; Green and Flowerdew 1996). This problem has two aspects: scale and zonation. The scale aspect of the MAUP refers to the variation in results obtained, based on the size of the size of the elemental unit. As recorded values are aggregated over larger and fewer units, variation is also lost. The zonation aspect of the MAUP refers to the variation in results obtained, depending on how the units are overlaid onto the space at the same scale. A slight shift can produce very different results. The MAUP problem exists for both discrete and gradual change over space, time, or space–time.

A variety of stochastic models have traditionally been applied to controlling or estimating the inexactness in spatial databases. However, these assume randomness in the variation of recorded values, which can limit their representational power. This approach, as well as others, were described by Goodchild and Gopal (1989).

To bring to bear specific knowledge about the behavior of the phenomenon represented, the probability vector approach has most commonly been used to represent uncertainty in spatial boundaries (Goodchild, Guoqing, and Shiren 1992; Mark and Csillag 1989). In this case, the steepness of the transition gradient from one area to another can be represented by the length and sequence of values contained within the probability vector. As noted by Goodchild et al. (1992), a similar approach has also been used in GIS for dealing with the MAUP in assigning values to spatial pixels containing mixed classifications (State of Maryland 1990; Tomlinson, Calkins, and Marble 1976). For example, a single pixel in a land use map may be assigned a vector of values, designating that 60% of the area in the pixel is residential, 30% is commercial, and 10% is parkland. In temporal databases, the probability vector approach was also used by Dyreson and Snodgrass (1993) in an extended tuple-timestamped relational model, to note potential times when the event might have occurred.

Another approach utilizes fuzzy set theory (Zadeh 1965). The fuzzy set representation provides a means to express degree of membership rather than probability of membership in a particular class or set. This approach has been used in dealing with classification of specific locations or specific objects (Burrough 1989; Burrough, MacMillan, and van Deursen 1992; Molenaar 1998; Wang and Hall 1996). Wang, Hall, and Subaryono (1990) also described a means of extending the relational model for the purpose of representing fuzzy category membership. The use of rough sets, another extension of classical set theory, has recently been proposed (Ahlqvist, Keukelaar, and Oukbir 2000). Rough sets are represented by the upper and lower approximation of an uncertain set. In using rough set theory, locations or objects can be designated as definitely within a given class (between the upper and lower bounds for the class), definitely outside, or possibly inside the class. A rule-based approach for handling uncertainty in spatial databases has also been proposed (Fisher 1989).

Very few researchers have addressed the problem of handling uncertainty in spatiotemporal databases. The work of Shibasaki (1994) represents an early attempt and describes some possible approaches at a general conceptual level. Work in this area has increased only in the past couple of years. Pfoser and Jensen (1999) described the use of interpolation for moving point objects. Pfoser and Tryfona (2000) explored the general conceptual differences in the use of fuzzy set theory and probability theory in the spatial–temporal domain.

On a level of practical implementation, linear interpolation has been the predominant approach used for dealing with inexactness in existing space–time databases. Dragicevic and Marceau (2000) described a fuzzy set approach that can be used to generate snapshots at times intermediate to those explicitly recorded within the database, if additional information about the overall nature of change is available from other sources. Using the likely trajectories of gradually progressing from one class to another (e.g., from forest to agricultural to urban) regions, the degree of membership in a specific class at a particular space–time location can be calculated on the basis of fuzzy set membership.

FORMAL ONTOLOGIES AS A BASIS FOR SHARED UNDERSTANDING

Research on ontology has received increasing attention over the past 10 years in a number of computer-related fields, particularly artificial intelligence (AI) and information systems. As I mentioned earlier, a distinction between ontology in the philosophical sense and what has become known as formal ontology has also arisen. This distinction between what some have called a *descriptive* ontology and a formal ontology is really one of representational level. A descriptive ontology is a cognitive-level representation, whereas a formal ontology utilizes the perspective of computer implementation. As described by researchers in formal ontology (Guarino 1998; Sowa 2000), a descriptive ontology denotes a particular system of concepts and categories that, taken together, form the basis for a particular world view. A formal ontology endeavors axiomatically to describe a set of concepts as elements, and the relationships between those elements. Gruber (1993) defined a formal ontology as "a specification of a conceptualization" (p. 1).

A descriptive ontology describes a particular reality at a very high level of generality. From a cognitive perspective, it is the basis for information sharing between the computer and the human users. A formal ontology can be viewed from a software-engineering perspective as the basis of a computer knowledge-base or semantic database design, as well as of the entire information system, influencing the design of retrieval capabilities, analytical capabilities, and the user interface.

A number of problems relating to formal ontologies remain research issues, however. Two key issues, described by Guarino (1998), are coarse versus fine-grained ontologies, and the ontology integration problem. Both issues relate to the fact that, on the one hand, any individual's ontological system—in terms of what exists—is highly complex and interrelated. On the other hand, the complexity and level of detail defined in a formal ontology should be kept to the minimum needed to enable shared understanding for the sake of storage efficiency and ease of maintenance.

Coarse-grained ontologies describe concepts in a general manner that may be fine for some applications and users, but not for others. For example, a formal ontology may contain the concept of "city," which is defined in part within the ontology as a geographic location, with a concentration of human dwellings and commercial activity. This would be sufficient for many applications, but other applications may require finer-grained distinctions. The question, then, is how to determine the minimum degree of fineness needed for a given application context.

The ontological integration problem involves finding the common ground between two distinct world views and how to organize representations from differing application perspectives that are still compatible with each other; it is a particularly important issue for shared databases when using a semantic approach. For example, what set of concepts and common definitions would be appropriate in a natural resources knowledge base for biologists, economists, and government policy analysts?

An additional problem with the geographic domain is that, unlike other application areas (e.g., medicine, banking, or civil engineering), there is little in the way of preexisting formalisms that can be translated into formal ontologies for database design. Definitions of concepts, and interrelationships between them, in the geographic domain remain largely informal and intuitive, or have a number of definitons (e.g., "region").

As a result, even though representational mechanisms exist for allowing better space–time database representations that coincide more closely with human cognition of geographic phenomena (e.g., frames, scripts, and semantic nets), populating such models with formalized ontologies that are detailed and extensive enough to be usable in a practical, real-world sense is problematic. No clear guidelines exist for the design and construction of formal ontologies in the face of these issues, although some criteria have been proposed. Gruber (1993) utilized a minimal commitment strategy, whereby the distinction between two concepts or types of entities should not be made unless needed. Borst, Akkermans, and Top (1997), in their emphasis on reusability and shared environments, utilized a hierarchical approach, with a view toward subdividing a generalized ontology into components, each as a derivative within a broader, metalevel ontology. Subsequently, Guarino and Welty (2000) have emphasized the need to revisit fundamental ontological notions from philosophy as guiding principles. This same principle was reinforced specifically within the context of geographic representations and the nature of geographic entities by Casati, Smith, and Varzi (1998).

Relative to the preceding discussion, Frank (1997) made an interesting assertion that building a single, unifying geographical ontology is impractical, and that it is therefore best to aim toward a "standard catalog of ontologies." Guarino (1997, 1998) has suggested that the single, unifying ontology is not really a limitation. Rather, that in recognizing that differing views of reality are inherent in the way people organize and use knowledge, a few kinds of ontologies quickly present themselves: top-level ontologies, domain and task ontologies, and application ontologies. Top-level ontologies describe very general concepts, such as "space," "time," and "object." It therefore seems reasonable to assume that such ontologies would be applicable in almost all cases over a broad range of contexts. They form a common basis for all other types of ontologies.

Although as discussed in Part I, there are indeed differing points of view that can be taken even at this most general level, it is these differences and variations (e.g., absolute vs. relative space) that are the best known and documented. Given this, it is reasonable to assume that a few top-level ontologies can be derived directly and refined through use. This is an important start. The more detailed the formal ontology needs to be, the more difficult it is to derive. It also needs to be recognized that, like the evolution of data models in the early days of GIS, the process of simply jumping in, building, and refining ontologies for advanced geographic representations is in itself a productive approach. There is no need to wait for formalized methodologies as a means to start deriving ontologies. Indeed, simply jumping in and building ontologies through a try-and-see approach can help to provide both a start in the shorter term, and through the process, provide formalization of ideas from a disciplinary perspective (in essence, we have already been doing this). Of course, an important part of the process of deriving and using ontologies needs to be recognition of the danger of falling into the same recursive, cognitive-formalized representation trap that for so long befell researchers with regard to raster/vector data models. The view of reality represented in the derived formal ontologies should not limit subsequent developments.

ARTIFICIAL CREATIVITY?

As I also discussed in Part I, the use of metaphor and other figurative devices is fundamental to human thought and learning. Metaphor provides mappings across different knowledge domains and is thus responsible on an intuitive level for what makes the unfamiliar or abstract more understandable via such mappings to something more familiar or concrete. Metaphor is also associated with what are considered creative or imaginative behaviors. Implementing the mechanism of metaphor at least to some degree in the computer would be an extremely powerful tool, extending the computer beyond the realm of the literal. Not only could this assist greatly in autonomous knowledge discovery for very large, heterogeneous databases, but it could also potentially enhance visualization capabilities by associating different sensory perceptions (e.g., utilizing visual displays to convey tactile information) to help in interactive data exploration and knowledge discovery (Marks 1996).

The current limitations in extending computer capabilities into the realm of metaphor and what can be generally described as creativity are threefold. The first is the need for formal ontologies and the development of algorithmic solutions that capture the functioning of metaphor. The problems in developing formal ontologies were just addressed. The second limitation is in developing algorithmic solutions that function in the manner of metaphor on stored knowledge, that find associations and linkages between what, in a literal interpretation, are dissimilar knowledge domains. The third limitation is in developing types of knowledge representation models that can support these algorithmic solutions.

Like formal ontology, work in the research community in the area of formal models of creative behavior has been increasing across a broad group of disciplines, particularly in AI and cognitive science, as well as in logic, mathematics, sociology, and software design (Dartnall 1994a). Creativity has historically been viewed as perhaps the most mysterious of human cognitive powers, and it has often been described as coming about through inspiration—ideas that simply "pop into the head." As such, it has also been widely assumed that computers cannot be creative, because they are deterministic state machines that can only follow the instructions that have been programmed into them. As discussed in Chapter 6, Finke asserted that novelty, incongruity, abstraction, and ambiguity seem to spark imagination in people. But can a computer be programmed to purposefully *generate* ambiguity or novelty?

Some general clues emerge from the examination of how imagery sparks the imagination, and from interpretation of linguistic metaphors. Creativity involves generative as well as exploratory processes. Generative processes modify the representational forms (i.e., the rules and pathways of the knowledge structures themselves), and this seems to be the key to knowledge discovery. Boden (1994) goes even further and argues that creativity *is* the mapping, exploration, and transformation of conceptual spaces.

Borrowing heavily from Karmiloff-Smith's representational redescription hypothesis (Clark and Karmiloff-Smith 1993; Karmiloff-Smith 1986, 1990, 1992), Dartnall (1994b), in describing the essential distinction between creatures who possess creativity and those that do not, asserts that those that do are rule *users*, and those that do not are, at best, rule *followers*. He uses the example of the beaver. Such a creature has both declarative and procedural knowledge, as do humans. As discussed in Chapter 3, declarative knowledge is knowledge about specific objects and locations, and their interrelationships. This kind of knowledge, then, corresponds to an ontology. Procedural knowledge focuses on how to perform specific tasks. The beaver can thus recognize trees, streams, and dams, and also knows how to build a dam. However, the beaver is simply following a procedure and will build a dam the same way every time. Although the beaver has a knowledge representation of how to build a dam, it has no *access* to that representational structure. Humans, on the other hand, posses a higher level of knowledge that provides a kind of metaknowledge that goes beyond domain-specific situations in how various kinds of knowledge interrelate.

Like the beaver, computers are rule followers, not rule users. In light of the representational redescription hypothesis and its development perspective, the question is whether computers can be programmed to exhibit behaviors that we would normally associate with creativity. Indeed, beavers cannot be "reprogrammed" to do things in a fundamentally different way, but computers can. Within languages and programming techniques developed in the field of AI, there are many examples of computer programs—particularly those called expert systems—that have access to their own knowledge base as it applies to a particular domain and can modify the knowledge structure itself. Even though this type of software still follows rules, the critical difference stems from the incorporation of different *levels* of rules that operate independently of each other: one level that describes a particular domain (i.e., the knowledge base), and another level that incorporates a set of context-free rules. The latter set of rules is usually referred to as an "inference engine." This consists of classical inductive and deductive logical procedures integrated with a collection of heuristics designed to traverse the knowledge base as a search space, in order to answer a specific question the way that a human expert within the given domain would answer the same question.

So, while the basic programming techniques are available to make computers rule users and not mere rule followers, examples of such rule using in computers are limited to a very specific and narrowly defined knowledge and task domain, such as medical diagnoses, legal reasoning, or minerals exploration. Even then, these programs do not come close to the performance of a human expert—even with their decades-long history of developing such computer-based systems. They simply do not have the ability to integrate broadly and bring knowledge to bear across varying domains. The approach used may itself have been self-limiting, in that expert systems historically have been focused on solving problems within specific knowledge domains. Usually, these systems are also intentionally deterministic, because part of the purpose behind such systems is to be able to replicate the result and to document the line of reasoning, whether to instruct medical students, to help safeguard the health of patients, or otherwise to assist in complex decision making. The fundamental limitation encountered with this type of program is a representational one.

Many of the classic representational forms developed in a cognitive context, particularly propositions, frames, conceptual graphs, scripts, and neural networks, have also been implemented on computers in order to represent knowledge, but each for specific types of applications. Considering the two basic types of formalized knowledge representations described in Chapter 8, propositions, frames, conceptual graphs, semantic nets, and scripts are symbolic (as opposed to connectionist) in nature, with propositions being the simplest and most clearly symbolic.

The basic proposition contains an ontology of elements, and the relationships among those elements. These components, as a whole, comprise a function that is either true or false. The algebraic form of a proposition consists of arguments and a predicate that asserts a relationship among those elements. Frames, as described by Minsky (1975) and Fillmore (1982), denote individual nodes in a network organization, with branches designating is-a, part-of, and other types of explicit and labeled linkages. Conceptual graphs (Burks 1960; Sowa, Foo, and Rao 1990), semantic nets (Quillian 1968), and scripts (Schank 1976) are in a general sense similar in nature. Minsky's frames, however, allow default attribute values and behavioral rules to be included. This allows "nontypical" examples and situations to be identified, and supports inexact (i.e., fuzzy) reasoning. All of these representational forms are designed to be able to designate a range of relationships. Scripts are specifically designed to represent *sequences* of events in time. Frames perhaps depart most from the discrete and deterministic nature usually attributed to symbolic forms. Thus, although each has its own functional strengths, symbolic forms in general have proved successful in dealing with structured reasoning tasks, and have subsequently been associated with "high-level" cognitive tasks.

Neural nets are the dominant connectionist representational form. This structure is characterized as a set of highly interconnected nodes and often has stochastic mechanisms for determining the state of connectedness for any given node. This has proven advantageous for simulating "low-level" cognitive tasks, such as vision processing and pattern recognition. Given the nature of the connections and how they are determined, neural nets are viewed as weakly structured. These are problems that require a high degree of parallelism and adaptability to be efficient.

The differing functional strengths of these two representational approaches have therefore led to a wide gap in capabilities. Despite the range of capabilities represented in these various forms, however, none employs the "imaginative" structuring principles of metaphorical mapping, and as such, all fall short of coming close to duplicating human knowledge or use of that knowledge. It is metaphorical mapping that is needed to bridge domains.

With the goal of developing a unified knowledge model, there has recently been a high level of activity, in the AI community in particular, in development of *hybrid* models that can support symbolic, structured reasoning, as well as parallel and nonconstrained problem solving. A number of classification schemes for such hybrid models have been proposed (Frasconi 1998; Sun and Bookman 1995; Wermter 2000). These reveal a continuum of forms that has evolved, incorporating connectionist and symbolic forms to varying degrees and in various ways.

Such hybrid or combined structures offer the possibility for supporting metaphoric mapping. Many existing computational implementations to model analogical and metaphorical reasoning utilize a knowledge representation that at least to some degree are hybrids. For example, Copycat (Hofstadter and Mitchell 1994) is intended to be a computational model of analogy making in a cognitive as well as computational sense, mapping functions onto conceptual spaces, and also modifying and creating these conceptual spaces as it goes along. ACME (the Analogical Constraint Mapping Engine) (Holyoak and Thagard 1989) and STAR (Structured Tensor Analogical Reasoning) (Wiles et al. 1994) are two other existing computational implementations of analogy that utilize a hybrid knowledge representation.

Computers are far from equaling humans in exhibiting behaviors associated with imagination and creativity: the mapping exploration and transformation of conceptual spaces, and finding novel associations. Computer capabilities in this area, however, continue slowly to increase. Although these capabilities are potentially very useful for helping to find meaningful patterns and associations in large, complex databases and for facilitating human users in complex analysis tasks, it must always be remembered that—unlike some computers portrayed in science fiction—while computers can have access to their own store of knowledge, they can never be "aware" of that knowledge. Computers will still only be able to perform tasks programmed into them. It is just that the tasks (and the representations that enable those tasks) will in the future be much more complex than those they are commonly capable of today.

THE MODELER'S DILEMMA

In moving forward, toward more advanced, human-centered, geographic database representations, it is interesting to note the tension within the DBMS research community between the extended relational approach and the object-oriented approach. The extended relational approach represents an adherence to the rigor, and indeed conceptual "neatness," that a unified conceptual framework based on mathematical theory provides. Extensions to the basic relational model are seen by its supporters as a means to represent more complex entities, with more complex interrelationships, while still taking advantage of a unified (and at this point also widely used and familiar) theoretical framework. The object-oriented approach, on the other hand, seen by its supporters as a more flexible means for modeling complex and nested entities, can result in representations that both better resemble how people cognitively represent such entities, and take advantage of the extended implementation capabilities built into modern programming languages.

As mentioned in the discussion of the development of DBMS technology in Chapter 11, object-oriented DBMS have been heralded by some as the next generation of DBMS and scorned by others as abandoning theory and taking a giant step backward to ad hoc systems. The developments in temporal DBMS are telling with respect to this debate. The relational model, with its conceptual foundation based on unordered sets, can be (and indeed has been) extended to a certain degree to handle a wider range of data types. SQL has similarly been extended to handle a wider range of data types and relational operators.

At a conceptual level, however, the relational model and SQL, by themselves, have no means for representing the complexities of multidimensional data and their inherent interrelationships. Interdependencies also go against one of the basic notions of the relational model. Data normalization, and what are known as first, second, and third normal forms, minimize interdependency of the data at increasing levels. The relational model itself is based on tables that contain scalar values.

The power of the relational model and the relational algebra comes from their conceptual simplicity, as well as from the minimized interdependencies dictated within the data represented by the relational model and the logical coherence of the relational algebra. Both have been extended somewhat, but there comes a point when adding data types and relational operators to accommodate complex data becomes counterproductive. Any design can also become ad hoc if extended significantly beyond the original principles. The absence of any explicit indexing in the relational database model also means that, at an implementation level, it does not have the performance efficiency required for very large database environments.

Commercial database software products that have tried to add object representation on top of the relational model have been likened to a fish with feathers—a creature that can neither swim nor fly very well (Celko and Celko 1997). Databases that have been built within the DBMS community for specialized applications involving space–time data have been forced to abandon the relational model for effective representational solutions. This leaves DBMS researchers with the seeming dilemma of having to develop representations with either a solid theoretical basis, but inherently limited to the kinds of things that can be represented, or "ad hoc" models that also provide representational freedom. Does this mean that those who need to deal with multidimensional or other forms of complex data must forever dwell in the realm of the ad hoc? Definitely not. It does seem to mean, however, that the deterministic and closed-world comfort of a *mathematical* basis for representation is insufficient. In general, object-based representations along with geographic and space–time representations, are currently ad hoc, but this is gradually changing. Cognitive principles must be incorporated as an important part of this. The resulting theory will be less closed and deterministic, and more open, flexible, and, ultimately, more robust.

This debate in the DBMS community seems to echo the representational debate that occurred in the GIS community, and a similar debate that has also been put to rest in the cognitive psychology community, in relation to knowledge representation. The resolution is the development of hybrid or multirepresentational forms that retain the advantages of each precedent form for a particular range of applications. Since various applications are best served with differing representations, development of a hybrid or multi-representational form becomes increasingly critical as the variety of tasks increases. Nevertheless, in extending the power of representational models, the cost will always be an increase in complexity. This has been the case with nonspatial database models, spatial database models, and knowledge representations—thus the famous adage: There is no such thing as the perfect model, except the thing you are modeling itself.

NOTE

1. This model is bitemporal in that it is designed to record valid, or world, time and transaction time. Valid time is the time of actual occurrence, and transaction time is the time at which the occurrence is recorded.

CHAPTER 15

EPILOGUE

Moving Forward

That the design of digital geographic representation models is best informed by human cognition is the basic premise of this book. It has been widely recognized that current representational techniques for geographic databases, in and of themselves, do not currently have the representational robustness needed in the task of unlocking and utilizing the vast wealth of environmental information stored in the worldwide network of computers that now exists. And, particularly for novice users of geographic information systems and related software, it often seems that the machines are winning in what does indeed seem to be conflicting views of the world between human and machine. This is a key issue as digital geographic databases and the software to utilize them become more an element of everyday life and an essential resource in the "information revolution."

From a computing standpoint, the human mind with its phenomenal capacity, flexibility, and power to represent and utilize information about the environment, provides a powerful model for building advanced and complex database and analysis tools for geographic data. From a cognitive standpoint, GIS and other forms of spatial information systems need to work *with* humans as a data-exploration and analysis aid in knowledge–discovery and problem solving.

The nature of space and time, as understood by humans, and how people cognitively represent geographic knowledge have fascinated humans for millennia. Nevertheless, many questions remain unanswered. We are at a disadvantage in studying cognition, because it is impossible simply to look inside our skulls and inspect how we accumulate, arrange, and utilize knowledge about our environment. This in itself would make the development of a systematic framework for computer-based space–time representation that appropriately reflects human cognition a difficult challenge. As we try to apply what is known about environmental cognition to geographic database representation and use, there is the additional problem that computer storage is a radically different external medium for storing geographic information than any other previously known.

While this combination of factors may constitute an insurmountable complex of problems in ultimately developing anything even close to a complete simulation or imitation of human environmental cognition within the computer, this is not what the real goal should be. One of the things that 30 years of experience with GIS, and with computers in general, has shown is that we need a theoretical framework for human–computer representation and interaction. Both of these must certainly be informed by human cognition, but in doing so, they also must allow the functional advantages of both the computer and the human to be utilized, without one overly hindering the capabilities of the other. The true power of GIS lies in *combining* the great speed, nonforgetfulness, and untiring accuracy of computers with the visual pattern recognition, information filtration, and generalization capabilities, insight, and imagination of human intelligence.

While there are limitations in the extent to which we can and should go in developing a framework for cognitively informed geographic database representation, there is much to be gained from some very basic insights. For approximately 20 years, approaches for digital representation have imitated previous representational approaches, particularly that of cartography, or borrowed from other computer-related fields, such as DBMS. As mentioned in Chapter 1, not until the late 1980s and early 1990s was the need for a human-centered theoretical framework generally recognized. This realization, in itself, quickly resulted in a new perspective for geographic database representation (Couclelis 1993; Hirtle 1994; Medyckyj-Scott and Blades 1992; Peuquet 1988a, 1988b, 1994).

Progress has been slow in the decade since this initial realization, and this in large part can be attributed to the simple fact that such a fundamental change in direction must first overcome inertia in adapting to new ways of thinking. In the case of geographic representation, this change of thinking involves the abandonment of world views that were adaptations designed to deal with the particular characteristics of the available representational medium (lines on paper, matrices, etc.). With modern computing technology, we are free of those constraints—free to utilize a more "natural" and intuitive world view. We can thereby tilt the balance more in favor of the human side of the equation in human–computer interaction.

So, what are our prospects for the next 10 years and beyond? As shown in this book, there are some well-known and widely accepted principles concerning human spatial cognition that correspond to well-known approaches for increasing the representational power and the efficiency of spatial data representations. These include the separation of different types of stored information and hierarchical approaches.

A number of characteristics of human cognition have historically been recognized as intrinsic shortcomings of spatial data models, particularly representation of uncertainty and inexactness, the use of context-dependent relationships, and the adaptive storage of meaningful groupings (i.e., categories) of real-world entities at varying levels of abstraction. However, methodologies for incorporating such characteristics in formalized representations have also been well-known for some time.

While a number of the experimental examples developed exhibit one or more of these properties, these efforts need to be extended and refined, utilizing what is known about these characteristics in the human context as guidance. How humans develop, refine, and utilize categories, for example, has been extensively studied. There has recently been much activity in both the nonspatial and spatial database research communities into the development and use of formal ontologies as a means of dealing with varying world views and application contexts. There have also been insights developed into the realm of human creativity and insight, but this knowledge has not yet been applied in an explicitly geographic context.

The challenge that lies ahead is to find how these various aspects can be drawn together to derive a more advanced framework for space–time database representation and to translate that framework into practice. I hope this book has provided some progress in this direction.

REFERENCES

Abel, D. J. (1988). Relational data management facilities for spatial information systems. *Proceedings, Third International Symposium on Spatial Data Handling, Sydney, Australia* (pp. 9–18).

Abel, D. J., and J. L. Smith. (1983). A data structure and algorithm based on a linear key for a rectangle retrieval problem. *Computer Vision, Graphics, and Image Processing, 24,* 4–14.

Adolph, K. E., M. A. Eppler, and E. J. Gibson. (1993). Development of perception of affordances. In C. Rovee-Collier and L. P. Lipsett (Eds.), *Advances in infancy research* (Vol. 8, pp. 51–98). Norwood, NJ: Ablex.

Adriaans, P., and D. Azntinge. (1996). *Data mining.* Harlow, UK: Addison-Wesley.

Afsarmanesh, H., D. McLeod, and A. Parker. (1985). An extensible object-oriented approach to databases for VLSI/CAD. *Proceedings, Eleventh International Conference on Very Large Data Bases, Institute of Electrical and Electronics Engineers, Stockholm* (pp. 13–24).

Ahlqvist, O., J. Keukelaar, and K. Oukbir. (2000). Rough classification and accuracy assessment. *International Journal of Geographical Information Science, 14*(5), 475–496.

Ahuja, N. (1983). On approaches to polygonal decomposition for hierarchical image decomposition. *Computer Vision, Graphics, and Image Processing, 24,* 200–214.

Akhundov, M. D. (1986). *Conceptions of space and time: Sources, evolution, directions.* Cambridge, MA: MIT Press.

Allen, J. F. (1983). Maintaining knowledge about temporal intervals. *Communications of the ACM, 26*(11), 832–843.

Al-Taha, K. K. (1992). *Temporal reasoning in cadastral systems.* Doctoral dissertation, Orono, University of Maine.

Andersen, P. B., B. Holmqvist, and J. F. Jensen (Eds.). (1993). *The computer as medium.* Cambridge, UK: Cambridge University Press.

Anderson, J. R. (1978). Arguments concerning representations for mental imagery. *Psychological Review, 85*(4), 249–276.

Anderson, R. A. (1987). Inferior parietal lobule function in spatial perception and visuomotor integration. In F. Plum and V. B. Mountcastle (Eds.), *Handbook of physiology* (pp. 483–518). Rockville, MD: American Physiological Society.

Andrews, J. H. (1990). Map and language: A metaphor extended. *Cartographica, 27*(1), 1–19.

Anselin, L., and A. Getis. (1992). Spatial statistical analysis and geographic information systems. *Annals of Regional Science, 26*(1), 19–33.

Aristotle. (1961). *Aristotle's physics*. Lincoln: University of Nebraska Press.

Aristotle. (1984a). Categories. In J. Barnes (Ed.), *The complete works of Aristotle*. Oxford, UK: Oxford University Press.

Aristotle. (1984b). Poetics. In J. Barnes (Ed.), *The complete works of Aristotle*. Oxford, UK: Oxford University Press.

Armstrong, S. L., L. R. Gleitman, and H. Gleitman. (1983). What some concepts might not be. *Cognition, 13*, 263–308.

Ashmead, D. H., R. S. Wall, K. A. Ebinger, M. Hill, X. Yang, and S. Eaton. (1998). Spatial hearing in children with visual disabilities. *Perception, 27*, 105–122.

Attneave, F. (1971). Multistability in perception. *Scientific American, 225*, 62–71.

Austin, J. L. (1961). *Philosophical papers*. Oxford, UK: Oxford University Press.

Aveni, A. F. (1989). *Empires of time: Calendars, clocks, and cultures*. New York: Basic Books.

Baillargeon, R., E. S. Spelke, and S. Wasserman. (1985). Object permanence in five-month-old infants. *Cognition, 20*, 191–208.

Ballard, D. (1981). Strip trees: A hierarchical representation for curves. *Communications of the ACM, 24*, 310–321.

Ballard, D., and C. Brown. (1982). *Computer vision*. Englewood Cliffs, NJ: Prentice-Hall.

Bang, J. (1993). The meaning of plot and narrative. In P. Andersen, B. Holmqvist, and J. Fensen (Eds.), *The computer as medium* (pp. 209–221). Cambridge, UK: Cambridge University Press.

Barber, P., and C. Board. (1993). *Tales from the map room: Fact and fiction about maps and their makers*. London: BBC Books.

Barbic, F., and B. Percini. (1985). Time modeling in office information systems. *Proceedings, SIGMOD '85*. Austin, TX: ACM Press.

Barr, A., and E. Feigenbaum. (1981). *The handbook of artificial intelligence*. Los Altos, CA: Kaufman.

Barsalou, L. W. (1983). Ad hoc categories. *Memory and Cognition, 11*(3), 211–227.

Bartels, D. (1982). Geography: Paradigmatic change or functional recovery?: A view from West Germany. In P. Gould and G. Olsson (Eds.), *A search for common ground* (pp. 24–33). London: Pion.

Bartholdi, John J., III, and P. Doldsman. (2001). Continuous indexing of hierarchical subdivisions of the globe. *International Journal of Geographical Information Science, 15*(6), 489–522.

Bartlett, F. C. (1932). *Remembering.* Cambridge, UK: Cambridge University Press.

Bean, K. L. (1938). An approach to the reading of music. *Psychological Monographs, 226,* 1–80.

Bechtel, W., and A. Abrahamsen. (2001). *Connectionism and mind: Parallel processing, dynamics, and evolution in networks* (2nd ed.). Oxford, UK: Blackwell.

Benedict, R. O. G. (1991). *Imagined communities: Reflections on the origin and spread of nationalism.* London: Verso.

Bengtsson, B., and S. Nordbeck. (1964). Construction of isorithms and isorithmic maps by computers. *Nordisk Tidschrift for Informations-Behandling, 4,* 87–105.

Berkeley, G. (1965). An essay towards a new theory of vision. In D. M. Armstrong (Ed.), *Berkeley's philosophical writings* (pp. 274–352). New York: Macmillan.

Berry, B. J. L. (1964). Approaches to regional analysis: A synthesis. *Annals of the Association of American Geographers, 54*(1), 2–11.

Berlin, B., and P. Kay. (1969). *Basic color terms: Their universality and evolution.* Berkeley: University of California Press.

Bertin, J. (1981). *Graphics and graphic information-processing.* Berlin: de Gruyter. (French edition, 1977)

Bertin, J. (1983). *Semiology of graphics.* Madison: University of Wisconsin Press. (French edition, 1967)

Biederman, I. (1987). Recognition-by-components: A theory of human image understanding. *Psychological Review, 94*(2), 115–147.

Biederman, I. (1990). Higher-level vision. In E. N. Osherson and S. M. Kosslyn (Eds.), *An invitation to cognitive science* (pp. 41–72). Cambridge, MA: MIT Press.

Black, M. (1969). *Models and metaphors: Studies in language and philosophy.* Ithaca, NY: Cornell University Press.

Black, M. (1993). More about metaphor. In A. Ortony (Ed.), *Metaphor and thought* (pp. 19–41). Cambridge, UK: Cambridge University Press.

Blades, M., and C. Spencer. (1986). The implications of psychological theory and methodology for cognitive cartography. *Cartographica, 23*(4), 1–13.

Blaser, A., M. Sester, and M. Egenhofer. (2000). Visualization in an early stage of the problem solving process. *Computers and Geosciences, 26*(1), 57–66.

Blaut, J. M., and D. Stea. (1971). Studies of geographic learning. *Annals of the Association of American Geographers, 61*(2), 387–393.

Boden, M. (1979). *Jean Piaget.* New York: Viking Press.

Boden, M. A. (1994). Précis of *The creative mind: Myths and mechanisms* commentary. *Behavioural and Brain Sciences, 17*(3), 519–570.

Bogartz, R. S., J. L. Shinskey, and T. Schilling. (2000). Object permanence in five-and-a-half month old infants? *Infancy, 1*(4), 403–428.

Bogartz, R. S., J. L. Shinskey, and C. J. Speaker. (1997). (1997). Interpreting infant looking: The event set x event set design. *Developmental Psychology, 33*(3), 408–422.

Böhlen, M. (1995). Temporal database system implementations. *ACM SIGMOD Record, 24*(4), 53–60.

Bohr, N. (1913). On the constitution of atoms and molecules. *Philosophical Magazine, 26*, 1–15.

Boots, B. N. (1979). Weighting Thiessen polygons. *Economic Geography, 56*(3), 248–259.

Boring, E. G. (1930). A new ambiguous figure. *American Journal of Psychology, 42*, 444–445.

Boroditsky, L. (2000). Metaphoric structuring: Understanding time through spatial metaphors. *Cognition, 75*, 1–28.

Borst, P., H. Akkermans, and J. Top. (1997). Engineering ontologies. *International Journal of Human Computer Studies, 46*(2/3), 365–406.

Bowerman, M. (1989). Learning a semantic system: What role does cognitive predispositions play? In M. L. Rice and R. C. Schiefenbusch (Eds.), *The teachability of language* (pp. 329–363). Baltimore: Brookes.

Brainerd, C. J. (1978). The stage question in cognitive-developmental theory. *Behavioural and Brain Sciences, 2*, 173–213.

Brann, E. T. H. (1991). *The world of the imagination: Sum and substance.* Savage, MD: Rowman & Littlefield.

Bryant, D. J., B. Tversky, and N. Franklin. (1992). Internal and external spatial frameworks for representing described scenes. *Jornal of Memory and Language, 31*(1), 74–98.

Bullinger, A.,, and J.-F. Chatillon. (1983). Recent theory and research of the geneval school. In P. H. Mussen (Ed.), *Child psychology* (pp. 231–262). New York: Wiley.

Bülthoff, H. H., and S. Edelman. (1992). Psychophysical support for a two-dimensional view interpolation theory of object-recognition. *Proceedings of the National Academy of Science, 89*, 60–64.

Bunge, W. (1962). *Theoretical geography* (pp. 38–88). Lund, Sweden: C.W.K. Gleerup.

Burks, A. W. (Ed.). (1960). *Collected papers of Charles Sanders Peirce.* Cambridge, MA: Harvard University Press.

Burrough, P. A. (1989). Fuzzy mathematical methods for soil survey and land evaluation. *Journal of Soil Science, 40*, 477–492.

Burrough, P. A. (1992). Are GIS data structures too simple minded? *Computers and Geosciences, 19*(4), 395–400.

Burrough, P. A., and A. U. Frank. (1995). Concepts and paradigms in spatial information: Are current geographical information systems truly generic? *International Journal of Geographical Information Systems, 9*(2), 101–116.

Burrough, P. A., and A. U. Frank. (Eds.). (1996). *Geographic objects with indeterminate boundaries*. London: Taylor & Francis.

Burrough, P. A., R. A. MacMillan, and W. van Deursen. (1992). Fuzzy classification methods for determining land suitability from soil profile observations and topography. *Journal of Soil Science, 43*, 193–210.

Burton, I., and R. W. Kates. (1964). The perception of natural hazards in resource management. *Natural Resources Journal, 3*, 412–441.

Bushnell, E. W. (1986). Cross-modal functioning in infancy: The basis of infant visual–tactual functioning: Amodal dimensions or multimodal compounds? In C. Rovee-Collier (Ed.), *Advances in infancy research* (pp. 182–194). Norwood, NJ: Ablex.

Byrne, R. W., and E. Salter. (1983). Distances and directions in cognitive maps of the blind. *Canadian Journal of Psychology, 37*, 293–299.

Caivano, J. L. (1990). Visual texture as a semiotic system. *Semiotica, 80*(3/4), 239–252.

Card, S. K., J. D. Mackinlay, and B. Shneiderman. (1999). Information visualization. In S. K. Card, J. D. Mackinlay, and B. Shneiderman (Eds.), *Readings in information visualization: Using vision to think* (pp. 1–32). San Francisco: Kaufmann.

Cartwright, W., J. Crampton, G. Gartner, S. Miller, K. Mitchell, E. Siekierska, and J. Wood. (2000). Geospatial information visualization user interface issues. *Cartography and Geographic Information Science, 28*(1), 45–60.

Casati, R., B. Smith, and A. C. Varzi. (1998). Ontological tools for geographic representation. In N. Guarino (Ed.), *Formal ontology in information systems* (pp. 77–85). Amsterdam: IOS Press.

Cassirer, E. (1944). *An essay on man: An introduction of the philosophy of human culture*. New Haven, CT: Yale University Press.

Cassirer, E. (1953–1996). *The philosophy of symbolic forms*. New Haven, CT: Yale University Press.

Cassirer, E. (1964). *The philosophy of symbolic forms I: Language*. New Haven, CT: Yale University Press.

Cassirer, E. (1966). *The philosophy of symbolic forms III: The metaphysics of symbolic forms*. New Haven, CT: Yale University Press.

Cassirer, E. (1973). *The philosophy of symbolic forms IV: The phenomenology of knowledge*. New Haven, CT: Yale Univeristy Press.

Celko, J., and J. Celko. (1997). Debunking object–database myths. *Byte, 22*(10), 101–106.

Chalmers, D. J., R. M. French, and D. R. Hofstadter. (1992). High-level perception, representation, and analogy: A critique of artificial intelligence methodology. *Journal of Experimental and Theoretical Artificial Intelligence, 4*(2), 185–211.

Chamberlin, D. D. (1976). Sequel 2: A unified approach to data definition, manipulation, and control. *IBM Journal of Research and Development, 20*(6), 560–575.

Chamberlin, D. D., and R. F. Boyce. (1974). Sequel: A structured English query language. *Proceedings, ACM SIGMOD Workshop on Data Description, Access, and Control, Ann Arbor, MI* (pp. 249–264).

Chan, T. C., and M. T. Turvey. (1991). Perceiving the vertical distances of surfaces by means of a hand-held probe. *Journal of Experimental Psychology: Human Perception and Performance, 17,* 347–358.

Chase, W. G., and H. A. Simon. (1973). The mind's eye in chess. In W. G. Chase (Ed.), *Visual information processing* (pp. 215–281). New York: Academic Press.

Chen, P. P.-S. (1976). The entity–relationship model—toward a unified view of data. *ACM Transactions on Database Systems, 1*(1), 9–36.

Choi, S., and M. Bowerman. (1991). Learning to express motion events in English and Korean: The influence of language-specific lexicalization patterns. *Cognition, 41,* 83–121.

Chomsky, N. (1957). *Syntactic structures.* The Hague: Mouton.

Chomsky, N. (1965). *Aspects of the theory of syntax.* Cambridge, MA: MIT Press.

Chown, E., S. Kaplan, and D. Kortenkamp. (1995). Prototypes, location, and associative networks (plan): Towards a unified theory of cognitive mapping. *Cognitive Science, 19,* 1–51.

Chrisman, N. R. (1974). The impact of data structure on geographic information processing. *Proceedings, Auto-Carto 1, Reston, VA* (pp. 165–177).

Chrisman, N. R. (Ed.). (1978). *Concepts of space as a guide to cartographic data structures.* Reading, MA: Addison-Wesley.

Chrisman, N. (1987). Fundamental principles of geographic information systems. *Proceedings, Auto-Carto 8* (pp. 32–41). Baltimore, MD: American Society for Photogrammetry and Remote Sensing/American Congress of Surveying and Mapping.

Christaller, W. (1933). *Die zentralen orte in süddeutschland.* Jena, Germany: Gustav Fischer. (Translated, 1966, by Charlisle W. Baskin, as *Central places in southern Germany*). New York: Prentice-Hall.

Churchland, P. S. (1992). *The computational brain.* Cambridge, MA: MIT Press.

Claramunt, C., and M. Thériault. (1995). Managing time in GIS: An event-oriented approach. In J. Clifford and A. Tuzhilin (Eds.), *Recent advances in temporal databases* (pp. 23–42). Berlin: Springer-Verlag.

Claramunt, C., and M. Thériault. (1996). Toward semantics for modelling spatiotemporal processes within GIS. In M. Kraak and M. Molenaar (Eds.), *Advances in GIS research II: Proceedings of of the Seventh International Symposium on Spatial Data Handling* (pp. 47–63). London: Taylor & Francis.

Clark, A., and A. Karmiloff-Smith. (1993). The cognizer's innards: A psychological and philosophical perspective on the development of thought. *Mind and Language, 8*(4), 520–530.

Clark, E. V. (1973). What's in a word?: On the child's acquisition of semantics in his first language. In T. E. Moore (Ed.), *Cognitive development and the acquisition of language* (pp. 65–110). New York: Academic Press.

Clark, H. H. (1973). Space, time, semantics, and the child. In T. E. Moore (Ed.), *Cognitive development and the acquisition of language* (pp. 28–64). New York: Academic Press.

Clowes, M. (1969). Transformational grammars and the organization of pictures. In A. Graselli (Ed.), *Automatic interpretation and the organization of pictures* (pp. 43–77). Orlando, FL: Academic Press.

Codd, E. F. (1970). A relational model of data for large shared data banks. *Communications of the ACM, 13*, 377–387.

Codd, E. F. (1972). Relational completeness of data base sublanguages. In R. J. Rustin (Ed.), *Data base systems* (Courant Computer Science Symposium, 6th). Englewood Cliffs, NJ: Prentice-Hall.

Codd, E. F. (1979). Extending the database relational model to capture more meaning. *ACM Transactions on Database Systems, 4*(4), 397–434.

Cook, D. (1997). *Immersed in statistics: Your worst nightmare or your wildest dream!* Presentation at the Joint Statistical Meeting, Anaheim, CA.

Cooper, L., and R. Shepard. (1973). Chronometric studies of the rotation of mental images. In W. Chase (Ed.), *Visual information processing* (pp. 175–176). New York: Academic Press.

Cooper, L. A. (1990). Mental representation of three-dimensional objects in visual problem solving and recognition. *Journal of Experimental Psychology: Learning, Memory and Cognition, 16*, 1097–1106.

Corbett, J. (1979). *Topological principles in cartography*. Washington, DC: U.S. Bureau of the Census.

Corner, J. (1999). The agency of mapping: Speculation, critique and invention. In D. Cosgrove (Ed.), *Mappings* (pp. 213–252). London: Reaktion Books.

Cornoldi, C., and R. H. Logie. (1996). Counterpoints in perception and mental imagery: Introduction. In C. Cornoldi, R. H. Logie, M. A. Brandimonte, G. Kaufmann, and D. Reisberg (Eds.), *Stretching the imagination: Representation and transformation in mental imagery* (pp. 3–30). Oxford, UK: Oxford University Press.

Cosgrove, D. (1999). Introduction: Mapping meaning. In D. Cosgrove (Ed.), *Mappings* (pp. 1–23). London: Reaktion Books.

Couclelis, H. (1993). People manipulate objects (but cultivate fields): Beyond the raster–vector debate in GIS. *Proceedings, International GIS Conference—From Space to Territory: Theories and Methods of Spatio-Temporal Reasoning, Pisa, Italy* (pp. 65–77). Berlin: Springer-Verlag.

Couclelis, H. (1999). Space, time, geography. In P. A. Longley, M. F. Goodchild, D. J. Maguire, and D. W. Rhind (Eds.), *Geographical information systems* (pp. 29–38). New York: Wiley.

Couclelis, H., R. Golledge, N. Gale, and W. Tobler. (1987). Exploring the anchor-point hypothesis of spatial cognition. *Journal of Environmental Psychology, 7*(2), 99–122.

Coulson, M. R. C. (1991). *Progress in creating tactile maps from geographic infor-*

mation systems (G.I.S.) output. Fifteenth Conference of the International Cartographic Association, Bournemouth, UK, International Cartographic Association.

Coveney, P., and R. Highfield. (1990). *The arrow of time: A voyage through science to solve time's greatest mystery.* New York: Fawcett Columbine.

Coxeter, H. S. M. (1973). *Regular polytopes.* New York: Dover.

Dacey, M. F. (1970). The syntax of a triangle and some other figures. *Pattern Recognition, 2,* 11–31.

Dangermond, J. (1982). A classification of software components commonly used in geographic information systems. *Proceedings, United States–Australia Workshop on the Design and Implementation of Computer-Based Geographic Information Systems, Honolulu, Hawaii.*

Darken, R., and H. Cevik. (1999). Map usage in virtual environments: Orientation issues. *Proceedings, Virtual Reality '99, Institute of Electrical and Electronics Engineers, Houston, TX.*

Darken, R. P., T. Allard, and L. Achille. (1998). Spatial orientation and wayfinding in large-scale virtual spaces: An introduction. *Presence: Teleoperators and Virtual Environments, 7,* 101–107.

Dartnall, T. (Ed.). (1994a). *Artificial intelligence and creativity.* Dordrecht: Kluwer.

Dartnall, T. (1994b). Creativity, thought and representational redescription. In T. Dartnall (Ed.), *Artificial intelligence and creativity* (pp. 43–62). Dordrecht: Kluwer.

Darwen, H., and C. J. Date. (1998). *Foundation for object/relational databases: The third manifesto.* Reading, MA: Addison-Wesley.

Date, C. J. (1995). *Relational database writings 1991–1994.* Reading, MA: Addison-Wesley.

Date, C. J. (2000). *An introduction to database systems* (7th ed.). Reading, MA: Addison-Wesley.

Davies, C., and D. Medyckyj-Scott. (1996). GIS users observed. *International Journal of Gegraphical Information Science, 10*(4), 363–384.

Dayal, U., and J. M. Smith. (1986). A knowledge-oriented database management system. In M. L. Brodie and J. Mylopoulos (Eds.), *On knowledge base management systems* (pp. 227–258). New York: Springer-Verlag.

de Groot, A. D. (1965). *Thought and choice in chess.* The Hague: Mouton.

Denis, M. (1990). Imagery and thinking. In C. Cornoldi and M. McDaniel (Eds.), *Imagery and cognition* (pp. 103–131). New York: Springer-Verlag.

Denis, M., and M. Cocude. (1992). Structural properties of visual images constructed from poorly or well-structured verbal descriptions. *Memory and Cognition, 20*(5), 497–506.

Denis, M., and G. Denhiere. (1990). Comprehension and recall of spatial descriptions. *European Bulletin of Cognitive Psychology, 10,* 115–143.

Dennett, D. C. (1969). *Content and consciousness.* London: Routledge & Kegan Paul.

Deregowski, J. B. (1990). On two distinct and quintessential kinds of pictorial representation. In K. Landwehr (Ed.), *Ecological perception research, visual communication, and aesthetics* (pp. 29–42). New York: Springer-Verlag.

Derthick, M., J. Kolojejchick, and S. F. Roth. (1997). An interactive visualization environment for data exploration. *Proceedings, KDD '97, Third ACM SIGKDD International Conference on Knowledge Discovery and Data Mining, Newport Beach, CA* (pp. 2–9).

Descartes, R. (1980). *Discourse on the method of rightly conducting one's reason and of seeking truth in the sciences.* Indianapolis: Hackett.

Descartes, R. (1983). *Principles of philosophy.* Dordrecht: Kluwer.

Descartes, R. (1996). *Meditations on first philosophy* (edited and translated by J. Cottingham). New York: Cambridge University Press.

de Vega, M., and M. Marschark. (1996). Visuospatial cognition: An historical and theoretical introduction. In M. de Vega, M. J. Intons-Peterson, P. Johnson-Laird, M. Denis, and M. Marschark (Eds.), *Models of visuospatial cognition* (pp. 3–19). Oxford, UK: Oxford University Press.

DiBiase, D., A. MacEachren, J. Krygier, and C. Reeves. (1992). Animation and the role of map design in scientific visualization. *Cartography and GIS, 19*(4), 201–214.

Dodd, B. (1977). The role of vision in the perception of speech. *Perception, 6,* 31–40.

Donald, M. (1991). *Origins of the modern mind: Three stages in the evolution of culture and cognition.* Cambridge, MA: Harvard University Press.

Downs, R. (1970). The cognitive structure of an urban shopping center. *Environment and Behavior, 2,* 13–39.

Downs, R. (1981a). Maps and mappings as metaphors for spatial representation. In L. Liben, A. Patterson, and N. Newcombe (Eds.), *Spatial representation and behavior across the life span* (pp. 143–166). New York: Academic Press.

Downs, R. (1981b). Maps and metaphors. *Professional Geographer, 33*(3), 287–293.

Downs, R. (1985). The representation of space: Its development in children and in cartography. In R. Cohen (Ed.), *The development of spatial cognition* (pp. 323–344). Hillsdale, NJ: Erlbaum.

Downs, R., and D. Stea. (1973). *Image and environment.* Chicago: Aldine.

Dragicevic, S., and D. J. Marceau. (2000). A fuzzy set approach for modelling time in GIS. *International Journal of Geographical Information Science, 14*(3), 225–245.

Drysdale, R., and D. Lee. (1978). Generalized Voronoi diagram in the plane. *Proceedings, 16th Annual Allerton Conference on Communications Control and Computers* (pp. 833–839).

Duncan, E. M., and T. Bourg. (1983). An examination of the effects of encoding and decision processes on the rate of mental rotation. *Journal of Mental Imagery, 7,* 33–56.

Duncan, J. (1993). Sites of representation: Place, time and the discourse of the other. In J. Duncan and D. Ley (Eds.), *Place/culture/representation* (pp. 39–56). London: Routledge.

Dutton, G. (1996). Improving locational specificity of map data—a multi-resolution metadata-driven approach and notation. *International Journal of Geographical Information Systems, 10*(3), 253–268.

Dutton, G. (1999). *A hierarchical coordinate system for geoprocessing and cartography.* Berlin: Springer.

Dutton, G. (2000). *Universal geospatial data exchange via global hierarchical coordinates.* International Conference on Discrete Global Grids, Santa Barbara, CA.

Dutton, J. (1983). Geodesic modelling of planetary relief. *Proceedings, Auto-Carto 6, Ottawa.*

Dykes, J. A., K. E. Moore, and J. D. Wood. (1999). Virtual environments for student fieldwork using networked components. *International Journal of Geographical Information Science, 13*(4), 397–416.

Dyreson, C. E., and R. T. Snodgrass. (1993). Valid-time indeterminacy. *Proceedings, Ninth International Conference on Data Engineering, Vienna* (pp. 335–343).

Eastman, J. R. (1985). Graphic organization and memory structures for map learning. *Cartographica, 22*(1), 1–20.

Easton, R. D., and B. L. Bentzen. (1999). The effect of extended acoustic training on spatial updating in adults who are congenitally blind. *Journal of Visual Impairment and Blindness, 93*, 405–415.

Edsall, R., and D. Peuquet. (1997). A graphical user interface for the integration of time into GIS. *Proceedings, Auto-Carto 13, American Society for Photogrammetry and Remote Sensing/American Congress of Surveying and Mapping, Seattle, WA* (pp. 182–189).

Edsall, R. M., M.-J. Kraak, A. MacEachren and D. J. Peuquet. (1997). Assessing the effectiveness of temporal legends in environmental visualization. *Proceedings, GIS/LIS '97, American Society for Photogrammetry and Remote Sensing/American Congress of Surveying and Mapping, Cincinnati, OH* (pp. 677–685).

Egenhofer, M. J. (1989). A spatial SQL dialect. Technical Report, Department of Surveying Engineering, University of Maine, Orono.

Egenhofer, M. J. (1991). Extending SQL for graphical display. *Cartography and Geographic Information Systems, 18*(4), 230–245.

Egenhofer, M. J. (1992). Why not SQL! *International Journal of Geographical Information Systems, 6*(2), 71–85.

Egenhofer, M. J., and J. R. Richards. (1993). *The geographer's desktop: A direct-manipulation user interface for map overlay.* Auto Carto-11, Minneapolis, MN, pp. 63–71.

Ehrlich, K., and P. N. Johnson-Laird. (1982). Spatial descriptions and referential continuity. *Journal of Verbal Learning & Verbal Behavior, 21*, 296–306.

Eliasmith, C. (1997). Structure without symbols: Providing a distributed account of high-level cognition. *Proceedings, Southern Society for Philosophy and Psychology Conference, Atlanta, GA.*

Erwig, M., R. H. Güting, M. Schneider, and M. Vazirgiannis. (1998). Abstract and discrete modeling of spatio-temporal data types. *Proceedings, ACM GIS '98, Washington, DC* (pp. 131–136).

Erwig, M., R. H. Güting, M. Schneider, and M. Vazirgiannis. (1999). Spatio-temporal data types: An approach to modeling and querying moving objects in databases. *GeoInformatica, 3*(3), 269–296.

Estes, W. K. (1993). Models of categorization and category learning. In G. V. Nakamura, R. Taraban, and D. L. Medin (Eds.). *Categorization by humans and machines.* San Diego: Academic Press.

Evans, G. W., and K. Pezdek. (1980). Cognitive mapping: Knowledge of real-world distance and location information. *Journal of Experimental Psychology: Human Learning and Memory, 1*, 13–24.

Everett-Green, R. (1995). Year in review 1995: Computers-and-info-systems. Brittanica Online, November 29, 2000, http://members.eb.com/.

Farah, M. J. (1988). Is visual imagery really visual?: Overlooked evidence from neuropsychology. *Psychological Review, 95*, 307–317.

Farah, M. J., K. M. Hammond, D. N. Levine, and R. Calvanio. (1988). Visual and spatial mental imagery: Dissociable systems of representation. *Cognitive Psychology, 20*, 439–462.

Farah, M. J., F. Peronnet, M. A. Gono, and M. H. Giard. (1988). Electrophysiological evidence for a shared representational medium for visual images and visual percepts. *Journal of Experimental Psychology: General, 117*, 248–257.

Fauconnier, G. (1994). *Mental spaces: Aspects of meaning construction in natural language* (2nd ed.). Cambridge, UK: Cambridge University Press.

Faust, N. (1998). Raster based GIS. In T. W. Foresman (Ed.), *The history of geography information systems* (pp. 59–72). Upper Saddle River, NJ: Prentice-Hall.

Fayyad, U. M., G. Piatetsky-Shapiro, P. Smyth, and R. Uthurusamy. (Eds.). (1996). *Advances in knowledge discovery and data mining.* Menlo Park, CA: AAAI Press.

Fekete, G. (1990). Rendering and managing spherical data with sphere quadtrees. *Proceedings, Visualization '90* (pp. 176–186). San Francisco: Institute of Electrical and Electronics Engineers Computer Society Press.

Fekete, G., and L. S. Davis. (1984). Property spheres: A new representation for 3-D object recognition. *Workshop on Computer Vision Representation and Control* (pp. 192–201). Annapolis, MD: Institute of Electrical and Electronics Engineers Computer Society Press.

Ferguson, E., and M. Hegarty. (1994). Properties of cognitive maps constructed from texts. *Memory and Cognition, 22*(4), 455–473.

Fillmore, C. (1982). Frame semantics. In Linguistic Society of Korea (Ed.), *Linguistics in the morning calm* (pp. 111–137). Seoul: Hanshin.

Finke, R. A. (1989). *Principles of mental imagery*. Cambridge, MA: MIT Press.

Finke, R. A., S. Pinker, and M. J. Farah. (1989). Reinterpreting visual patterns in mental imagery. *Cognitive Science, 13*, 51–78.

Finke, R. A., and R. N. Shepard (1986). Visual functions of mental imagery. In K. R. Boff, L. Kaufman, and J. Thomas (Eds.), *Handbook of perception and human performance* (pp. 37–1 to 37–55). New York: Wiley.

Finke, R. A., and K. Slayton. (1988). Explorations of creative visual synthesis in mental imagery. *Memory and Cognition, 16*, 252–257.

Finke, R. A., T. B. Ward, and S. M. Smith. (1992). *Creative cognition: Theory, research, and applications*. Cambridge, MA: MIT Press.

Finkel, R. A., and J. L. Bentley. (1974). Quad trees: A data structure for retrieval on composite keys. *Acta Informatica, 4*, 1–9.

Fisher, P. F. (1989). Knowledge-based approaches to determining and correcting areas of unreliability in geographic databases. In M. Goodchild and S. Gopal (Eds.), *Accuracy of spatial databases* (pp. 45–54). London: Taylor & Francis.

Fisher, P. F. (1999). Models of uncertainty in spatial data. In P. A. Longley, M. F. Goodchild, D. J. Maguire, and D. W. Rhind (Eds.), *Geographical information systems* (pp. 191–205). New York: Wiley.

Fletcher, J. F. (1980). Spatial representation in blind children: 1. Development compared to sighted children. *Visual Impairment and Blindness, 74*, 381–385.

Fodor, J. A. (1975). *The language of thought*. New York: Crowell.

Fodor, J. A., and B. McLaughlin. (1990). Connectionism and the problem of systematicity: Why Smolensky's solution doesn't work. *Cognition, 35*, 183–204.

Fodor, J. A., and Z. W. Pylyshyn. (1988). Connectionism and cognitive architecture: A critical analysis. *Cognition, 28*, 3–71.

Foley, J., and A. Cohen. (1984). Working mental representations of the environment. *Environment and Behavior, 16*, 713–729.

Fortune, S. (1986). A sweepline algorithm for voronoi diagrams. *Proceedings, 2nd ACM Symposium on Computational Geometry*.

Fotheringham, A. S., and D. W. S. Wong. (1991). The modifiable areal unit problem in multivariate statistical analysis. *Environment and Planning A, 23*, 1025–1044.

Frank, A. (1982). Mapquery: Database query language for retrieval of geometric data and their graphical representation. *Computer Graphics, 16*(3), 199–207.

Frank, A. U. (1997). Spatial ontology: A geographical point of view. In O. Stock (Ed.), *Spatial and temporal reasoning* (pp. 135–153). Dordrecht: Kluwer Academic.

Franklin, N. (1992). Spatial representation for described environments. *Geoforum, 23*(2), 165–174.

Franklin, N., and B. Tversky. (1992). Switching points of view in spatial mental models. *Memory and Cognition, 20*(5), 507–518.

Franklin, W. R. (1990). Calculating map overlay polygons' areas without explicitly calculating the polygons—implementation. *Proceedings, Fourth International Symposium on Spatial Data Handling, 1*, Zurich, Switzerland, pp. 151–160.

Frasconi, P. (1998). A general framework for adaptive processing of data structures. *Institute of Electrical and Electronics Engineers Transactions on Neural Networks, 9*(5), 768–786.

Frasconi, P., and A. Sperduti. (2001). Guest editors' introduction: Special section on connectionist models for learning in structured domains. *Institute of Electrical and Electronics Engineers Transactions on Knowledge and Data Engineering, 13*(2), 145–147.

Freeman, J. (1973). The modeling of spatial relations. *Computer Graphics and Image Processing, 4*, 156–171.

Frege, G. (1980). On sense and meaning. In P. Geach and M. Black (Eds.), *Translations from the philosophical writings of Gottlob Frege* (pp. 56–78). Totowa, NJ: Rowman & Littlefield.

Freksa, C. (1992). Temporal reasoning based on semi-intervals. *Artificial Intelligence, 54*, 199–227.

Freundschuh, S. M. (1992). Is there a relationship between spatial cognition and environmental patterns? *Proceedings, International GIS Conference—From Space to Territory: Theories and Methods of Spatio-Temporal Reasoning, Pisa, Italy* (pp. 288–304). Berlin: Springer-Verlag.

Fuhrmann, S., and W. Kuhn. (1999). Interface design issues for interactive animated maps. *Proceedings, 19th ICA/ACI International Cartographic Conference, Ottawa, Canada* (pp. 875–884).

Fuhrmann, S., and A. MacEachren. (2001). Navigation in desktop geovirtual environments: Usability assessment. *Proceedings, 20th International Cartographic Conference, Beijing, China*.

Galison, P. L. (1985). Minkowski's space–time: From visual thinking to the absolute world. *Historical Studies in the Physical Sciences, 10*, 85–121.

Gallistel, C. R. (1990). *The organization of learning*. Cambridge, MA: MIT Press.

Gardels, K. (1994). Sequoia 2000 and geographic information: The Guernewood geoprocessor. *Proceedings, Sixth International Symposium on Spatial Data Handling, International Geographical Union, Edinburgh, Scotland* (pp. 1072–1085).

Garling, T. (1985). Adults' memory representations of the spatial properties of their everyday physical environment. In R. Cohen (Ed.), *The development of spatial cognition* (pp. 141–184). Hillsdale, NJ: Erlbaum.

Garling, T., A. Book, and E. Lindberg. (1984). Cognitive mapping of large-scale environments. *Environment and Planning, 16*(1), 3–34.

Gatrell, A. C. (1991). Concepts of space and geographical data. In D. J. Maguire, M. F. Goodchild, and D. Rhind (Eds.), *Geographical information systems, principles and applications* (pp. 119–133). New York: Wiley.

Gero, J. S. (1994). Computational models of creative design. In T. Dartnall (Ed.), *Artificial intelligence and creativity* (pp. 269–281). Dordrecht: Kluwer.

Ghassemzadeh, H. (1999). Some reflections on metaphoric processing: A move toward a meta-sign formulation. *New Ideas in Psychology, 17*, 41–54.

Gibbs, R., and H. Colston. (1995). The cognitive psychological reality of image schemas and their transformations. *Cognitive Linguistics, 6*(4), 347–378.

Gibbs, R. W., Jr., and J. Bogdonovich. (1999). Mental imagery in interpreting poetic metaphor. *Metaphor and Symbol, 14*(1), 37–44.

Gibson, E. J. (1969). *Principles of perceptual learning and development.* New York: Prentice-Hall.

Gibson, E. J. (1993). Ontogenesis of the perceived self. In U. Neisser (Ed.), *The perceived self: Ecological and interpersonal sources of self-knowledge* (pp. 25–42). Cambridge, UK: Cambridge University Press.

Gibson, E. J., and E. S. Spelke. (1983). The development of perception. In J. H. Flavell and E. M. Markman (Eds.), *Handbook of child psychology: Cognitive development* (pp. 2–76). New York: Wiley.

Gibson, J. J. (1966). *The senses considered as perceptual systems.* Boston: Houghton Mifflin.

Gibson, J. J. (1979). *The ecological approach to visual perception.* Boston: Houghton Mifflin.

Gibson, W. (1984). *Neuromancer.* New York: Ace Books.

Gleick, J. (1987). *Chaos: Making a new science.* New York: Viking Press.

Gleitman, L. R., and J. Gillette. (1999). The role of syntax in verb learning. In W. C. Ritchie and T. K. Bhatia (Eds.), *Handbook of child language acquisition* (pp. 279–295). San Diego: Academic Press.

Glenberg, A., and M. McDaniel. (1992). Mental models, pictures, and text: Integration of spatial and verbal information. *Memory and Cognition, 20*(5), 458–460.

Glenberg, A. M., and T. Grimes. (1995). Memory and faces: Pictures help you remember who said what. *Personality and Social Psychology Bulletin, 21*(3), 196–206.

Golledge, R. (1988). Integrating spatial knowledge. *Proceedings, International Geographical Congress, Sydney, Australia.*

Golledge, R. (1991). Tactual strip maps as navigational aids. *Journal of Blindness and Vision Impairment, 87*(7), 296–301.

Golledge, R. (1992). Place recognition and wayfinding: Making sense of space. *Geoforum, 23*(2), 199–214.

Golledge, R. (1993). Geography and the disabled—a survey with special reference to vision impaired and blind populations. *Transactions of the Institute of British Geographers, 18*, 63–85.

Golledge, R. (2001). *Expanding computer interfaces beyond visualization.* Position paper prepared for the Workshop on Intersections between Geospatial Information and Information Technology project of the Computer Science and Telecommunications Board, National Research Council, Washington, DC.

Golledge, R., V. Dougherty, and S. Bell. (1995). Acquiring spatial knowledge: Survey versus route-based knowledge in unfamiliar environments. *Annals of the Association of American Geographers*, *85*(1), 134–158.

Golledge, R., R. L. Klatzky, and J. M. Loomis. (1996). Cognitive mapping and wayfinding by adults without vision. In J. Portugali (Ed.), *The construction of cognitive maps* (pp. 215–245). Dordrecht: Kluwer Academic.

Golledge, R., and G. Rushton. (1976). Fundamental conflicts and the search for geographical knowledge. In P. Gould and G. Olsson (Eds.), *A search for common ground* (pp. 11–23). London: Pion.

Golledge, R., and H. Timmermans. (1990). Applications of behavioural research on spatial problems I: Cognition. *Progress in Human Geography*, *14*(1), 57–99.

Gollin, E. S. (1960). Developmental studies of visual recognition of incomplete objects. *Perceptual and Motor Skills*, *11*, 289–298.

Goodchild, M. (1987). Towards an enumeration and classification of GIS functions. *Proceedings, International Geographic Information Systems (IGIS) Symposium: The Research Agenda, Arlington, VA, NASA*.

Goodchild, M. (1989). Optimal tiling for large cartographic databases. *Proceedings, Arto-Carto 9* (pp. 444–451). Baltimore: American Congress on Surveying and Mapping.

Goodchild, M. (1990). Spatial information science. *Proceedings, Fourth International Symposium on Spatial Data Handling, Zurich, Switzerland* (pp. 3–12).

Goodchild, M. (1992). Geographical information science. *International Journal of Geographical Information Systems*, *6*(1), 31–45.

Goodchild, M., and S. Gopal. (1989). *Accuracy of spatial databases*. London: Taylor & Francis.

Goodchild, M., S. Guoqing, and Y. Shiren. (1992). Development and test of an error model for categorical data. *International Journal of Geographical Information Systems*, *6*(2), 87–104.

Goodchild, M., and Y. Shiren. (1992). A hierarchical spatial data structure for global geographic information systems. *CVGIP: Graphical Models and Image Processing*, *54*(1), 31–44.

Goonatilake, S., and S. Khebbal (1995). *Intelligent hybrid systems*. Chichester, UK: Wiley.

Gould, P. (1963). Man against his environment: A game theoretic framework. *Annals of the Association of American Geographers*, *53*(3), 290–297.

Gould, P. (1966). *On mental maps*. Ann Arbor: University of Michigan Press.

Green, M., and R. Flowerdew. (1996). New evidence on the modifiable areal unit problem. In P. Longley and M. Batty (Eds.), *Spatial analysis modelling in a GIS environment* (pp. 41–54). New York: Wiley.

Gruber, T. R. (1993). Toward principles for the design of ontologies used for knowledge sharing. *International Journal of Human and Computer Studies*, *43*(5/6), 907–928.

Guarino, N. (1997). Understanding, building, and using ontologies. *International Journal of Human and Computer Studies*, *46*(2/3), 293–310.

Guarino, N. (1998). Formal ontology and information systems. *Proceedings, FOIS '98* (pp. 3–15). Trento, Italy: IOS Press.

Guarino, N., and C. Welty. (2000). A formal ontology of properties. *Proceedings, 12th International Conference on Knowledge Engineering and Knowledge Management, Lecture Notes on Computer Science* (pp. 97–112), Juan-les-Pim, France: Springer-Verlag.

Guibas, L., and J. Stofoli. (1985). Primitives for the manipulations of general subdivisions and the computation of voronoi diagrams. *ACM Transactions on Graphics, 4*(2), 74–123.

Guptill, S. C., and M. Stonebraker. (1992). The Sequoia 2000 approach to managing large spatial object databases. *Proceedings, Fifth International Symposium on Spatial Data Handling, Charleston, SC* (pp. 642–651).

Güting, R. H. (1989). Gral: An extensible relational database system for geometric applications. *Proceedings, 15th International Conference on Very Large Data Bases, Amsterdam* (pp. 33–44).

Güting, R. H., M. H. Böhlen, M. Erwig, C. S. Hensen, N. A. Lorentzos, M. Schneider, and M. Vazirgiannis. (2000). A foundation for representing and querying moving objects. *ACM Transactions on Database Systems, 25*(1), 1–42.

Hägerstrand, T. (1967). *Innovation diffusion as a spatial process.* Chicago: University of Chicago Press.

Hägerstrand, T. (1970). What about people in regional science? *Papers of the Regional Science Association, 14*(7), 7–21.

Harley, J. B., and D. Woodward. (1987). *The history of cartography* (Vol. 1). Chicago: University of Chicago Press.

Hart, R. A., and G. T. Moore. (1973). The development of spatial cognition: A review. In R. M. Downs and D. Stea (Eds.), *Image and environment: Cognitive mapping and spatial behavior* (pp. 246–288). Chicago: Aldine.

Hartman, R. S. (1949). Cassirer's philosophy of symbolic forms. In P. A. Schilpp (Ed.), *The philosophy of Ernst Cassirer* (pp. 289–333). New York: Tudor.

Hartshorne, C., and P. Weiss. (Eds.). (1931–1958). *The collected papers of Charles Sanders Peirce.* Cambridge, MA: Harvard University Press.

Hayward, W. G., and M. J. Tarr. (1995). Spatial language and spatial representation. *Cognition, 55*(1), 39–84.

Hayward, W. G., and M. J. Tarr. (2000). Differing views on views: Comments on Biederman and Barr (1999). *Vision Research, 40*, 3895–3899.

Hayward, W. H., and P. Williams. (2000). Viewpoint dependence and object discriminability. *Psychological Science, 11*(1), 7–12.

Hazelton, N. W. J. (1991). *Integrating time, dynamic modelling and geographical information systems: Development of four-dimensional GIS.* Dissertation, Department of Surveying and Land Information, University of Melbourne, Melbourne, Australia.

Head, C. G. (1991). Mapping as language or semiotic system: Review and comment. In D. Mark and A. Frank (Eds.), *Cognitive and linguistic aspects of geographic space* (pp. 237–262). Dordrecht: Kluwer Academic.

Hebb, D. O. (1968). Concerning imagery. *Psychological Review, 75*, 466–477.

Hegarty, M., P. A. Carpenter, and M. A. Just. (1990). Diagrams in the comprehension of scientific text. In R. Barr, M. L. Kamil, P. Mosenthal, and P. D. Pearson (Eds.), *Handbook of reading research* (pp. 641–668). New York: Longman.

Hegel, G. W. F. (1990). *Encyclopedia of the philosophical sciences in outline and critical writings* (edited by Ernst Behler). New York: Continuum. (Original published 1817)

Held, G. D., M. R. Stonebraker and E. Wong (1975). INGRES—a relational data base system. *Proceedings, National Computer Conference 44.* Anaheim, CA: AFIPS Press.

Hernández, D. (1995). HCI aspects of a framework for the qualitative representation of space. In T. L. Nyerges et al. (Eds.), *Cognitive aspects of human–computer interaction for geographic information systems, Palma de Mallorca, Spain* (pp. 45–59). New York: Kluwer Academic Publishers.

Herring, J., R. Larsen, and J. Shivakumar. (1988). Extensions to the SQL language to support spatial analysis in a topological data base. *Proceedings, GIS/LIS '88, San Antonio, TX* (pp. 741–750).

Herring, J. R. (1992). TIGRIS: A data model for an object-oriented geographic information system. *Computers and Geosciences, 18*(4), 443–452.

Herskovits, A. (1985). Semantics and pragmatics of locative expressions. *Cognitive Science, 9*, 341–378.

Herskovits, A. (1986). *Language and spatial cognition: An interdisciplinary study of the preposition in English.* Cambridge, UK: Cambridge University Press.

Herskovits, A. (1998). Schematization. In P. Olivier and K.-P. Gapp (Eds.), *Representation and processing of spatial expressions* (pp. 149–162). Mahwah, NJ: Erlbaum.

Hesiod. (1999). *Theogony, works and days* (translated by M. L. West). Oxford, UK: Oxford University Press.

Heylighen, F., C. Joslyn, and V. Turchin. (1993). Metaphysics. *Principia Cybernetica Web*, February 5, 2001, http://pespmc1.vub.ac.be/METAPHYS.htm.

Hill, C. (1982). Up/down, front/back, left/right: A contrastive study of Hausa and English. In J. Weissenborn and W. Klein (Eds.), *Here and there: Cross-linguistic studies on deixis and demonstration* (pp. 11–42). Amsterdam: Benjamins.

Hirtle, S., and J. Jonides. (1985). Evidence of hierarchies in cognitive maps. *Memory and Cognition, 13*(3), 208–217.

Hirtle, S. C. (1994). Towards a cognitive GIS. In M. Molenaar and S. de Hoop (Eds.), *Advanced geographic data modelling: Spatial data modelling and query languages for 2d and 3d applications* (pp. 217–227). Delft, The Netherlands: Netherlands Geodetic Commission.

Hochberg, J. (1968). In the mind's eye. In R. N. Haber (Ed.), *Contemporary theory and research in visual perception* (pp. 309–331). New York: Holt, Rinehart & Winston.

Hofstadter, D. R. (1979). *Gödel, Escher, Bach: An eternal golden braid.* New York: Vintage Books.

Hofstadter, D. R., and M. Mitchell. (1994). The copycat project: A model of mental fluidity and analogy-making. In K. J. Holyoak and J. A. Barnden (Eds.), *Analogical connections* (pp. 31–73). Norwood, NJ: Ablex.

Hollins, M. (1989). *Understanding blindness.* Hillsdale, NJ: Erlbaum.

Holmes, N. (1992). *Pictorial maps.* London: Herbert Press.

Holyoak, K. J., and P. Thagard. (1989). Analogical mapping by constraint satisfaction. *Cognitive Science, 13*(3), 295–355.

Hooker, C. A. (Ed.). (1973a). Contemporary research in the foundations and philosophy of quantum theory. *Proceedings of a conference held at the University of Western Ontario, London, Canada.* Dordrecht, The Netherlands: D. Reidel.

Hooker, C. A. (1973b). Metaphysics and modern physics: A prolegomenon to the understanding of quantum theory. In C. A. Hooker (Ed.), *Contemporary Research in the Foundations and Philosophy of Quantum Theory* (pp. 174–304). London, Ontario, Canada: D. Reidel.

Hooker, C. A. (1995). *Reason, regulation, and realism: Toward a regulatory systems theory of reason and evolutionary epistemology.* Albany: State University of New York Press.

Hoppe, H. (1998). Smooth view-dependent level-of-detail control and its application to terrain rendering. *Proceedings, Visualization '98, Institute of Electrical and Electronics Engineers* (pp. 35–42).

Hornsby, K., and M. J. Egenhofer. (2000). Identity-based change: A foundation for spatio-temporal knowledge representation. *International Journal of Geographical Information Science, 14*(3), 207–224.

Hubel, D. H., and T. N. Wiesel. (1965). Receptive fields of single neurons in two nonstriate visual areas (18 and 19) of the cat. *Journal of Neurophysiology, 28,* 229–289.

Hunter, G. M. (1978). *Efficient computation and data structures for graphics.* Princeton, NJ: Princeton University Press.

Ingram, K., and W. Phillips (1987). Geographic information processing using a SQL-based query language. *Proceedings, Auto-Carto 8, American Society for Photogrammetry and Remote Sensing, Baltimore, MD* (pp. 326–335).

Intons-Peterson, M., W. Russell, and S. Dressel. (1992). The role of pitch in auditory imagery. *Journal of Experimental Psychology: Human Perception and Performance, 18,* 233–240.

Ioerger, T. R. (1994). The manipulation of images to handle indeterminacy in spatial reasoning. *Cognitive Science, 18,* 551–593.

Ivic, M. (1965). *Trends in linguistics.* London: Mouton.

Jackendoff, R. (1983). *Semantics and cognition.* Cambridge, MA: MIT Press.

Jackendoff, R. (1987). *Consciousness and the computational mind*. Cambridge, MA: MIT Press.

Jackendoff, R. (1992). *Languages of the mind: Essays on mental representation*. Cambridge, MA: MIT Press.

Jackendoff, R. (1996). The architecture of the linguistic–spatial interface. In P. Bloom, M. A. Peterson, L. Nadel, and M. F. Garrett (Eds.), *Language and space* (pp. 1–30). Cambridge, MA: MIT Press.

Jackendoff, R., and B. Landau. (1991). Spatial language and spatial cognition. In D. J. Napoli and J. A. Kegl (Eds.), *Bridges between psychology and linguistics: A Swarthmore festschrift for Lila Gleitman*. Hillsdale, NJ: Erlbaum.

Jackins, C. L., and S. L. Tanimoto. (1980). Oct-trees and their use in representing three-dimensional objects. *Computer Graphics and Image Processing, 14*(3), 249–270.

Jackins, C. L., and S. L. Tanimoto. (1983). Quad-trees, oct-trees, and k-trees: A generalized approach to recursive decomposition of Euclidean space. *IEEE Transactions on Pattern Analysis and Machine Intelligence, 5*(5), 533–539.

Jacob, C. (1999). Mapping in the mind: The earth from ancient Alexandria. In D. Cosgrove (Ed.), *Mappings* (pp. 24–49). London: Reaktion.

Jensen, C. S., and R. T. Snodgrass. (1996). Semantics of time-varying information. *Information Systems, 21*(4), 311–352.

Jensen, C. S., and R. Snodgrass. (1999). Temporal data management. *IEEE Transactions on Knowledge and Data Engineering, 11*(1), 36–44.

Jensen, J. R. (1996). *Introductory digital image processing: A remote sensing perspective*. Upper Saddle River, NJ: Prentice-Hall.

Johansson, G. (1973). Visual perception of biological motion and a model for its analysis. *Perception and Psychophysics, 14*(2), 201–211.

Johnson, M. (1987). *The body in the mind: The bodily basis of meaning, imagination, and reason*. Chicago: University of Chicago Press.

Johnson, M. (1991). Knowing through the body. *Philosophical Psychology, 4*(1), 3–18.

Johnson-Laird, P. N. (1983). *Mental models*. Cambridge, UK: Cambridge University Press.

Johnson-Laird, P. (1996). Images, models and propositional representations. In M. de Vega, M. Intons-Peterson, P. Johnson-Laird, M. Denis, and M. Marschark (Eds.), *Models of visuospatial cognition* (pp. 90–127). Oxford, UK: Oxford University Press.

Jolicoeur, P., D. Snow, and J. Murray. (1987). The time to identify disoriented letters: Effects of practice and font. *Canadian Journal of Psychology, 41*, 303–316.

Jones, B., and S. O'Neill. (1985). Combining vision and touch in texture perception. *Perception and Psychophysics, 37*, 66–72.

Jordan, K., and L. Huntsman. (1990). Image rotation of misoriented letter strings: Effects of orientation cuing and repetition. *Perception and Psychophysics, 48*, 363–374.

Jungert, E. (1992). The observer's point of view: An extension of symbolic projections. *International Conference GIS—From Space to Territory: Theories and Methods of Spatio-Temporal Reasoning* (pp. 179–195). Pisa, Italy: Springer-Verlag.

Kainz, W. (1987). A classification of digital map data models. *Proceedings, Euro-Carto 6, Brno, Czechoslovakia* (pp. 105–113).

Kant, I. (1950). *Critique of pure reason.* New York: Humanities Press.

Kant, I. (1955). *Kant's Prolegomena to any future metaphysics* (P. Carus, Ed.). LaSalle, IL: Open Court.

Karmiloff-Smith, A. (1986). From metaprocess to conscious access: Evidence from children's metalinguistic and repair data. *Cognition, 23,* 95–147.

Karmiloff-Smith, A. (1990). Constraints on representational change: Evidence from children's drawing. *Cognition, 34,* 57–83.

Karmiloff-Smith, A. (1992). *Beyond modularity: A developmental perspective on cognitive science.* Cambridge, MA: MIT Press.

Keil, F. C., and N. Batterman. (1984). A characteristic-to-defining shift in the development of word meaning. *Journal of Verbal Learning and Verbal Behavior, 23,* 221–236.

Kelmelis, J. (1991). *Time and space in geographic information: Toward a four-dimensional spatio-temporal data model.* Doctoral dissertation, Pennsylvania State University, University Park, PA.

Kennedy, J. M., and P. Gabias. (1985). Metaphoric devices in drawings of motion mean the same to the blind and sighted. *Perception, 14,* 189–195.

Kennedy, J. M., P. Gabias, and M. A. Heller. (1992). Space, haptics and the blind. *Geoforum, 23*(2), 175–189.

Kerr, N. H. (1983). The role of vision in visual imagery experiments: Evidence from the congenitally blind. *Journal of Experimental Psychology: General, 112,* 265–277.

Klatzky, R. L., R. G. Golledge, J. M. Loomis, J. G. Cicinelli, and J. W. Pellegrino. (1995). Performance of blind and sighted persons on spatial tasks. *Journal of Visual Impairment and Blindness, 89,* 70–82.

Klinger, A. (1971). Patterns and search statistics. In J. S. Rustagi (Ed.), *Optimizing methods in statistics* (pp. 303–337). New York: Academic Press.

Klinger, A., and C. Dyer. (1976). Experiments on picture representation using regular decomposition. *Computer Graphics and Image Processing, 5,* 68–105.

Knox, P. (1991). The restless urban landscape: Economic and sociocultural change and the transformation of metropolitan Washington, DC. *Annals of the Association of American Geographers, 81*(2), 181–209.

Knuth, D. (1998). *The art of computer programming.* Reading, MA: Addison-Wesley.

Koh, B., and T. Chen. (1999). *Progressive browsing of 3-D models.* Workshop on Multimedia Signal Processing. Copenhagen: IEEE Signal Processing Society.

Kopec, R. R. (1963). An alternative method for the construction of Thiessen polygons. *Professional Geographer, 15*(5), 24–26.

Kosslyn, S. (1975). Information representation in visual images. *Cognitive Psychology, 7,* 341–370.

Kosslyn, S. (1980). *Image and mind.* Cambridge, MA: Harvard University Press.

Kosslyn, S. (1987). Seeing and imagining in the cerebral hemispheres: A computational approach. *Psychological Review, 94,* 148–175.

Kosslyn, S., T. Ball, and B. Reiser. (1978). Visual images preserve metric spatial information: Evidence from studies of image scanning. *Journal of Experimental Psychology: Human Perception and Performance, 4,* 47–60.

Kosslyn, S., C. Chabris, C. Marsolek, and O. Koenig. (1992). Categorical versus coordinate spatial relations: Computational analyses and computer simulations. *Journal of Experimental Psychology: Human Perception and Performance, 18*(2), 562–577.

Kosslyn, S. M., and O. Koenig. (1992). *Wet mind: The new cognitive neuroscience.* New York: Free Press.

Kosslyn, S., O. Koenig, A. Barrett, C. B. Cave, J. Tang, and J. D. E. Gabrieli. (1989). Evidence for two types of spatial representations: Hemispheric specialization of categorical and coordinate relations. *Journal of Experimental Psychology: Human Perception and Performance, 15*(4), 723–735.

Kosslyn, S., and J. Pomerantz. (1977). Imagery, propositions, and the form of internal representations. *Cognitive Psychology, 9,* 52–76.

Kraak, M.-J. (1999). Visualising spatial distributions. In P. A. Longley, M. F. Goodchild, D. J. Maguire, and D. W. Rhind (Eds.), *Geographical information systems* (pp. 157–173). New York: Wiley.

Kraak, M.-J., and R. van Driel. (1997). Principles of hypermaps. *Computers and Geosciences, 23*(4), 457–464.

Kroes, P. A. (1988). *Newton's mathematization of physics in retrospect.* Dordrecht: Kluwer Academic.

Kuhn, T. S. (1996). *The structure of scientific revolutions* (3rd ed.). Chicago: University of Chicago Press.

Kuipers, B. (1978). Modeling spatial knowledge. *Cognitive Science, 2,* 129–153.

Kuipers, B. (1982). The "map in the head" metaphor. *Environment and Behavior, 14*(2), 202–220.

Lakoff, G. (1987). *Women, fire, and dangerous things: What categories reveal about the mind.* Chicago: University of Chicago Press.

Lakoff, G. (1990). The invariance hypothesis: Is abstract reason based on image-schemas? *Cognitive Linguistics, 1*(1), 39–74.

Lakoff, G. (1993). The contemporary theory of metaphor. In A. Ortony (Ed.), *Metaphor and thought* (pp. 202–251). Cambridge, UK: Cambridge University Press.

Lakoff, G., and M. Johnson. (1980). *Metaphors we live by.* Chicago: University of Chicago Press.

Lakoff, G., and M. Turner. (1989). *More than cool reason: A field guide to poetic metaphor*. Chicago: University of Chicago Press.

Landau, B. (1986). Early map use as an unlearned ability. *Cognition, 22*(3), 201–223.

Landau, B., and R. Jackendoff. (1993). "What" and "where" in spatial language and spatial cognition. *Behavioral and Brain Sciences, 16*(2), 217–265.

Landau, B., and E. Munnich. (1998). The representation of space and spatial language: Challenges for cognitive science. In P. Olivier and K.-P. Gapp (Eds.), *Representation and processing of spatial expressions* (pp. 263–272). Mahwah, NJ: Erlbaum.

Langran, G. (1992). *Time in geographic information systems*. London: Taylor & Francis.

Langran, G., and N. R. Chrisman. (1988). A framework for temporal geographic information. *Cartographica, 25*(3), 1–14.

Larkin, J. H., and H. A. Simon. (1987). Why a diagram is (sometimes) worth ten thousand words. *Cognitive Science, 11*, 65–99.

Lauzon, J. P., D. Mark, L. Kikuchi, and J. A. Guevara. (1985). Two-dimensional run-encoding for quadtree representation. *Computer Vision, Graphics and Image Processing, 30*(1), 56–69.

Lederer, A., H. Gleitman, and L. Gleitman. (1995). Verbs of a feather flock together: Semantic information in the structure of maternal speech. In M. Tomasello and W. E. Merriman (Eds.), *Beyond names for things: Young children's acquisition of verbs* (pp. 277–297). Hillsdale, NJ: Erlbaum.

Lederman, S. J., and S. G. Abbott. (1981). Texture perception: Studies of intersensory organization using a discrepancy paradigm and visual vs. tactual psychophysics. *Journal of Experimental Psychology: Human Perception and Performance, 47*, 54–64.

Lee, H.-Y., H.-L. Ong, and L.-H. Quek. (1995). Exploiting visualization in knowledge discovery. *Proceedings, KDD '95, First ACM SIGKDD International Conference on Knowledge Discovery and Data Mining, Montréal, Québec* (pp. 198–203).

Lefebvre, H. (1992). *The production of space*. Oxford, UK: Blackwell.

Levine, D., J. Warach, and M. Farah. (1985). Two visual systems in mental imagery: Dissociation of "what" and "where" in imagery disorders due to bilateral posterior cerebral lesions. *Neurology, 35*, 1010–1018.

Levinson, S. C. (1996). Frames of reference and Molyneux's question: Cross-linguistic evidence. In P. Bloom, M. A. Peterson, L. Nadel, and M. F. Garrett (Eds.), *Language and space* (pp. 109–169). Cambridge, MA: MIT Press.

Levy, R. (1973). *Tahitians: Mind and experience in the society islands*. Chicago: University of Chicago Press.

Ley, D. (1974). *The black inner city as frontier outpost* (AAG Monograph Series No. 7). Washington, DC: Association of American Geographers.

Liben, L. (1981). Spatial representation and behavior: Multiple perspectives. In L. Liben, A. Patterson, and N. Newcombe (Eds.), *Spatial representation and behavior across the life span* (pp. 3–36). New York: Academic Press.

Liben, L. S. (1988). Conceptual issues in the development of spatial cognition. In J. Stiles-Davis, M. Kritchevsky, and U. Bellugi (Eds.), *Spatial cognition: Brain bases and development* (pp. 167–194). Hillsdale, NJ: Erlbaum.

Livingstone, D. N., and R. T. Harrison. (1981). Meaning through metaphor: Analogy as epistemology. *Annals of the Association of American Geographers*, *71*(1), 95–107.

Lloyd, R. (1989). Cognitive maps: Encoding and decoding information. *Annals of the Association of American Geographers*, *79*(1), 101–124.

Lloyd, R., and C. Heivly. (1987). Systematic distortions in urban cognitive maps. *Annals of the Association of American Geographers*, *77*(2), 191–207.

Lloyd, R., D. Patton, and R. Cammack. (1996). Basic-level categories. *Professional Geographer*, *48*(2), 181–194.

Logan, G. D. (1995). Linguistic and conceptual control of visual spatial attention. *Cognitive Psychology*, *28*(2), 103–174.

Logan, G. D., and D. D. Sadler. (1996). A computational analysis of the apprehension of spatial relations. In P. Bloom and M. A. Peterson (Eds.), *Language and space: Language, speech and communication* (pp. 493–529). Cambridge, MA: MIT Press.

Lolonis, P., and M. Armstrong. (1993). *Temporal information in spatial decision support systems*. Minneapolis: GIS/LIS.

Loomis, J. M., R. L. Klatzky, R. G. Golledge, J. G. Cicinelli, J. W. Pellegrino, and P. A. Fry. (1993). Nonvisual navigation by blind and sighted: Assessment of path integration ability. *Journal of Experimental Psychology: General*, *122*(1), 73–91.

Lounsbury, F. (1964). A formal account of the Crow- and Omaha-type kinship terminologies. In W. H. Goodenough (Ed.), *Explorations in cultural anthropology* (pp. 351–394). New York: McGraw-Hill.

Luchins, A. S., and E. H. Luchins. (1959). *Rigidity of behavior*. Eugene: University of Oregon Press.

Lynch, K. (1960). *The image of the city*. Cambridge, MA: MIT Press.

MacEachren, A. (1992). Learning spatial information from maps: Can orientation-specificity be overcome? *Professional Geographer*, *44*(4), 431–443.

MacEachren, A. (1995). *How maps work: Representation, visualization, and design*. New York: Guilford Press.

MacEachren, A., C. Brewer, and E. Steiner. (2001). Geovisualization to mediate collaborative work: Tools to support different-place knowledge construction and decision-making. *Proceedings, 20th International Cartographic Conference, Bejing, China*.

MacEachren, A. M., R. Edsall, D. Haug, R. Baxter, G. Otto, R. Masters, S.

Fuhrmann, and L. Qian. (1999). Virtual environments for geographic visualization: Potential and challenges. *Proceedings, ACM Workshop on New Paradigms in Information Visualization and Manipulation, Kansas City, KS* (pp. 35–40).

MacEachren, A. M., and M.-J. Kraak. (2001). Research challenges in geovisualization. *Cartography and Geographic Information Science, 28*(1), 3–12.

MacEachren, A. M., M. Wachowicz, D. Haug, R. Edsall, and R. Masters. (1999). Constructing knowledge from multivariate spatiotemporal data: Integrating geographic visualization with knowledge discovery in database methods. *International Journal of Geographic Information Science, 13*(4), 311–334.

Maguire, D., G. Stickler, and G. Browning. (1992). Handling complex objects in geo-relational GIS. *Proceedings, Fifth International Symposium on Spatial Data Handling, Charleston, SC* (pp. 652–661).

Maki, R. H. (1981). Categorization and distance effects with spatial linear orders. *Journal of Experimental Psychology: Human Learning and Memory, 7*, 15–32.

Mandl, H., and J. R. Levin (Eds.). (1989). *Knowledge acquisition from text and pictures*. Amsterdam: North-Holland.

Manovich, L. (2000). *The language of new media*. Cambridge, MA: MIT Press.

Mark, D., and F. Csillag. (1989). The nature of boundaries on "area-class" maps. *Cartographica, 26*(1), 65–78.

Mark, D. M. (1975). Computer analysis of topography: A comparison of terrain storage methods. *Geografiska Annaler, 57a*, 179–188.

Mark, D. M., A. Frank, and M. Egenhofer. (1989). *Languages of spatial relations*. Initiative two specialist meeting report (National Center for Geographic Information and Analysis Technical Paper No. 89-2), Santa Barbara, CA.

Mark, D. M., C. Freksa, S. C. Hirtle, R. Lloyd, and B. Tversky. (1999). Cognitive models of geographical space. *International Journal of Gegraphical Information Science, 13*(8), 747–774.

Mark, D. M., and S. M. Freundschuh. (1995). Spatial concepts and cognitive models for geographic information use. In T. L. Nyerges et al. (Eds.), *Cognitive aspects of human-computer interaction for geographic information systems* (pp. 21–28). Palma de Mallorca, Spain: Kluwer Academic.

Markman, E. M., and M. A. Callanan. (1984). An analysis of hierarchical classification. In R. Sternberg (Ed.), *Advances in the psychology of human intelligence* (pp. 325–365). Hillsdale, NJ: Erlbaum.

Marks, L. E. (1996). On perceptual metaphors. *Metaphor and Symbolic Activity, 11*(1), 39–66.

Marr, D. (1982). *Vision*. New York: Freeman.

Martin, J. (1977). *Computer data-base organization*. Englewood Cliffs, NJ: Prentice-Hall.

Massaro, D. W., and D. Friedman. (1990). Models of integration given multiple sources of information. *Psychological Review, 97*, 225–252.

Matlin, M. (1998). *Cognition*. Fort Worth, TX: Harcourt Brace.

Mayer, R. E., and J. K. Gallini. (1991). When is an illustration worth ten thousand words? *Journal of Educational Psychology, 82*(4), 715–726.

McClelland, J., D. Rumelhart, and G. Hinton. (1986). The appeal of parallel distributed processing. In D. Rumelhart and J. McClelland (Eds.), *Parallel distributed processing, explorations in the microstructures of cognition* (pp. 3–44). Cambridge, MA: MIT Press.

McCormick, B. H., T. A. DeFanti, and M. D. Brown. (1987). Visualization in scientific computing [Special issue]. *ACM Siggraph, 21*(6), 1–14.

McGurk, H., and J. MacDonald (1976). Hearing lips and seeing voices. *Nature, 264,* 746–748.

McNamara, T. P. (1992). Spatial representation. *Geoforum, 23*(2), 139–150.

Medsker, L. R. (1995). *Hybrid intelligent systems*. Boston: Kluwer.

Medyckyj-Scott, D., and M. Blades. (1992). Human spatial cognition: Its relevance to the design and use of spatial information systems. *Geoforum, 23*(2), 215–226.

Meier, A., and M. Ilg. (1986). Consistent operations on a spatial data structure. *IEEE Transactions on Pattern Analysis and Machine Intelligence, 8*(4), 532–538.

Meinig, D. (1979). Symbolic landscapes: Models of American community. In D. Meinig (Ed.), *The interpretation of ordinary landscapes: Geographical essays* (pp. 164–192). New York: Oxford University Press.

Meltzoff, A. N., and M. K. Moore. (1998). Object representation, identity, and the paradox of early permanence: Steps toward a new framework. *Infant Behavior and Development, 21,* 201–235.

Mennis, J., D. J. Peuquet, and L. Qian. (2000). A conceptual framework for incorporating cognitive principles into geographical database representation. *International Journal of Geographical Information Science, 14*(6), 501–520.

Metzler, J., and R. N. Shepard. (1982). Transformational studies of the internal representations of three-dimensional objects. In R. N. Shepart and L. A. Cooper (Eds.), *Mental images and their transformations* (pp. 25–71). Cambridge, MA: MIT Press.

Millar, S. (1972). The effects of interpolated tasks on latency and accuracy of intramodal and crossmodal shape recognition by children. *Journal of Experimental Child Psychology, 16,* 170–175.

Millar, S. (1975a). Effects of input variables on visual and kinaesthetic matching by children within and across modalities. *Journal of Experimental Child Psychology, 19,* 63–78.

Millar, S. (1975b). Visual experience or translation rules?: Drawing the human figure by blind and sighted children. *Perception, 4,* 363–371.

Millar, S. (1985). The perception of complex patterns by touch. *Perception, 14,* 293–303.

Millar, S. (1988). Models of sensory deprivation: The nature/nurture dichotomy and spatial representation in the blind. *International Journal of Behavioural Development, 11,* 69–87.

Millar, S. (1994). *Understanding and representing space: Theory and evidence from studies with blind and sighted children.* Oxford, UK: Clarendon Press.

Miller, A. I. (1996a). *Insights of genius: Imagery and creativity in science and art.* New York: Springer-Verlag.

Miller, A. I. (1996b). Metaphors in creative scientific thought. *Creativity Research Journal, 9*(2/3), 113–130.

Minkowski, H. (1953). *Geometrie der Zahlen.* New York: Chelsea. [Reprint of the original edition, published in 2 parts, 1896–1910]

Minsky, M. (1975). A framework for representing knowledge. In P. H. Winston (Ed.), *The psychology of computer vision* (pp. 211–277). New York: McGraw-Hill.

Molenaar, M. (1998). *An introduction to the theory of spatial object modelling for GIS.* London: Taylor & Francis.

Monmonier, M., and H. J. de Blij. (1991). *How to lie with maps.* Chicago: University of Chicago Press.

Montanari, A., and B. Pernici. (1993). Towards a temporal logic reconstruction of temporal databases. *Proceedings, International Workshop on an Infrastructure for Temporal Databases, Arlington, TX* (pp. BB1–BB12).

Montello, D. (1992). The geometry of environmental knowledge. In A. U. Frank, I. Campari, and U. Formentini (Eds.), *International Conference GIS—From Space to Territory: Theories and methods of spatio-temporal reasoning: Lecture notes in Computer Science 639* (pp. 136–152). Pisa, Italy: Springer-Verlag.

Montello, D. (1993). Scale and multiple psychologies of space. In A. U. Frank and I. Campari (Eds.), *Spatial information theory: A theoretical basis for GIS: Lecture notes in computer science 716* (pp. 312–321). Berlin: Springer-Verlag.

Moore, K., J. Dykes, and J. Wood. (1999). Using JAVA to interact with geo-referenced VRML within a virtual field course. *Computers and Geosciences, 25*(10), 1125–1136.

Morris, C. (1938). Foundations of the theory of signs. In O. Neurath (Ed.), *International encyclopedia of unified science* (Vol. 1, No. 2). Chicago: University of Chicago Press.

Morrison, J. L. (1974). A theoretical framework for cartographic generalization with the emphasis on the process of symbolization. *International Yearbook of Cartography, 14,* 115–127.

Morton, G. M. (1966). *A computer oriented geodetic data base; and a new technique in file sequencing.* Ottawa, Canada: IBM.

Moyer, R. S. (1973). Comparing objects in memory: Evidence suggesting an internal psychophysics. *Perception and Psychophysics, 13,* 180–184.

Munnich, E., B. Landau, and B. A. Dosher. (2001). Spatial language and spatial representation: A cross-linguistic comparison. *Cognition, 81,* 171–207.

Murphy, G. L., and E. J. Wisniewski. (1989). Categorizing objects in isolation and in scenes: What a superordinate is good for. *Journal of Experimental Psychology: Learning, Memory, and Cognition*, 15(4), 572–586.

National Research Council. (1999). *Distributed geolibraries: Spatial information resources*. Washington, DC: National Academy Press.

Neisser, U. (1967). *Cognitive psychology*. New York: Appleton–Century–Crofts.

Neisser, U. (1976). *Cognition and reality*. New York: Freeman.

Neisser, U. (Ed.). (1987a). *Concepts and conceptual development: Ecological and intellectual factors in categorization*. Cambridge, UK: Cambridge University Press.

Neisser, U. (1987). From direct perception to conceptual structure. In U. Neisser (Ed.), *Concepts and conceptual development: Ecological and intellectual factors in categorization* (pp. 11–24). Cambridge, UK: Cambridge University Press.

Neisser, U. (1989). Direct perception and recognition as distinct perceptual systems. *Proceedings, Annual Conference of the Cognitive Science Society, Ann Arbor, MI*.

Neisser, U. (1993). *Distinct systems for "where" and "what": Reconciling the ecological and representational views of perception*. San Diego: American Psychological Society.

Neves, N., J. P. Silva, P. Gonçalves, J. Muchaxo, J. M. Lilva, and A. Câmara. (1997). Cognitive spaces and metaphors: A solution for interacting with spatial data. *Computers and Geosciences*, 23(4), 483–488.

Newman, S. S. (1933). Further experiments in phonetic symbolism. *American Journal of Psychology*, 45, 53–75.

Newton, S. I. (1962). *Mathematical principles of natural philosophy and his system of the world* (translated by Andrew Motte in 1729, revised by Florian Cajori). New York: Greenwood Press.

North, C., and F. Korn. (1996). Browsing anatomical image databases: A case study of the visible human (Conference Companion and Video). *Proceedings, ACM CHI '96, Human Factors in Computing Systems, Association for Computing Machinery, Vancouver, Canada* (pp. 414–415).

North, C., and B. Shneiderman. (2000). Snap-together visualization: Can users construct and operate coordinated views? *International Journal of Human–Computer Studies*, 53(5), 715–739.

Nuti, L. (1999). Mapping places: Chorography and vision in the renaissance. In C. Cosgrove (Ed.), *Mappings* (pp. 71–108). London: Reaktion.

Nyerges, T. L. (1991). Geographic information abstractions: Conceptual clarity for geographic modeling. *Environment and Planning A*, 23, 1483–1499.

Nyerges, T. L., D. M. Mark, R. Laurini, and M. J. Egenhofer. (Eds.). (1995). *Cognitive aspects of human–computer interaction for geographic information systems*. Dordrecht: Kluwer.

Nystuen, J. (1963). Identification of some fundamental spatial concepts. *Papers of the Michigan Academy of Science, Arts, and Letters*, 48, 373–384.

Oakhill, J. V., and P. N. Johnson-Laird. (1985). Representation of spatial descriptions in working memory. *Current Psychological Research and Reviews, 3*(1), 52–62.

O'Keefe, J., and L. Nadel. (1978). *The hippocampus as a cognitive map.* Oxford, UK: Clarendon Press.

Olson, J., and C. Brewer. (1997). An evaluation of color selections to accommodate maps users with color-vision impairments. *Annals of the Association of American Geographers, 87*(1), 103–134.

Ooi, B., R. Sacks-Davis, and K. McDonell. (1989). Extending a DBMS for geographic applications. *Proceedings, IEEE Fifth International Conference on Data Engineering, Los Angeles, CA* (pp. 590–597).

Openshaw, S., M. Charlton, C. Wymer, and A. Craft. (1987). A Mark I geographical analysis machine for the automated analysis of point data sets. *International Journal of Gegraphical Information Systems, 1*(4), 335–358.

Ortony, A. (1979). Beyond literal similarity. *Psychological Review, 86,* 161–180.

Otoo, E. J., and H. Zhu. (1993). Indexing on spherical surfaces using semi-quadcodes. In D. Abel and B. C. Ooi (Eds.), *Advances in Spatial Databases— Third Annual Symposium, SSD '93: Lecture notes in Computer Science 692* (pp. 510–529). Singapore: Springer-Verlag.

Oviatt, S. (1996). Multimodal interfaces for dynamic interactive maps. *Proceedings, CHI '96, Human Factors in Computing Systems. Association for Computing Machinery, Vancouver, Canada* (pp. 95–102).

Paivio, A. (1971). *Imagery and verbal processes.* New York: Holt, Reinehart & Winston.

Paivio, A. (1978). On exploring visual knowledge. In B. S. Randhawa and W. E. Coffman (Eds.), *Visual learning, thinking and communication* (pp. 113–132). New York: Academic Press.

Paivio, A. (1983). The empirical case for dual coding. In J. C. Yuille (Ed.), *Imagery, memory and cognition* (pp. 307–332). Hillsdale, NJ: Erlbaum.

Paivio, A. (1986). *Mental representations: A dual coding approach.* Oxford, UK: Clarendon Press.

Palmer, S. E. (1978). Fundamental aspects of cognitive representation. In E. Rosch and B. B. Lloyd (Eds.), *Cognition and categorization* (pp. 259–303). New York: Wiley.

Passini, R., G. Prouix, and C. Rainville. (1990). The spatio-cognitive abilities of the visually impaired population. *Environment and Behavior, 22*(1), 91–118.

Peano, G. (1973). *Selected works* (Ed., H. C. Kennedy). Toronto: Toronto University Press.

Peirce, C. S. (1868). On a new list of categories. *Proceedings, American Academy of Arts and Sciences.*

Peterson, M. P. (1995). *Interactive and animated cartography.* Englewood Cliffs, NJ: Prentice-Hall.

Petrie, H. G. (1993). Metaphor and learning. In A. Ortony (Ed.), *Metaphor and thought* (pp. 579–609). Cambridge, UK: Cambridge University Press.

Peucker, T. K., and N. Chrisman. (1975). Cartographic data structures. *American Cartographer*, 2(1), 55–69.

Peuquet, D. J. (1978). Raster data handling in geographic information systems. In J. Dutton (Ed.), *Harvard papers on geographic information systems* (pp. 68–76). Reading, MA: Addison-Wesley.

Peuquet, D. J. (1979). Raster processing: An alternative approach to automated cartographic data handling. *American Cartographer*, 6(2), 129–139.

Peuquet, D. J. (1981a). An examination of techniques for reformatting digital cartographic data: Part I. The raster-to-vector process. *Cartographica*, 18(1), 34–48.

Peuquet, D. J. (1981b). An examination of techniques for reformatting digital cartographic data: Part II. The raster-to-vector process. *Cartographica*, 18(3), 21–33.

Peuquet, D. J. (1983). A hybrid structure for the storage and manipulation of very large spatial data sets. *Computer Vision, Graphics, and Image Processing*, 24, 14–27.

Peuquet, D. J. (1984). A conceptual framework and comparison of spatial data models. *Cartographica*, 21(4), 66–113.

Peuquet, D. J. (1988a). Representations of geographic space: Toward a conceptual synthesis. *Annals of the Association of American Geographers*, 78(3), 375–394.

Peuquet, D. J. (1988b). Toward the definition and use of complex spatial relationships. *Proceedings, Third International Symposium on Spatial Data Handling, Sydney, Australia* (pp. 211–223).

Peuquet, D. J. (1992). An algorithm for calculating minimum Euclidean distance between two geographic features. *Computers and Geosciences*, 18(8), 989–1001.

Peuquet, D. J. (1994). It's about time: A conceptual framework for the representation of temporal dynamics in geographic information systems. *Annals of the Association of American Geographers*, 84(3), 441–461.

Peuquet, D. J. (2000). Space–time representation: An overview. In L. Heres (Ed.), *Time in GIS: Issues in spatio-temporal modelling* (pp. 3–12). Apeldoorn, The Netherlands: Netherlands Geodetic Commission Publications on Geodesy, No. 47.

Peuquet, D. J. (2001). Making space for time: Issues in space–time data representation. *Geoinformatica*, 5(1), 11–32.

Peuquet, D. J., and N. Duan. (1995). An event-based spatiotemporal data model (ESTDM) for temporal analysis of geographical data. *International Journal of Geographical Information Systems*, 9(1), 7–24.

Peuquet, D. J., and M.-J. Kraak. (2002). Geobrowsing: Creative thinking and knowledge discovery using geographic visualization. *Information Visualization*, 1(1).

Peuquet, D. J., and L. Qian. (1996). *An integrated database design for temporal GIS*. Seventh International Symposium on Spatial Data Handling, Delft, Holland: Taylor & Francis, pp. 2-1–2-11.

Peuquet, D. J., and Z. C. Xiang. (1987). An algorithm to determine the directional relationship between arbitrarily shaped polygons. *Pattern Recognition, 20*(1), 65–74.

Pfoser, D., and C. S. Jensen. (1999). Capturing the uncertainty of moving-object representations. *Proceedings, Sixth International Symposium on the Advances in Spatial Databases—SSD '99, Hong Kong* (pp. 111–132).

Pfoser, D., and N. Tryfona. (2000). Fuzziness and uncertainty in spatiotemporal applications: Chorochronos technical report, CH-00-4.

Piaget, J. (1963). *The origins of intelligence in children.* New York: Norton.

Piaget, J. (1969a). *The child's conception of time.* New York: Basic Books.

Piaget, J. (1969b). *The psychology of the child.* New York: Basic Books.

Piaget, J., and B. Inhelder. (1956). *The child's conception of space.* London: Routledge & Kegan Paul.

Pick, H. L., and J. J. Lockman. (1981). From frames of reference to spatial representations. In L. S. Liben, A. H. Patterson, and N. Newcombe (Eds.), *Spatial representation and behavior across the life span: Theory and application* (pp. 39–61). New York: Academic Press.

Pigot, S., and B. Hazelton. (1992). The fundamentals of a topological model for a four-dimensional GIS. *Proceedings, Fifth International Symposium on Spatial Data Handling, Charleston, SC* (pp. 580–591).

Pinker, S., and R. A. Finke. (1980). Emergent two-dimensional patterns in images rotated in depth. *Journal of Experimental Psychology: Human Perception and Performance, 6*, 224–264.

Plaisant, C., D. Carr, and B. Shneiderman. (1995). Image browsers: Taxonomy, guidelings, and informal specifications. *IEEE Software, 12*(2), 21–32.

Plato. (1949). *Timaeus.* Indianapolis: Bobbs-Merrill.

Portugali, J. (1992). Geography, environment and cognition: An introduction. *Geoforum, 23*(2), 107–109.

Portugali, J. (Ed.). (1996). *The construction of cognitive maps.* Dordrecht: Kluwer Academic.

Pos, H. J. (1958). Recollections of Ernst Cassirer. In P. Schilpp (Ed.), *The philosophy of Ernst Cassirer* (pp. 61–72). New York: Tudor.

Presson, C., N. DeLange, and M. Hazelrigg. (1989). Orientation-specificity in spatial memory: What makes a path different from a map of a path? *Journal of Experimental Psychology: Learning, Memory, and Cognition, 15*, 887–897.

Presson, C., and M. Hazelrigg. (1984). Building spatial representations through primary and secondary learning. *Journal of Experimental Psychology: Learning, Memory, and Cognition, 10*, 716–772.

Pulaski, M. A. S. (1980). *Understanding Piaget: An introduction to children's cognitive development.* New York: Harper & Row.

Pylyshyn, Z. (1978). Imagery and artificial intelligence. In C. W. Savage (Ed.), *Perception and cognition issues in the foundations of psychology* (pp. 19–56). Minneapolis: University of Minnesota Press.

Pylyshyn, Z. (1981). The imagery debate: Analogue media versus tacit knowledge. *Psychological Review, 88*(1), 16–45.

Pylyshyn, Z. (1984). *Computation and cognition.* Cambridge, MA: MIT Press.

Pylyshyn, Z. (1989). The role of location indexes in spatial perception: A sketch of the FINST spatial-index model. *Cognition, 32,* 65–97.

Qian, L. (2000). *A visual query language for geographic information systems.* Dissertation, Pennsylvania State University, University Park, PA.

Qian, L., and D. Peuquet. (1998). Design of a visual query language for GIS. *Proceedings, Eighth International Symposium on Spatial Data, Handling, Vancouver, Canada* (pp. 592–602).

Qualcomm. (2001). Omnitracs: System overview. www.qualcomm.com/qwbs/products/omnitracs-overview.html.

Quillian, M. R. (1968). Semantic memory. In M. Minsky (Ed.), *Semantic information processing* (pp. 227–270). Cambridge, MA: MIT Press.

Rafaat, H. M., Z. Yang, and D. Gauthier. (1994). Relational spatial topologies for historical geographical information. *International Journal of Geographical Information Science, 8*(2), 163–173.

Raisz, E. (1956). *Mapping the world.* New York: Abelard-Schuman.

Raper, J., and D. Livingstone. (1995). Development of a geomorphological spatial model using object-oriented design. *International Journal of Geographical Information Systems, 9*(4), 359–383.

Raper, J. F., and D. J. Maguire. (1992). Design models and functionality in GIS. *Computers and Geosciences, 18*(4), 387–394.

Rauschecker, J. P. (1995). Compensatory plasticity and sensory substitution in the cerebral cortex. *Trends in the Neurosciences, 18,* 36–43.

Reason, J. T., and D. Lucas. (1984). Using cognitive diaries to investigate naturally occurring memory blocks. In J. Harris and P. E. Morris (Eds.), *Everyday memory, actions, and absent mindedness* (pp. 53–70). New York: Academic Press.

Reddy, D. R., and S. Rubin. (1978). *Representation of three-dimensional objects.* Report No. CMU-CS-78-113, Computer Science Department, Carnegie-Mellon University.

Reddy, M. (1993). The conduit metaphor: A case of frame conflict in our language about language. In A. Ortony (Ed.), *Metaphor and thought* (pp. 164–201). Cambridge, UK: Cambridge University Press.

Revesz, G. (1950). *Psychology and art of the blind.* London: Longman.

Rhynsburger, D. (1973). Analytic delineation of Thiessen polygons. *Geographical Analysis, 5*(2), 133–144.

Rips, L. (1975). Inductive judgments about natural categories. *Journal of Verbal Learning and Verbal Behavior, 14,* 665–681.

Robinson, A. H., and B. B. Petchenik. (1976). *The nature of maps: Essays toward understanding maps and mapping.* Chicago: University of Chicago Press.

Rock, I. (1973). *Orientation and form.* New York: Academic Press.

Roddick, J. F., and J. D. Patrick. (1992). Temporal semantics in information systems—a survey. *Information Systems, 17*, 249–267.

Rosch, E. (1973a). Natural categories. *Cognitive Psychology, 4*, 328–350.

Rosch. E. (1973b). On the internal structure of perceptual and semantic categories. In T. M. Moore (Ed.), *Cognitive development and the acquisition of language*. Academic Press.

Rosch, E. (1978). Principles of categorization. In E. Rosch and B. Lloyd (Eds.), *Cognition and categorization* (pp. 27–77). Hillsdale, NJ: Erlbaum.

Rosenfeld, A., and A. Kak. (1976). *Picture processing*. New York: Academic Press.

Ross, D. (1993). *Metaphor, meaning and cognition*. New York: Peter Lang.

Roth, S. F., M. C. Chuah, S. Kerpedjiev, J. Kolojejchick, and P. Lucas. (1997). Towards an information visualization workspace: Combining multiple means of expression. *Human–Computer Interaction Journal, 12*(1/2), 131–185.

Roussopoulos, N. (1984). *An introduction to PSQL: A pictorial structures query language*. Institute of Electrical and Electronics Engineers Computer Society Workshop on Visual Languages. Hiroshima, Japan: Computer Society Press, pp. 77–123.

Rudel, R. G., and H. L. Teuber. (1964). Crossmodal transfer of shape discrimination by children. *Neurophychologica, 2*, 1–8.

Rueckl, J., K. Cave, and S. Kosslyn. (1989). Why are "what" and "where" processed by separate cortical visual systems": A computational investigation. *Journal of Cognitive Neuroscience, 1*(2), 171–186.

Rumelhart, D., and J. McClelland. (1986). *Parallel distributed processing: Explorations in the microstructures of cognition*. Cambridge, MA: MIT Press.

Rumelhart, D. E. (1975). Notes on a schema for stories. In D. B. Bobrow and A. Collins (Eds.), *Representation and understanding* (pp. 211–236). New York: Academic Press.

Sack, R. (1980). *Conceptions of space in social thought: A geographic perspective*. Minneapolis: University of Minnesota Press.

Saffran, J., E. L. Newport, and R. N. Aslin. (1996). Word segmentation: The role of distributional clues. *Journal of Memory and Language, 35*(4), 606–621.

Samet, H. (1990a). *Applications of spatial data structures: Computer graphics, image processing and GIS*. Reading, MA: Addison-Wesley.

Samet, H. (1990b). *The design and analysis of spatial data structures*. Reading, MA: Addison-Wesley.

Samet, H., A. Rosenfeld, C. A. Shaffer, and R. E. Webber. (1984). Use of heirarchical data structures in geographical information systems. *Proceedings of the International Symposium on Spatial Data Handling, 2*, 392–411.

Sanders, A. F., and J. J. F. Schroots. (1969). Cognitive categories and memory span: III. Effects of similarity on recall. *Quarterly Journal of Experimental Psychology, 21*, 21–28.

Saussure, F. D. (1986). *Cours de linguistique générale* (25th ed.). Paris: Payot.

Scafi, A. (1999). Mapping Eden: Cartographies of the earthly paradise. In D. Cosgrove (Ed.), *Mappings* (pp. 50–70). London: Reaktion.

Schaefer, F. (1953). Exceptionalism in geography: A methodological examination. *Annals of the Association of American Geographers, 43*(3), 226–249.

Schank, R. C. (1976). The role of memory in language processing. In C. N. Coffer (Ed.), *The structure of human memory* (pp. 162–189). San Francisco: Freeman.

Schank, R. C. (1988). *The creative attitude: Learning to ask and answer the right questions*. New York: Macmillan.

Schenkelaars, V. (1994). Query classification, a first step towards a graphical interaction language. *Proceedings, Advanced Data Modeling Symposium, Delft, The Netherlands* (pp. 53–65).

Schiel, U. (1983). An abstract introduction to the temporal–hierarchic data model. *Proceedings, Ninth International Conference on Very Large Data Bases, New York, Institute of Electrical and Electronics Engineers* (pp. 322–330).

Schlaegel, T. F. (1953). The dominant method of imagery in blind compared to sighted adolescents. *Journal of Genetic Psychology, 83*, 265–277.

Schober, M. F. (1998). How addresses affect spatial perspective. In P. Olivier and K.-P. Gapp (Eds.), *Representation and processing of spatial expressions* (pp. 231–245). Mahwah, NJ: Erlbaum.

Schön, D. A. (1993). Generative metaphor: A perspective on problem-setting in social policy. In A. Ortony (Ed.), *Metaphor and thought* (pp. 137–163). Cambridge, UK: Cambridge University Press.

Scutenaire, L. (1948). *René Magritte*. Brussels: Libraire Sélection.

Sellis, T. (1999). Research issues in spatio-temporal database systems. *Proceedings, Advances in Spatial Databases: 6th International Symposium, SSD '99* (pp. 5–11). Hong Kong, China: Springer.

Seltz, O. (1927). The revision of the fudamental conceptions of intellectual processes. In J. M. Mandler and G. Mandler (Eds.), *Thinking: From association to gestalt* (pp. 225–234). New York: Wiley.

Sergent, J. (1991). Judgements of relative position and distance on representations of spatial relations. *Journal of Experimental Psychology: Human Perception and Performance, 17*(3), 762–780.

Shamos, M. (1978). *Computational geometry*. Dissertation, Yale University, New Haven, CT.

Shamos, M., and J. Bently. (1978). Optimal algorithms for structuring geographic data. *Proceedings, First International Advanced Study Symposium on Topological Data Structures for Geographic Information Systems, Boston, MA* (pp. 43–51).

Shapiro, L. G., and R. Haralick. (1982). An experimental relational database system for cartographic applications. *Proceedings, American Society for Photogrammetry and Remote Sensing/American Congress of Surveying and Mapping, Denver, CO*.

Sharma, R., I. Poddar, E. Ozyildiz, S. Kettebekov, H. Kim, and T. S. Huang. (1999). Toward interpretation of natural speech/gesture: Spatial planning on a virtual map. *Proceedings, ARL Advanced Displays Annual Symposium, Adelphi, MD* (pp. 35–39).

Shemyakin, F. N. (1962). General problems of orientation in space and space representations. In B. G. Ananyev (Ed.), *Psychological science in the USSR* (pp. 186–255). Washington, DC: U.S. Office of Technical Reports. (National Technical Information Service, No. TT62–11083)

Shepard, R. (1984). Ecological constraints on internal representation: Resonant kinematics of perceiving, imagining, thinking, and dreaming. *Psychological Review, 91*(4), 417–447.

Shepard, R. (1990). *Mind sights*. New York: Freeman.

Shepard, R., and C. W. Cermak. (1973). Perceptual-cognitive explorations of a toroidal set of free-form stimuli. *Cognitive Psychology, 4*, 351–377.

Shepard, R., and S. Chipman. (1970). Second-order isomorphism of internal representation: Shapes of states. *Cognitive Psychology, 1*, 1–17.

Shepard, R., and J. Metzler. (1971). Mental rotation of three-dimensional objects. *Science, 171*, 701–713.

Shepherd, I. D. H. (1994). Multi-sensory GIS: Mapping out the research frontier. *Proceedings, Sixth International Symposium on Spatial Data Handling, International Geographical Union, Edinburgh, Scotland* (pp. 356–390).

Shibasaki, R. (1994). Handling spatio-temporal uncertainties of geo-objects for dynamic update of GIS databases from multi-source data. *Proceedings, Advanced Geographic Data Modelling, Netherlands Geodetic Commission, Delft, The Netherlands* (pp. 228–242).

Shipley, T. F. (1991). Perception of a unified world: The role of discontinuities. In D. J. Napoli and J. A. Kegl (Eds.), *Bridges between psychology and linguistics: A Swarthmore festschrift for Lila Gleitman* (pp. 55–88). Hillsdale, NJ: Erlbaum.

Shneiderman, B. (2000). Universal usability. *Communications of the ACM, 43*(5), 84–91.

Shu, N. C. (1999). Visual programming: Perspectives and approaches. *IBM Systems Journal, 38*(2/3), 199–221. (Reprinted from *IBM Systems Journal, 28*(4), 1989)

Sider, D. (1981). *The fragments of Anaxagoras*. Meisenheim am Glan: Hain.

Siegel, A. W. (1981). The externalization of cognitive maps by children and adults: In search of ways to ask better questions. In L. S. Liben, A. H. Patterson, and N. Newcombe (Eds.), *Spatial representation and behavior across the life span: Theory and application* (pp. 167–194). New York: Academic Press.

Siegel, A. W., and S. H. White. (1975). The development of spatial representations of large-scale environments. In H. W. Reese (Ed.), *Advances in child development and behavior* (pp. 9–55). New York: Academic Press.

Siegel, L., and C. Brainerd (Eds.). (1978). *Alternatives to Piaget: Critical essays on the theory.* New York: Academic Press.

Simons, D. J., and F. C. Keil. (1995). An abstract to concrete shift in the development of biological thought: The insides story. *Cognition, 56,* 129–163.

Sinclair-deZwart, H. (1973). Language acquisition and cognitive development. In T. E. Moore (Ed.), *Cognitive development and the acquisition of language* (pp. 9–26). New York: Academic Press.

Sinton, D. (Ed.). (1978). *The inherent structure of information as a constraint to analysis: Mapped thematic data as a case study.* Reading, MA: Addison-Wesley.

Sloman, A. (1971). Interactions between philosophy and artificial intelligence: The role of intuition and non-logical reasoning in intelligence. *Artificial Intelligence, 2,* 209–225.

Sloman, A. (1975). Afterthoughts on analogical representations. *Proceedings, First Workshop on Theoretical Issues in Natural Language Processing (TINLAP-1), Cambridge, MA* (pp. 164–168).

Smith, S. M., and S. E. Blankenship. (1991). Incubation and the persistence of fixation in problem solving. *American Journal of Psychology, 104,* 61–87.

Smith, T., D. Peuquet, S. Menon, and P. Agarwal. (1987). KBGIS-II. A knowledge-based geographical information system. *Internation Journal of Geographical Information Systems, 1*(2), 149–172.

Smith, T. R., J. W. Pellegrino, and R. G. Golledge. (1982). Computational process modeling of spatial cognition and behavior. *Geographical Analysis, 14*(4), 305–325.

Snodgrass, R. T. (Ed.). (1995). *The TSQL2 temporal query language.* Boston: Kluwer.

Snyder, J. P. (1984). *Computer-assisted map projection research.* Reston, VA: U.S. Geological Survey Bulletin No. 1629.

Sowa, J. F. (2000). *Knowledge representation: Logical, philosophical, and computational foundations.* Pacific Grove, CA: Brooks/Cole.

Sowa, J. F., N. Y. Foo, and A. S. Rao. (1990). *Conceptual graphs for knowledge systems.* Reading, MA: Addison-Wesley.

Spencer, C., K. Morsley, S. Ungar, E. Pike, and M. Blades. (1992). Developing the blind child's cognition of the environment: The role of direct and map-given experience. *Geoforum, 23*(2), 191–197.

Starkey, P., E. S. Spelke, and R. Gelman. (1983). Detection of intermodal numerical correspondences by human infants. *Science, 222,* 179–181.

Starkey, P., E. S. Spelke, and R. Gelman. (1990). Numerical abstraction by human infants. *Cognition, 36*(2), 97–127.

State of Maryland. (1990). MAGI: Maryland Automated Geographic Information System. In D. J. Peuquet and D. F. Marble (Eds.), *Introductory readings in geographic information systems* (pp. 65–89). London: Taylor & Francis.

Stea, D., J. M. Blaut, and J. Stephens. (1996). Mapping as a cultural universal. In J. Portugali (Ed.), *The construction of cognitive maps* (pp. 345–360). Dordrecht: Kluwer Academic.

Stevens, A., and P. Coupe. (1978). Distortions in judged spatial relations. *Cognitive Psychology, 10,* 422–437.

Stevens, R. J., A. F. Lehar, and F. H. Preston. (1983). Manipulation and presentation of multidimensional image data using the Peano scan. *IEEE Transactions on Pattern Analysis and Machine Intelligence, 5,* 520–526.

Stonebraker, M. (1986). Triggers and inference in database systems. In M. L. Brodie and J. Mylopoulos (Eds.), *On knowledge base management systems* (pp. 222–231). New York: Springer-Verlag.

Stonebraker, M., L. A. Rowe, B. G. Lindsay, J. Gray, M. J. Carey, M. L. Brodie, P. A. Bernstein, and D. Beech. (1990). Third-generation database system manifesto. *ACM SIGMOD Record, 19*(3), 31–44.

Stonebraker, M., E. Wong, P. Kreps, and G. Held. (1976). The design and implementation of INGRES. *ACM Transactions on Database Systems, 1*(3), 189–222.

Strabo. (1917). *Geography.* New York: Putnam.

Streri, A. (1987). Tactile perception of shape and intermodal transfer in two- to three-month old infants. *British Journal of Developmental Psychology, 5,* 213–220.

Studer, R. (1986). Modeling time aspects of information systems. *Proceedings, International Conference on Data Engineering, Los Angeles, CA* (pp. 364–372).

Suchan, T. (1997). *The categories "urban" and "rural."* Dissertation, Pennsylvania State University, University Park, PA.

Sui, D. Z., and M. F. Goodchild. (2001). GIS as media? *International Journal of Gegraphical Information Science, 15*(5), 387–390.

Sun, R., and L. A. Bookman. (1995). *Computational architectures integrating neural and symbolic processes.* Boston: Kluwer.

Sutherland, N. S. (1968). Outlines of a theory of visual pattern recognition in animals and man. *Proceedings of the Royal Society, 171,* 297–317.

Suwa, M., B. Tversky, J. Gero, and T. Purcell. (2001). Seeing into sketches: Regrouping parts encourages new interpretations. *Proceedings, Visual and Spatial Reasoning in Design II, Key Centre of Design Computing and Cognition, University of Sydney, Australia* (pp. 207–219).

Talmy, L. (1978). Figure and ground in complex sentences. In J. Greenberg, C. Ferguson, and E. Moravcsik (Eds.), *Universals of human language* (pp. 625–649). Palo Alto, CA: Stanford University Press.

Talmy, L. (1983). How language structures space. In H. L. Pick and L. P. Acredolo (Eds.), *Spatial orientation: Theory, research, and application* (pp. 225–319). New York: Plenum Press.

Talmy, L. (1987). The relation of grammar to cognition. In B. Rudzka-Ostyn (Ed.), *Topics in cognitive linguistics* (pp. 165–205). Amsterdam: Benjamins.

Talmy, L. (1988). Force dynamics in language and cognition. *Cognitive Science, 12,* 49–100.

Tanaka, J. W., and M. Taylor. (1991). Object categories and expertise: Is the basic level in the eye of the beholder? *Cognitive Psychology, 23,* 457–482.

Tansel, A. U., J. Clifford, S. Gadia, S. Jajodia, A. Segev, and R. Snadgrass. (Eds.). (1993). *Temporal databases: Theory, design and application.* Redwood City, CA: Benjamin/Cummings.

Tarr, M. J. (in press). Visual object recognition: Can a single mechanism suffice? In M. A. Peterson and G. Rhodes (Eds.), *Analytic and holistic processes in the perception of faces, objects, and scenes.* JAI/Ablex.

Tarr, M. J., and S. Pinker. (1989). Mental rotation and orientation-dependence on shape recognition. *Cognitive Psychology, 21*(28), 233–282.

Tarr, M. J., and S. Pinker. (1990). When does human object recognition use a viewer-centered reference frame? *Psychological Science, 1*(4), 253–256.

Tatham, A. F. (1991). The design of tactile maps: Theoretical and practical consiserations. *Proceedings, 15th Annual Conference of the International Cartographic Association, Bournemouth, UK* (pp. 157–166).

Taylor, H., and B. Tversky. (1992a). Descriptions and depictions of environments. *Memory and Cognition, 20*(5), 483–496.

Taylor, H., and B. Tversky. (1992b). Spatial mental models derived from survey and route descriptions. *Journal of Memory and Language, 31,* 261–292.

Thiessen, A. H. (1911). Precipitation averages for large areas. *Monthly Weather Review, 39,* 1082–1084.

Thomas, N. J. T. (1999). Are theories of imagery theories of imagination?: An active perception approach to conscious mental content. *Cognitive Science, 23*(2), 207–245.

Thorndyke, P. (1984). Applications of schema theory in cognitive research. In J. Anderson and S. Kosslyn (Eds.), *Tutorials in learning and memory: Essays in honor of Gordon Bower* (pp. 167–192). San Francisco: Freeman.

Thrower, N. J. W. (1996). *Maps and civilization: Cartography in culture and society.* Chicago: University of Chicago Press.

Tobler, W. (1959). Automation and cartography. *Geographical Review, 49*(4), 526–534.

Tobler, W. (1961). *Map transformations of geographic space.* Department of Geography, University of Washington, Seattle, WA.

Tobler, W. (1963). Geographic area and map projections. *Geographical Review, 53*(1), 59–78.

Tobler, W. R., and Z.-T. Chen. (1986). A quadtree for global information storage. *Geographical Analysis, 18*(4), 360–371.

Tolman, E. C. (1948). Cognitive maps in rats and men. *Psychological Review, 55,* 189–208.

Tomlin, C. D. (1983). *Digital cartographic modeling techniques in environmental planning.* New Haven, CT: Yale University, Graduate School.

Tomlin, C. D. (1990). *Geographic information systems and cartographic modeling.* Englewood Cliffs, NJ: Prentice-Hall.

Tomlinson, R. F. (1973). A technical description of the Canada geographic information system. Lands Directorate, Environment Canada, Ottawa.

Tomlinson, R. F., H. W. Calkins, and D. F. Marble. (1976). *Computer handling of geographical data: An examination of selected geographical information systems.* Paris, UNESCO, Natural Resource Series No. 8.

Torp, K., C. S. Jensen, and R. T. Snodgrass. (1998). *Stratum approaches to temporal DBMS implementation.* International Database Engineering and Applications Symposium (IDEAS 1998), Cardiff, Wales, Institute of Electrical and Electronics Engineers Computer Society.

Torpf, H., and H. Herzog. (1981). Multidimensional range search in dynamically balanced trees. *Angewandte Informatik, 2,* 71–77.

Tuan, Y.-F. (1977). *Space and place: The perspective of experience.* Minneapolis: University of Minnesota Press.

Tuan, Y.-F. (1978). Sign and metaphor. *Annals of the Association of American Geographers, 68*(3), 362–372.

Tulving, E. (1972). Episodic and semantic memory. In E. Tulving and W. Donaldson (Eds.), *Organization of memory* (pp. 382–403). New York: Academic Press.

Turner, M. (1987). *Death is the mother of beauty: Mind, metaphor, criticism.* Chicago: University of Chicago Press.

Tversky, A. (1977). Features of similarity. *Psychological Review, 84*(4), 327–352.

Tversky, B. (1992). Distortions in cognitive maps. *Geoforum, 23*(2), 131–138.

Tversky, B. (1993). Cognitive maps, cognitive collages, and spatial mental models. In A. Frank and I. Campari (Eds.), *Spatial information theory: A theoretical basis for GIS* (pp. 14–24). New York: Springer-Verlag.

Tversky, B. (2000). Some ways that maps and diagrams communicate. In C. Freksa, C. Habel, and K. F. Wender (Eds.), *Spatial cognition II: Integrating abstract theories, empirical studies, formal methods, and practical applications* (pp. 72–79). Berlin: Springer-Verlag.

Tversky, B., and K. Hemenway. (1983). Categories of environmental scenes. *Cognitive Psychology, 15,* 121–149.

Tversky, B., and K. Hemenway. (1984). Objects, parts, and categories. *Journal of Experimental Psychology: General, 113*(2), 169–193.

U.S. Census Bureau. (1969). The DIME geocoding system. *Report No. 4: Census use study.* Washington, DC: Author.

Ungar, S., M. Blades, and C. Spencer. (1996). The construction of cognitive maps by children with visual impairments. In J. Portugali (Ed.), *The construction of cognitive maps* (pp. 247–273). Dordrecht: Kluwer.

Ungerleider, L. G., and M. Mishkin. (1982). Two cortical visual systems. In D. J. Ingle, M. A. Goodale, and R. J. W. Mansfield (Eds.), *Analysis of visual behavior* (pp. 549–586). Cambridge, MA: MIT Press.

Urban, S. D., and L. Delcambre. (1986). An analysis of the structural, dynamic and temporal aspects of semantic data models. *Proceedings, International Conference on Data Engineering, New York, Institute of Electrical and Electronics Engineers* (pp. 382–389).

Usery, E. L. (1993). Category theory and the structure of features in geographic information systems. *Cartography and Geographic Information Systems*, *20*(1), 5–12.

van Oosterom, P., and T. Vijlbrief. (1991). Building a GIS on top of the open DBMS "postgres." *Proceedings, EGIS '91, Brussels, Belgium* (pp. 775–787).

van Roessel, J. W. (1987). Design of a spatial data structure using the relational normal forms. *International Journal of Geographical Information Systems*, *1*(1), 33–50.

van Selst, M., and P. Jolicoeur (1994). Can mental rotation occur before the dual-task bottleneck? *Journal of Experimental Psychology: Human Perception and Performance*, *20*, 905–921.

Verbree, E., G. van Maren, R. Germs, F. Jansen, and M.-J. Kraak. (1999). Interaction in virtual world views—linking 3D GIS with VR. *International Journal of Geographical Information Science*, *13*(4), 385–396.

von Senden, M. (1932). *Raum und gestalt: Auffassung bei operierten blindgeborenen vor und nach der operation*. Leipzig: Barth.

Wachowicz, M. (1999). *Object-oriented design for temporal GIS*. London: Taylor & Francis.

Wagman, M. (1991). *Cognitive science and concepts of mind*. New York: Praeger.

Wallace, K. D. (1974). *Epistemological basis for Ernst Cassirer's philosophy of science and its application to spacial theory*. Dissertation, Fordham University, New York, NY.

Wang, F., and G. B. Hall. (1996). Fuzzy representation of geographical boundaries in GIS. *International Journal of Geographical Information Systems*, *10*(5), 573–590.

Wang, F., G. B. Hall, and S. Subaryono. (1990). Fuzzy information representation and processing in conventional GIS software: Database design and application. *International Journal of Geographical Information Systems*, *4*(3), 261–283.

Warren, D. H. (1974). Early vs. late vision: The role of early vision in spatial reference systems. *New Outlook for the Blind*, *68*, 157–162.

Warren, D. H., and M. J. Rossano. (1991). Intermodality relations: Vision and touch. In M. A. Heller and W. Schiff (Eds.), *Psychology of touch* (pp. 119–137). Hillsdale, NJ: Erlbaum.

Waugh, T. C., and R. G. Healey. (1987). The geoview design: A relational data base approach to geographical data handling. *International Journal of Geographical Information Systems*, *1*(2), 101–118.

Way, E. C. (1991). *Knowledge representation and metaphor*. Boston: Kluwer Academic.

Wermter, S. (2000). The hybrid approach to artificial neural network-based language processing. In R. Dale, H. Moisl and H. Somers (Eds.), *A handbook of natural language processing* (pp. 823–846). New York: Marcel Dekker.

Wermter, S., and R. Sun (2000). *Hybrid neural systems. (Lecture notes in Computer Science 1778).* Berlin: Springer-Verlag.

Werner, H. (1948). *Comparative psychology of mental development.* New York: International Universities Press.

Werner, H. (1957). The concept of development from a comparative and organismic point of view. In D. B. Harris (Ed.), *The concept of development* (pp. 125–148). Minneapolis: University of Minnesota Press.

Wheless, G. H., C. M. Lascara, A. Valle-Levinson, and D. Sherman. (1996). The Chesapeake Bay virtual ecosystem initial results from the prototypical system. *International Journal of Supercomputer Applications and High Performance Computing, 10*(2).

White, M. (1979). A survey of the mathematics of maps. *Proceedings, Auto-Carto 4, Reston, VA* (pp. 82–96).

White, M. (1981). N-trees: Large ordered indexes for multi-dimensional space, Application Mathematics Research Staff, Statistical Research Division, U.S. Bureau of the Census.

White, M. (1984). Technical requirements and standards for a multipurpose geographic data system. *American Cartographer, 11*(1), 15–26.

Whorf, B. L. (1956). *Language, thought and reality.* New York: Technology Press/ Wiley.

Wiedel, J. (Ed.). (1983). *Proceedings of the first international symposium on maps and graphics for the visually handicapped.* Washington, DC: AAG.

Wiles, J., G. S. Halford, J. E. M. Stewart, M. S. Humphreys, J. D. Bain, and W. H. Wilson. (1994). Tensor models. In T. Dartnall (Ed.), *Artificial intelligence and creativity* (pp. 147–161). Dordrecht, The Netherlands: Kluwer.

Willmott, C. J., and G. L. Gaile. (1992). Reality, models, and knowledge. In R. F. Abler, M. G. Marcus, and J. M. Olson (Eds.), *Geography's inner worlds: Pervasive themes in contemporary American geography* (pp. 163–186). New Brunswick, NJ: Rutgers University Press.

Wilton, R., and P. File. (1975). Knowledge of spatial relations: A preliminary investigation. *Quarterly Journal of Experimental Psychology, 27,* 251–258.

Winograd, T., and F. Flores. (1986). *Understanding computers and cognition.* Reading, MA: Addison-Wesley.

Winston, P. H. (1992). *Artificial intelligence* (3rd ed.). Reading, MA: Addison-Wesley.

Wittgenstein, L. (1953). *Philosophical investigations.* New York: Macmillan.

Woldenberg, M. J., G. Cumming, K. Horsfield, K. Prowse, and S. Singhal. (1970). *Law and order in the human lung.* Cambridge, MA: Harvard Laboratory for Computer Graphics.

Wolf, T. (1976). A cognitive model of musical sight-reading. *Journal of Psycholinguistic Research*, 5(2), 143–171.

Wood, D. (1992). *The power of maps*. New York: Guilford Press.

Wood, D., and J. Fels. (1986). Designs on signs: Myth and meaning in maps. *Cartographica*, 23(3), 54–103.

Wood, M. (1994). The traditional map as a visualization technique. In H. Hearnshaw and D. Unwin (Eds.), *Visualisation in geographical information systems* (pp. 9–17). Chichester, UK: Wiley.

Woodward, D. (1985). Reality, symbolism, time and space in medieval world maps. *Annals of the Association of American Geographers*, 75(4), 510–521.

Woodward, D. (1992). Representations of the world. In R. F. Abler, M. G. Marcus, and J. M. Olson (Eds.), *Geography's inner worlds: Pervasive themes in contemporary American geography* (pp. 50–72). New Brunswick, NJ: Rutgers University Press.

Woodward, D., and G. M. Lewis. (Eds.). (1998). *Cartography in the traditional African, American, Arctic, Australian, and Pacific societies* (The History of Cartography, vol. 2, bk. 3). Chicago: University of Chicago Press.

Worboys, M. F. (1994). Object oriented approaches to geo-referenced information. *International Journal of Geographic Information Systems*, 8(4), 385–389.

Worchel, P. (1951). Space perception and orientation in the blind. *Psychological Monographs*, 65(No. 332), 1–27.

Young, M. (1988). *The metronomic society: Natural rhythms and human timetables*. Cambridge, MA: Harvard University Press.

Zadeh, L. A. (1965). Fuzzy sets. *Information and Control*, 8(8), 338–353.

Zimler, J., and J. M. Keenan. (1983). Imagery in the congenitally blind: How visual are visual images? *Journal of Experimental Psychology: Learning, Memory, and Cognition*, 9(2), 269–282.

INDEX

ABOUT THE AUTHOR

Donna J. Peuquet, PhD, is Professor of Geography at the Pennsylvania State University, University Park. Dr. Peuquet conducts research in the areas of geographic knowledge representation theory, spatiotemporal data models, spatial cognition, AI approaches to knowledge representation, and GIS design.